D0577757

Second Edition

Op-Amp Handbook

Fredrick W. Hughes
Electronics Training Consultant

Prentice-Hall, *Englewood Cliffs, New Jersey 07632*

Library of Congress Cataloging-in-Publication Data

HUGHES, FREDRICK W.
Op-amp handbook.

Includes index.
1. Operational amplifiers. 2. Integrated circuits. I. Title.
TK7871.58.06H83 1986 621.381′735 85-20998
ISBN 0-13-637315-1

Editorial/production supervision and
interior design: **Kathryn Pavelec**
Cover design: **Wanda Lubelska Design**
Manufacturing buyer: **Rhett Conklin**

This book is dedicated to my loving wife, Roberta, whose enormous energy, productivity, and confidence made it possible.

Printed in the United States of America

10 9 8 7 6 5 4 3 2

ISBN 0-13-637315-1 025

PRENTICE-HALL INTERNATIONAL (UK) LIMITED, *London*
PRENTICE-HALL OF AUSTRALIA PTY. LIMITED, *Sydney*
PRENTICE-HALL CANADA INC., *Toronto*
PRENTICE-HALL HISPANOAMERICANA, S.A., *Mexico*
PRENTICE-HALL OF INDIA PRIVATE LIMITED, *New Delhi*
PRENTICE-HALL OF JAPAN, INC., *Tokyo*
PRENTICE-HALL OF SOUTHEAST ASIA PTE. LTD., *Singapore*
EDITORA PRENTICE-HALL DO BRASIL, LTDA., *Rio de Janeiro*
WHITEHALL BOOKS LIMITED, *Wellington, New Zealand*

Contents

Preface

The First Edition of this book was widely accepted because its straightforward, practical approach enabled students to grasp the concepts necessary to understand and use the operational amplifier (op amp). Its popularity among schools was based on the consistent learning format which is found again in this Second Edition in Chapters 1 through 6. First, in each chapter, the text material is presented in the form of theory, functional diagrams, nomenclature, and basic practical circuits. These circuits have the component values given and can be constructed directly from the book, although Chapter 7 is dedicated to step-by-step method experiments. These experiments are keyed to their respective theory portions in the book. Second, the text is followed by Summary Points, which are a brief review of important points found in the chapter. Third, a Terminology Exercise is given to familiarize the reader with the language of op amps and associated circuits. Fourth, Problems and Exercises, rearranged for this Edition, enable the reader to gain a working knowledge of op amps. Last, a Self-Checking Quiz provides the reader with immediate follow-up for determining his or her mastery of the theory and circuit operations previously covered. This type of chapter format can provide instructors with student-centered instructional material that does not require preparation.

The prerequisites for using this book successfully go little beyond a knowledge of fundamental algebra and basic dc/ac circuit theory. However, bipolar transistor theory is helpful, and it is recommended that this book be used in post-solid-state-device courses as an interface between discrete devices and integrated circuits.

There are many fine books and manuals concerning op amps that are already in print. However, this book uses a direct, easy-to-read approach that develops the basic understanding and practical skills essential to working with op amps, for readers at various levels: the electronics student, the technician, the engineer, and the instructor.

What each chapter contains:

Chapter 1 explains the function of an op amp, its nomenclature, IC package pin identification, important characteristics, and parameters. References are made to the experiments in Chapter 7 to verify theory and to provide an instant hands-on approach to develop working skills with op-amp circuits. Some simple dual-voltage power supply circuits needed to operate op-amp circuits are shown in Section 1-4.

Chapter 2 describes the operation of basic op-amp circuits: voltage comparators, the inverting amplifier, the noninverting amplifier, voltage followers, summing amplifiers, and the difference amplifier.

Chapter 3 shows basic signal-processing circuits, such as the integrator, the differentiator, the low-pass filter, the high-pass filter, the bandpass filter, and the notch filter. This chapter provides the necessary formulas for determining output voltages, cut-off frequencies, and bandwidth capabilities.

Chapter 4 deals with oscillators and describes the square-wave generator, the sawtooth-wave generator, the triangle-wave generator, the sine-wave oscillator, the quadrature oscillator, and a basic function generator.

Chapter 5 shows op-amp applications to audio circuits in the form of voltage amplifiers, equalization preamplifiers, active-tone control circuits, basic audio mixers, and miscellaneous audio circuits.

Chapter 6 shows associated circuits found with op amps that provide protection and stability. This chapter also presents basic op-amp testing and troubleshooting techniques. Four op-amp tester circuits with construction hints are given to aid the reader with building projects.

Chapter 7 contains 20 easy-to-perform experiments to provide the reader with the basic skills necessary to troubleshoot and test op-amp circuits. Six new experiments provide op-amp applications for a voltage-level detector, a low-pass filter, a high-pass filter, a bandpass filter, a notch filter, and a square-wave generator.

Chapter 8 shows practical step-by-step design procedures for 20 basic op-amp circuits. These circuits, while easy enough to give the beginner a straightforward approach to designing, provide the experienced designer with instant, time-saving applications.

Chapter 9 is a collection of 62 practical op-amp circuits with a brief description of each, to furnish a reference for readers interested in constructing circuits or creating electronic systems.

The Appendix lists 11 op-amp manufacturers' specification sheets with information for further study of, or circuit construction of op-amp circuits presented in the text.

Although there are countless integrated circuit (IC) op amps available, the 741 op amp is used throughout this book because it is inexpensive, reliable, does not burn out easily, and can readily be found in most electronic supply houses.

This book will help the beginner with its easy-to-understand approach to learning op-amp circuits; it will make the instructor's job less difficult with its individualized-instruction format; and it will present different ideas and approaches to the experienced person.

FREDRICK W. HUGHES

Miami, Florida

Chapter 1

Operational Amplifier Functions and Characteristics

Operational amplifiers (op amps) are specially designed and packaged electronic circuits that can be used for various purposes with only a few external components. Until recently, op amps were constructed of discrete components within a sealed package and were so costly that few engineers and technicians ever became involved with them. Today however, with improved integrated-circuit (IC) technology, these inexpensive IC packages are found in nearly every aspect of electronics.

Originally, op amps were used for analog computing circuits, control circuits, and instrumentation. Their main function was to provide linear (voltage and current) mathematical operations, such as comparison, addition, subtraction, differentiation, integration, and amplification. Now they are found virtually everywhere—in audio reproduction, communication systems, digital processing systems, consumer electronics, and many unique hobbyist's devices.

Op-amp configurations may have a single input and single output, differential input and differential output, or differential input and single output. The latter configuration is the most widespread in the electronics industry and will be used as the basis of this book. Anyone involved in electronics should know how the op amp functions, its characteristics, and should be able to recognize and work with basic circuit configurations.

1-1 WHAT IS AN OP AMP?

The IC op amp is a solid-state device capable of sensing and amplifying DC and AC input signals.

A typical IC op amp consists of three basic circuits, a high-input impedance differential amplifier, a high-gain voltage amplifier, and a low-impedance output amplifier (usually a push-pull emitter follower). Figure 1-1 shows a block diagram of an op amp. Notice that it usually requires a positive and a negative power supply. This allows the output voltage to swing positive and negative with respect to ground.

The most important characteristics of an op amp are:

1. *Very high input impedance*, which produces negligible currents at the inputs.
2. *Very high open-loop gain.*
3. *Very low output impedance*, so as not to affect the output of the amplifier by loading.

The standard op amp schematic symbol is represented by a triangle, as illustrated in Figure 1-2. The input terminals are at the base of the triangle. The inverting input is represented by the minus sign. A DC voltage or AC signal placed on this input will be 180° out of phase at the output. The noninverting input is represented by the plus sign. A DC voltage or AC signal placed on this input will be in phase at the output. The output terminal is shown at the apex of the triangle.

Power-supply terminals and other leads for frequency compensation or null adjusting are shown extending above and below the triangle. These leads are not always shown in schematic diagrams, but are implied. Power connections are understood, whereas the other leads may not be used at all.

The type of op amp or manufacturer's part number is centered within the body of the triangle. A general circuit not indicating a specific op amp might use the symbol A_1, A_2, etc., or OP-1, OP-2, etc.

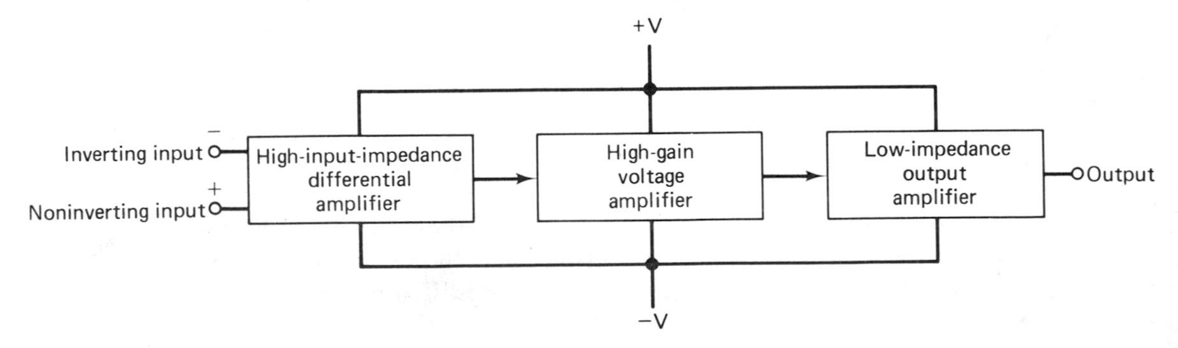

FIGURE 1-1 Block diagram of op amp.

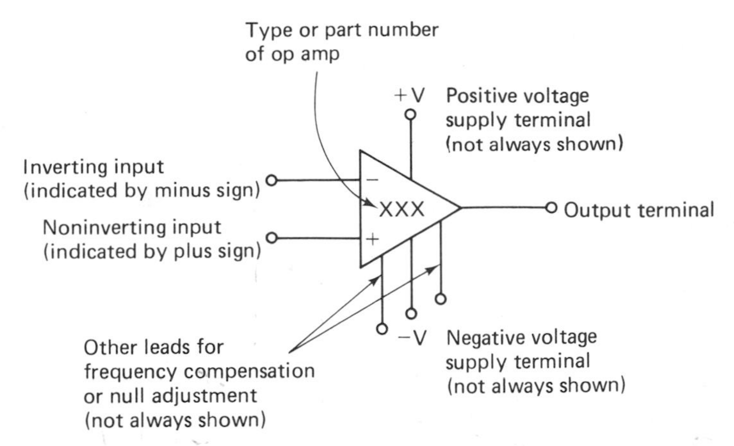

FIGURE 1-2 Standard op-amp schematic symbol.

Although we can use the op amp without knowing exactly what goes on inside it, we can better understand its operating characteristics by having some idea of its internal circuitry. Figure 1-3 shows the schematic diagram of the popular 741 op-amp IC. Other op amps are similar. Resistors and capacitors are held to an absolute minimum in IC design, using transistors wherever possible. No coupling capacitors are used, allowing the circuit to amplify DC as well as AC signals. The 30-pF capacitor shown provides internal frequency compensation, which will be discussed later in this chapter.

The op amp consists basically of three stages; a high-input-impedance differential amplifier, a high-gain voltage amplifier with a level shifter (permitting the output to swing positive and negative) and a low-impedance output amplifier.

The three stages are indicated. Transistors Q_1 and Q_2 are the differential inputs. Transistor Q_{16} is a high-gain Darlington driver. The output stage is complementary symmetry using transistors Q_{14} and Q_{20}. Output short-circuit protection is provided by current-limiting transistor Q_{15}. The rest of the components provide bias and amplification. An external potentiometer can have its ends connected to pins 1 and 5 (offset null) with its wiper connected to pin 4 ($-V_{cc}$) for nulling adjustments.

1-2 HOW AN OP AMP FUNCTIONS

Ideally, the gain of an op amp would be infinite; however, practically, the gain may exceed 200,000 in the open-loop mode. In the open-loop mode there is no feedback from the output to the inputs and the voltage gain (A_v) is maximum, as shown in Figure 1-4a. In a practical circuit, the slightest voltage difference at the inputs will cause the output voltage to attempt to

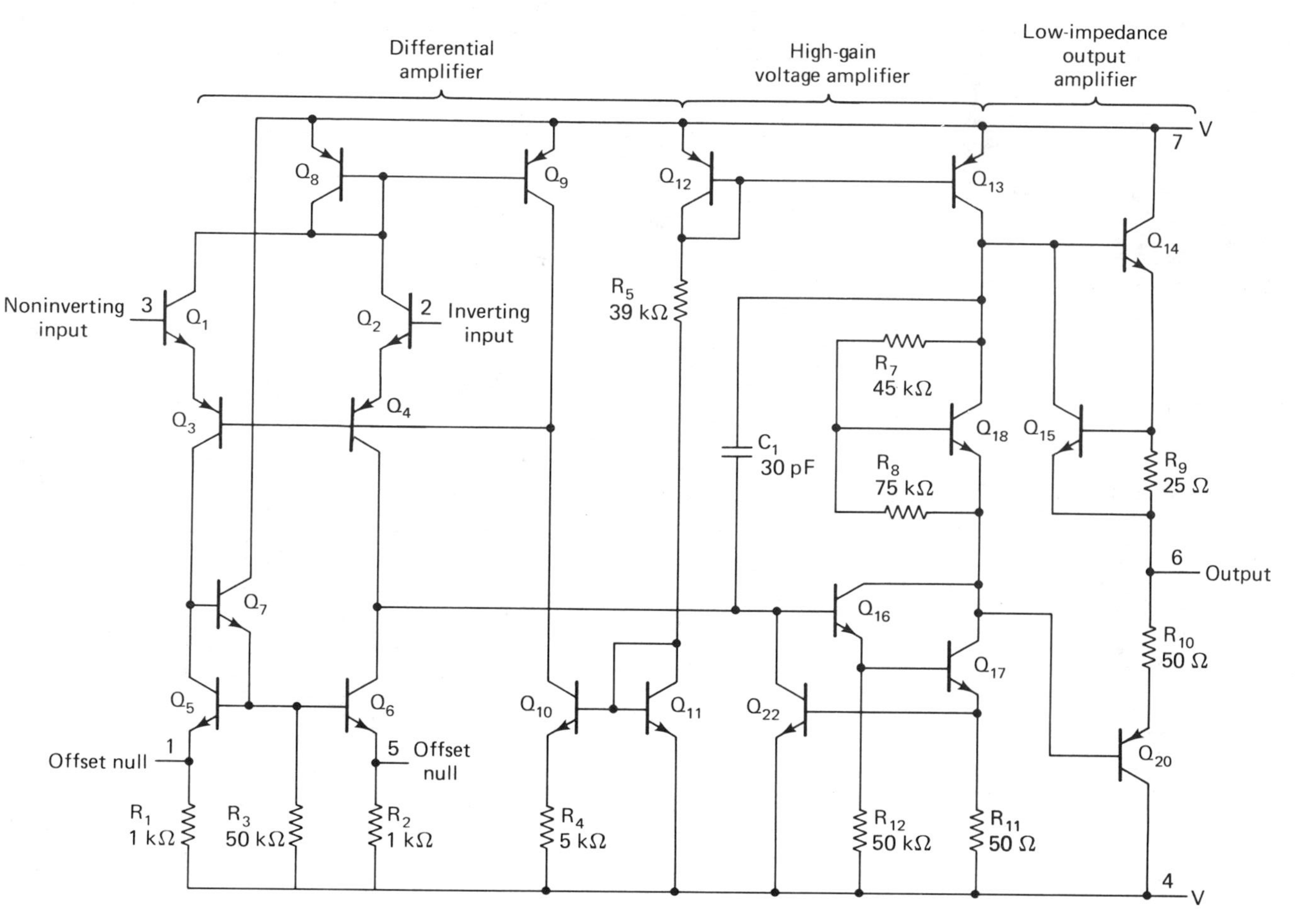

FIGURE 1-3 Typical op-amp schematic diagram.

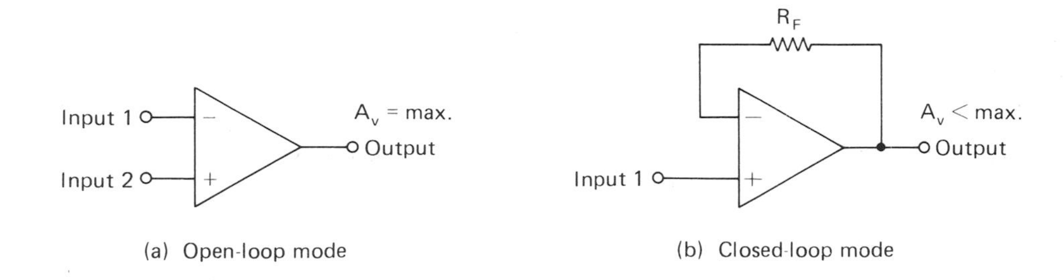

(a) Open-loop mode (b) Closed-loop mode

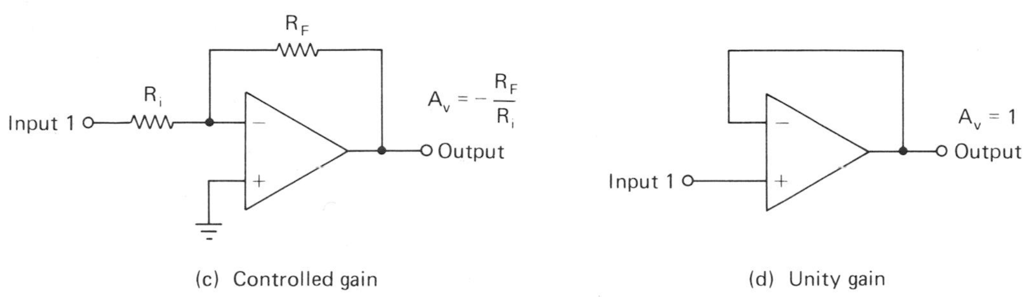

(c) Controlled gain (d) Unity gain

FIGURE 1-4 Op-amp gain.

swing to the maximum power-supply level. The maximum voltage at the output will be about 90% of the supply voltages, because of the internal voltage drops of the op amp. (Refer to Figure 1-3 and note components Q_{14}, R_9, R_{10}, and Q_{20}.) The output is said to be at saturation and can be represented (for either polarity) by $+V_{\text{sat}}$ and $-V_{\text{sat}}$. As an example, an op-amp circuit in the open-loop mode using a ±15-V supply would have its output swing from +13.5 V to −13.5 V. With this type of circuit the op amp is very unstable and the output will be 0 V for a 0-V difference between the inputs, or the output voltage will be at either extreme, with a slight voltage difference at the inputs. The open-loop mode is found primarily in voltage comparators and level-detector circuits.

The versatility of the op amp is demonstrated by the fact that it can be used in so many types of circuits in the closed-loop mode, as shown in Figure 1-4b. External components are used to feed back a portion of the output voltage to the inverting input. This feedback stabilizes most circuits and can reduce the noise level. The voltage gain (A_v) will be less ($<$) than maximum gain in the open-loop mode.

Closed-loop gain must be controlled to be of any value in a practical circuit. By adding a resistor R_{in} to the inverting input as shown in Figure 1-4c, the gain of the op amp can be controlled. The resistance ratio of R_F to R_{in}

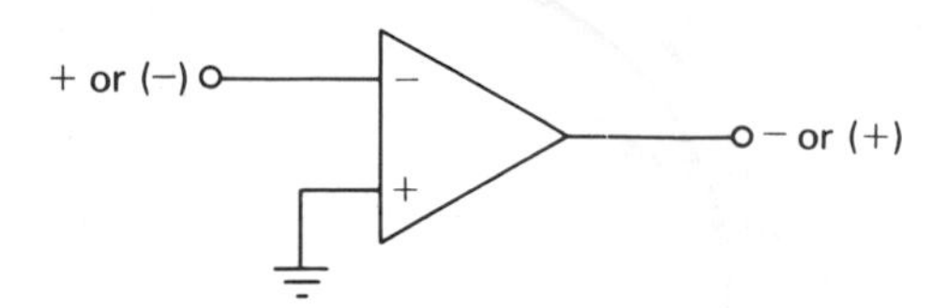

1. If − input is more positive than + input, output will be negative.
2. If − input is more negative than + input, output will be positive.

FIGURE 1-5 Input/output polarity relationship.

determines the voltage gain of the circuit and can be found by the formula

$$A_v = -\frac{R_F}{R_{\text{in}}}$$

The minus sign indicates that the op-amp circuit is in the inverting configuration and is disregarded for calculations. For example, if $R_{\text{in}} = 10\ \text{k}\Omega$ and $R_F = 100\ \text{k}\Omega$, then A_v would equal 10. An input voltage of 0.01 V would yield an output voltage of 0.1 V. If R_{in} were changed to 1 kΩ, the A_v would increase to 100. Now, an input voltage of 0.01 V would yield an output voltage of 1 V.

If both R_{in} and R_F are the same value, the A_v equals 1, or unity gain. A direct connection from output to input also results in unity gain, as shown in Figure 1-4d. In this noninverting configuration, the voltage out equals the voltage in and A_v equals +1.

These various types of gain will be used in basic circuits later in this book to better acquaint you with the functions of op amps. One important function to remember is the relationship of input polarity to output polarity. Simply stated, if the inverting input is more positive than the noninverting input, the output will be negative. Similarly, if the inverting input is more negative than the noninverting input, the output will be positive. Figure 1-5 shows this important function, where the noninverting input is at ground or zero volts. (See Experiment 7-1.)

1-3 OP-AMP CHARACTERISTICS AND PARAMETERS

Understanding the characteristics and parameters of an electronic device will enable you to better understand the circuit in which it is used. Knowing what to expect from op amps will aid you in servicing or designing circuits in which they are used. This section lists the pertinent information on the characteristics and parameters of op amps used in most circuits.

1-3-1 Input Impedance

Ideally, the input impedance of an op amp should be infinite, but in practice is about 1 MΩ or more. Some special op amps may have an input impedance of as much as 100 MΩ. The higher the input impedance, the better the op amp will perform. The input capacitance of an op amp may become important at high frequencies. Typically, this capacitance is less than 2 pF, when one input terminal is grounded. (See Experiment 7-9.)

1-3-2 Output Impedance

Ideally, the output impedance of an op amp should be zero. In actuality, each op amp is different and its output impedance may range from 25 to several thousand ohms. For most applications, the output is assumed to be zero and will function as a voltage source capable of providing current for a wide range of loads. With high input impedance and low output impedance, the op amp becomes an impedance-matching device. (See Experiment 7-10.)

1-3-3 Input Bias Current

Theoretically, the input impedance is infinite; therefore, there should not be any input current. However, there do exist small input currents, of the order microamperes down to picoamperes. The average of these two currents is termed the input bias current. This current can cause an unbalance in the op amp, which can affect the output. Generally, the lower the input bias current, the smaller the imbalance will be. Op amps using field-effect transistors (FETs) at the inputs have the least amount of input bias current.

1-3-4 Output-Offset Voltage

Output-offset voltage (error voltage) is caused by the input bias current. When both inputs are the same voltage the output of an op amp should be zero volts. Since this is seldom the case, a small voltage will usually appear at the output. This situation can be minimized or corrected somewhat by offset nulling, which is applying an input-offset voltage or current.

1-3-5 Input-Offset Current

Both input currents should be equal to obtain zero output voltage. However, this is impossible, and there will be an input offset current to maintain the output at zero volts. In other words, to set the output to zero volts, one input may require more current than the other. This offset current may range up to 20 mA. (See Experiment 7-11.)

1-3-6 Input-Offset Voltage

Ideally, the output voltage of an op amp should be zero when the voltage at both inputs is zero. However, owing to the high gain of an op amp, a slight circuit imbalance can cause an output voltage. By applying a small offset voltage at one of the inputs, the output voltage can be brought back to zero. (See Experiment 7-12.)

1-3-7 Offset Nulling

There are different methods to introduce an input-offset voltage to zero or null the output voltage. Op-amp manufacturers have taken this into account and their data sheets usually give the best recommendations for a particular op amp. Figure 1-6 shows how to null a typical op amp. Offset null terminals have been shown in Figures 1-2 and 1-3. The following procedure describes the steps for nulling the output voltage.

1. Make sure that the circuit has all the correct components, including the null circuit. (Null circuits are usually not shown on normal schematic diagrams.)
2. Reduce input signals to zero. If any series input resistor is about 1% larger than the signal source impedance, nothing else need be done at this point. If the series resistor is equal to or less than the source impedance, replace each source with a resistor equivalent to its impedance.
3. Connect the load to the output terminal.
4. Apply DC power and wait a few minutes for the circuit to settle down.
5. Connect a sensitive voltmeter (capable of reading a few millivolts)

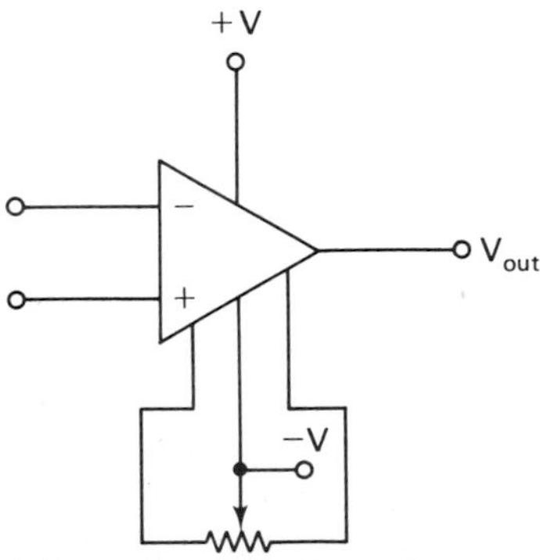

FIGURE 1-6 Offset nulling.

or a DC-coupled oscilloscope across the load to read the output voltage (V_{out}).

6. Adjust the variable resistor until V_{out} reads zero.
7. Remove any added components to the input and reconnect the source inputs, making sure not to touch the offset-voltage adjusting resistor. (See Experiments 7-12 to 7-14.)

1-3-8 Effects of Temperature

Changes in temperature affect all solid-state devices, and op amps are not immune to this problem. DC circuits using op amps tend to be more susceptible than do AC circuits. A change in temperature can cause a change in offset current and offset voltage and is termed drift. A drift caused by temperature will upset any adjusted op-amp imbalance and produce errors in the output voltage.

1-3-9 Frequency Compensation

Because of the op amp's high gain and the phase shift from one internal circuit to another, a point is reached at some high frequency where enough output signal could be fed back to the input and cause oscillations. Frequently, compensation capacitors are added to the op amp, either internally or externally, to prevent these oscillations, by decreasing the op amp's gain as frequency increases.

1-3-10 Slew Rate

Slew rate is the maximum rate of change of the op amp output voltage and can be stated thus:

$$\text{slew rate} = \frac{\text{maximum change in output voltage}}{\text{change in time}} = \frac{\Delta V_{out(max)}}{\Delta t}$$

The 741 general-purpose op amp has a slew rate of 0.5 V/μs, which means that the output voltage can change a maximum of 0.5 V in 1 μs. Capacitance limits this slewing ability and the output voltage will be delayed from the input voltage, as shown in Figure 1-7. Most often, the frequency-compensation capacitor, either internal or external, causes slew-rate limiting in an op amp. At high frequencies or high rates of signal change, slew-rate limiting becomes more pronounced. Slew rate is a large-signal-performance parameter. Slew rate is usually specified at unity gain. Op amps with higher slew rates have wider bandwidths. (See Experiment 7-15.)

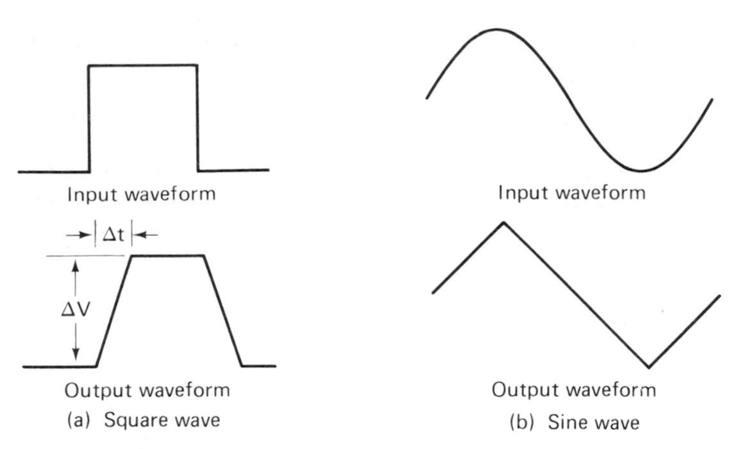

FIGURE 1-7 Example of slew-rate limiting on waveforms.

1-3-11 Frequency Response

The gain of an op amp decreases with an increase in frequency. The gain given by manufacturers is generally at zero hertz or DC. Figure 1-8 shows a voltage-gain versus frequency-response curve. In the open-loop mode, the gain falls off very rapidly as frequency increases. When the frequency increases tenfold, a tenfold decrease in gain results. The breakover point occurs at 70.7% of the maximum gain. The frequency bandwidth is normally considered at the point where the gain falls to 70.7% of maximum.

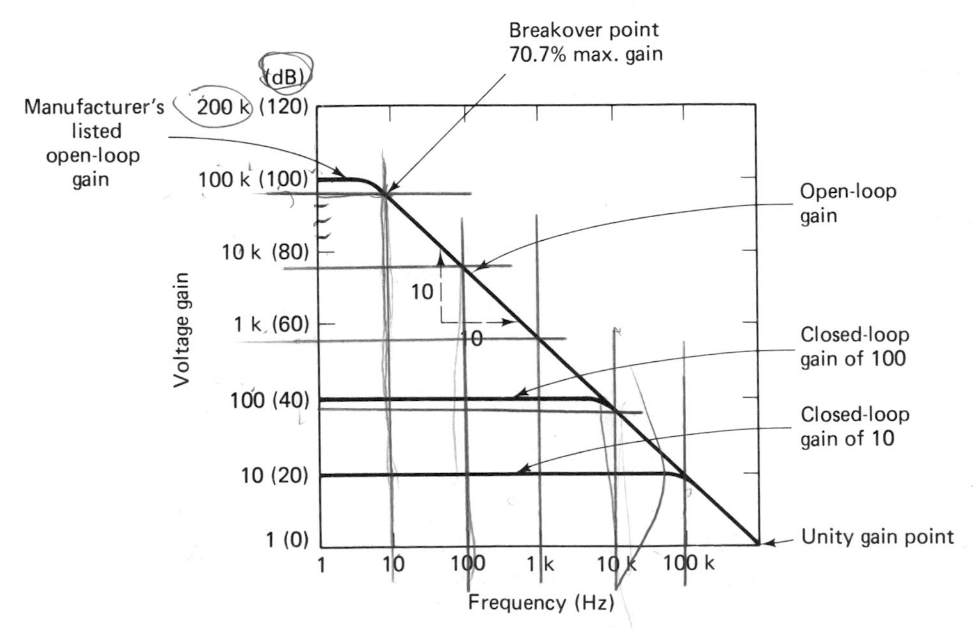

FIGURE 1-8 Voltage gain versus frequency.

Therefore, the open-loop bandwidth is about 10 Hz, for this example. Fortunately, op amps usually require degenerative feedback in amplifier circuits, and this feedback increases the bandwidth of the circuit. For a closed-loop gain of 100, the bandwidth has increased to about 10 kHz. Lowering the gain to 10 increases the bandwidth to about 100 kHz. The unity-gain point occurs at 1 MHz and is called the unity-gain frequency. The unity-gain frequency establishes the reference point at which many op amps are specified by manufacturers. (See Experiments 7-16 to 7-18.)

1-3-12 Gain-Bandwidth Product

The gain-bandwidth product is equal to the unity-gain frequency. It not only tells us the upper useful frequency of a circuit, but allows us to determine the bandwidth for a given gain. For example (referring to Figure 1-8, which shows a frequency-response curve for a frequency-compensated op amp, such as the 741), if you multiply the gain and bandwidth of a specific circuit, the product will equal the unity-gain frequency:

$$\text{gain-bandwidth product} = \text{gain} \times \text{bandwidth} = \text{unity-gain frequency}$$

$$(\text{GBP}) = 100 \times 10 \text{ kHz} = 1{,}000{,}000 \text{ Hz (1 MHz)}$$

or

$$\text{GBP} = 10 \times 100 \text{ kHz} = 1{,}000{,}000 \text{ Hz (1 MHz)}$$

Therefore, if we wanted to know the upper frequency limit or bandwidth of a circuit with a gain of 100, we would divide the unity-gain frequency by gain:

$$\text{bandwidth} = \frac{\text{unity-gain frequency}}{\text{gain}}$$

$$(\text{BW}) = \frac{1{,}000{,}000}{100} = 10 \text{ kHz}$$

1-3-13 Noise

Op amps are susceptible to noise as any other electronic circuit. External noise is generated by electrical devices and the inherent noise of electronic components (resistors, capacitors, etc.), ranging in frequency from 0.01 Hz to the megahertz range. Proper circuit construction techniques can minimize external noise. Internal noise of an op amp results from the internal components, bias current, and drift. Noise is also amplified by the op amp, just as offset voltage and signal voltage. The noise gain is expressed

$$\text{noise gain} = 1 + \frac{R_F}{R_{\text{in}}}$$

Internal noise can be minimized by keeping series input resistors and the feedback resistor as low in value as practically possible to satisfy circuit requirements. Also, bypassing the feedback resistor with a small capacitor ($\approx$3 pF) reduces the noise gain at high frequencies.

1-3-14 Common-Mode Rejection Ratio (CMRR)

Common-mode rejection is a feature associated with differential amplifiers. If the same in-phase voltage is applied to the inputs of the amplifier, the output will be zero. Only a difference of potential at the inputs will produce a voltage at the output. For instance, a 1020-Hz signal is applied to the inverting input of the op amp shown in Figure 1-9. The same frequency is applied to the noninverting input but is 180° out of phase. This is the differential signal. However, the 1020-Hz signal has picked up a 60-Hz hum. The 60 Hz is in phase at the two inputs and represents the common-mode signal. The differential amplifier will tend to reject the 60-Hz common-mode signal while amplifying the differential 1020-Hz signal. The ability of an op amp to amplify the differential signal while rejecting the common-mode signal is called the common-mode rejection ratio (CMRR). This ratio can be expressed as

$$\text{CMRR} = \frac{A_D}{A_{\text{cm}}}$$

where A_D is the differential gain and A_{cm} the common-mode gain. The CMRR is usually expressed in decibels, where the higher the rating, the better the rejection. (See Experiment 7-19.)

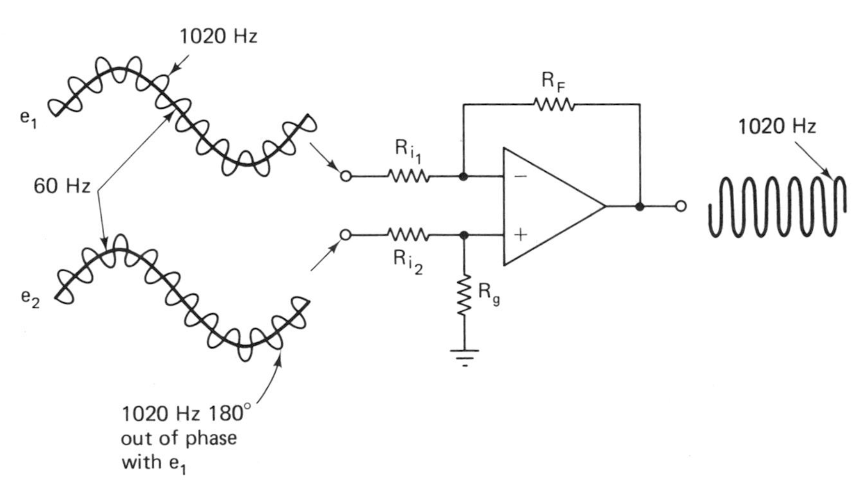

FIGURE 1-9 Common mode rejection.

1-3-15 Short-Circuit Protection

An op amp can produce damaging current if its output is shorted to ground, $+V_c$ supply or $-V_c$ supply, unless it is provided with short-circuit protection. Transistor Q_{15} shown in Figure 1-3 is a current-limiting device that provides this protection. Most newer-type op amps have this short-circuit protection built in, while some older types do not.

1-3-16 Electrical Limitations

Like all solid-state devices, op amps have electrical limitations which must not be exceeded, to ensure proper operation and to prevent possible destruction. These limitations are referred to as *absolute maximum ratings.*

Supply voltage $\pm V$. The safe maximum allowable voltage that can be applied to the device, including both positive and negative supplies.

Power dissipation. The safe amount of power the device is capable of dissipating on a continuous basis while operating within a specified temperature range.

Differential-input voltage. The safe maximum voltage that can be applied between the two inputs without excessive current flow.

Input voltage. The maximum voltage that can safely be applied between either input terminal and ground (circuit common). The magnitude of this input voltage should never exceed the magnitude of the supply voltage (typically 15 V).

Output short-circuit duration. The length of time that the op amp can withstand a direct short circuit from the output terminal to ground or either voltage supply terminal.

Operating-temperature range. The range of temperature over which the op amp will perform within the rated specifications. Commercial-grade devices operate from 0 to +70°C, industrial-grade devices operate from −25 to +85°C, and military-grade devices operate from −55 to +125°C.

Storage-temperature range. The safe range of temperature over which the device can be stored, typically −65 to +150°C.

Lead temperature. The temperature that the device will withstand for a specific period of time during the lead soldering process. This rating is typically 300°C for a period of 10 to 60 s.

Figure 1-10 shows a typical manufacturer's data sheet for a 741 op amp.

absolute maximum ratings

Supply Voltage LM747	±22V
LM747C	±18V
Power Dissipation (Note 1)	800 mW
Differential Input Voltage	±30V
Input Voltage (Note 2)	±15V
Output Short-Circuit Duration	Indefinite
Operating Temperature Range LM747	−55°C to 125°C
LM747C	0°C to 70°C
Storage Temperature Range	−65°C to 150°C
Lead Temperature (Soldering, 10 sec)	300°C

electrical characteristics (Note 3)

PARAMETER	CONDITIONS	LM747			LM747C			UNITS
		MIN	TYP	MAX	MIN	TYP	MAX	
Input Offset Voltage	$T_A = 25°C$, $R_S \leq 10\ k\Omega$		1.0	5.0		1.0	6.0	mV
Input Offset Current	$T_A = 25°C$		80	200		80	200	nA
Input Bias Current	$T_A = 25°C$		200	500		200	500	nA
Input Resistance	$T_A = 25°C$	0.3	1.0		0.3	1.0		MΩ
Supply Current Both Amplifiers	$T_A = 25°C$, $V_S = \pm 15V$		3.0	5.6		3.0	5.6	mA
Large Signal Voltage Gain	$T_A = 25°C$, $V_S = \pm 15V$ $V_{OUT} = \pm 10V$, $R_L \geq 2\ k\Omega$	50	160		50	160		V/mV
Input Offset Voltage	$R_S \leq 10\ k\Omega$			6.0			7.5	mV
Input Offset Current				500			300	nA
Input Bias Current				1.5			0.8	μA
Large Signal Voltage Gain	$V_S = \pm 15V$, $V_{OUT} = \pm 10V$ $R_L \geq 2\ k\Omega$	25			25			V/mV
Output Voltage Swing	$V_S = \pm 15V$, $R_L = 10\ k\Omega$	±12	±14		±12	±14		V
	$R_L = 2\ k\Omega$	±10	±13		±10	±13		V
Input Voltage Range	$V_S = \pm 15V$	±12			±12			V
Common Mode Rejection Ratio	$R_S \leq 10\ k\Omega$	70	90		70	90		dB
Supply Voltage Rejection Ratio	$R_S \leq 10\ k\Omega$	77	96		77	96		dB

Note 1: The maximum junction temperature of the LM747 is 150°C, while that of the LM747C is 100°C. For operating at elevated temperatures, devices in the TO-5 package must be derated based on a thermal resistance of 150°C/W, junction to ambient, or 45°C/W, junction to case. For the flat package, the derating is based on a thermal resistance of 185°C/W when mounted on a 1/16-inch-thick epoxy glass board with ten, 0.03-inch-wide, 2-ounce copper conductors. The thermal resistance of the dual-in-line package is 100°C/W, junction to ambient.

Note 2: For supply voltages less than ±15V, the absolute maximum input voltage is equal to the supply voltage.

Note 3: These specifications apply for $V_S = \pm 15V$ and $-55°C \leq T_A \leq 125°C$, unless otherwise specified. With the LM747C, however, all specifications are limited to $0°C \leq T_A \leq 70°C$ $V_S = \pm 15V$.

FIGURE 1-10 Typical data sheet. (Courtesy of National Semiconductor Corp.)

1-4 POWER-SUPPLY REQUIREMENTS FOR OP AMPS

Most op amps require a dual ± power supply for proper operation. Using this type of power source allows the output of the op amp to swing positive and negative with reference to ground. This feature is particularly useful in DC circuits and special audio applications.

The simplest power source is batteries, as shown in Figure 1-11a. Two 9-V dry-cell batteries can be connected in series, with the common connection being reference ground. The output will be a ±9-V power supply. Adding two more 9-V batteries in series will produce a ±18-V power supply, if needed or desired. Although this battery supply has portability, each battery must be fresh for proper circuit operation.

A single battery, such as an automobile 12-V rechargeable battery, can utilize a resistive voltage divider network to produce a ±6-V power supply, as shown in Figure 1-11b. If this circuit is used in an automotive application, remember that the ground reference for the op-amp circuitry is not the same as the battery ground for the vehicle.

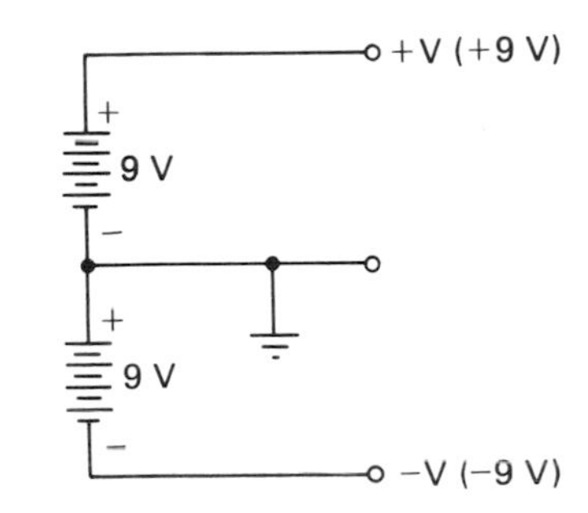

(a) ±9-volt battery power supply

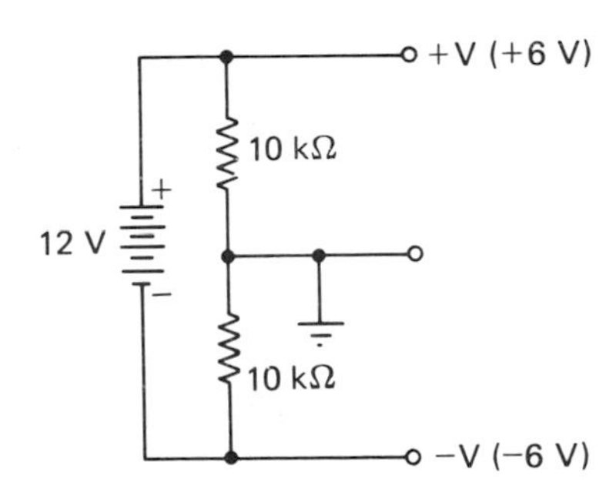

(b) ±6-volt battery power supply

FIGURE 1-11 Battery dual power supplies.

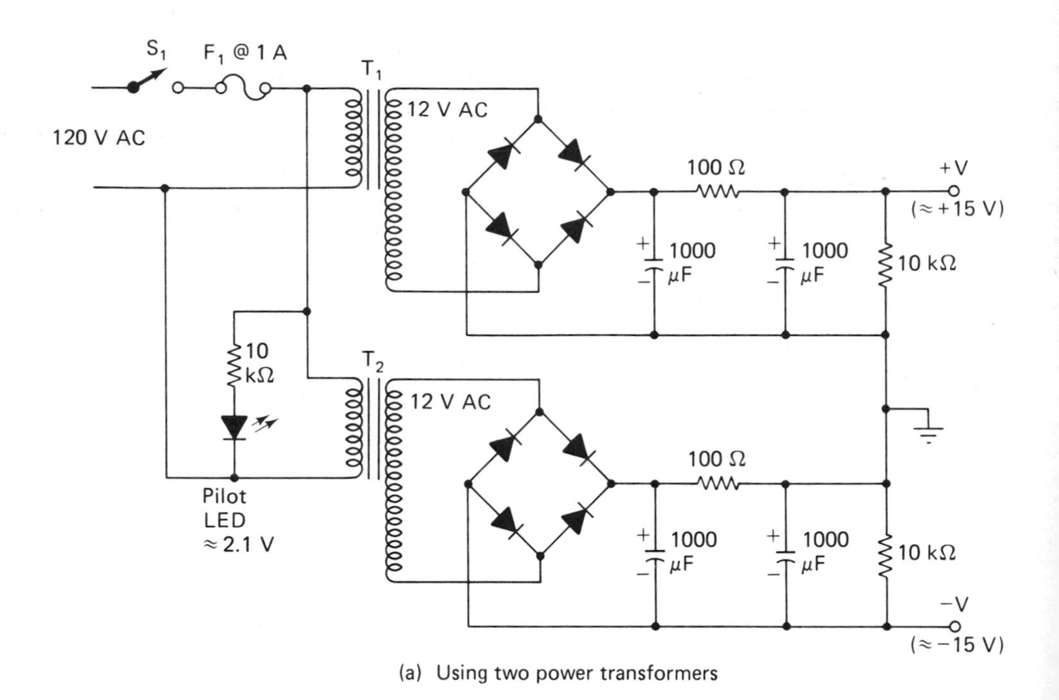

(a) Using two power transformers

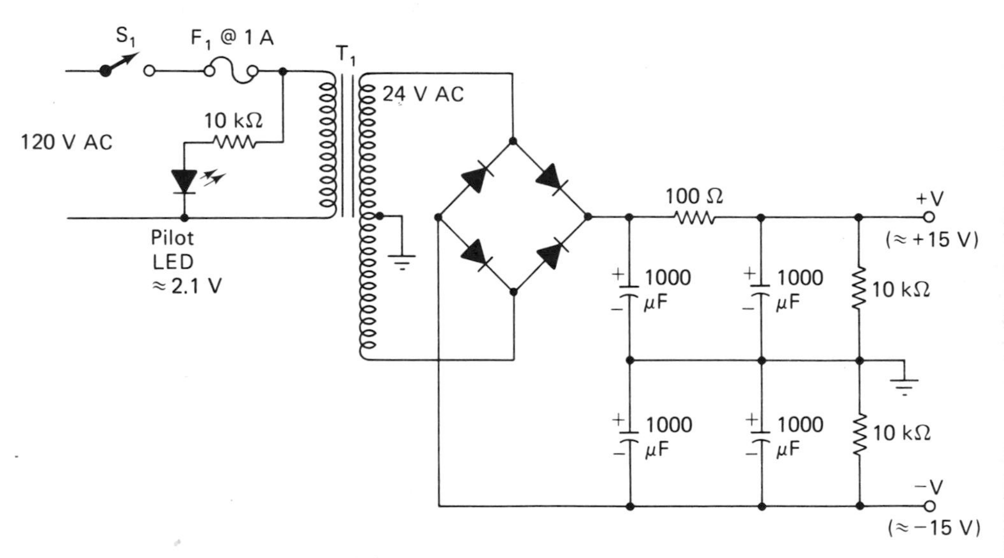

(b) Using a center-tapped transformer

FIGURE 1-12 Dual polarity power supplies.

Battery-operated power supplies have the disadvantage of maintaining the rated voltage up to maximum, either through recharging or replacement. Electronic dual power supplies operating from commercial electric service are shown in Figure 1-12. Two step-down 12-V transformers can be used to construct a ±15-V dual power supply with the ground reference connecting the two supplies together (Figure 1-12a).

The same dual power supply can be constructed more easily and is less bulky and less costly by using a 24-V center-tapped transformer, as shown in Figure 1-12b.

The power supplies discussed provide the basic voltages for op-amp operation. Many op-amp applications require these voltages to be regulated and as free as possible from noise. Basic zener-diode voltage regulation is shown in Figure 1-13a. It is good practice to have V_{in} from the rectifier about 2 V greater than the total regulated voltage. For example, a ±15-V supply would be 30 V from $-V$ to $+V$; therefore, V_{in} should be about 32 V.

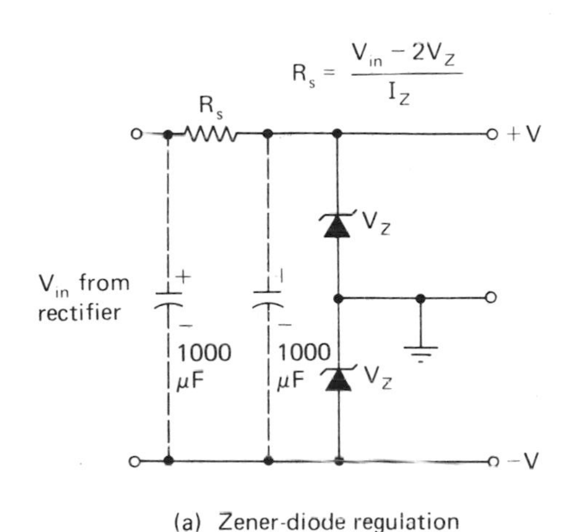

(a) Zener-diode regulation

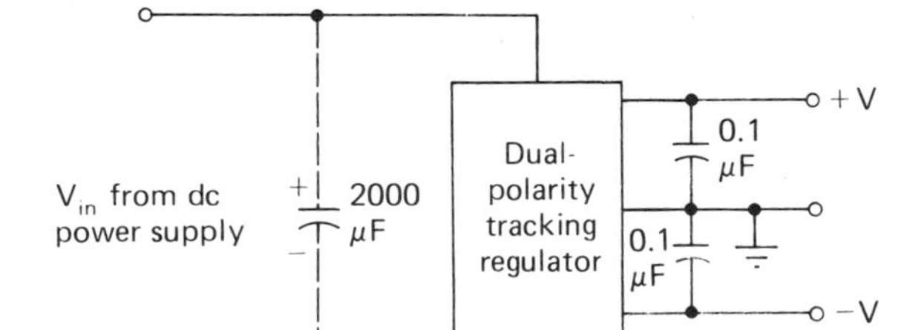

(b) Dual-polarity IC voltage regulator

FIGURE 1-13 Dual polarity voltage regulation.

The value of the limiting resistor R_s (the resistor of the pi-type filter in Figure 1-12) can be found by the formula

$$R_s = \frac{V_{in} - 2V_z}{I_z}$$

where V_z is the zener-regulated voltage of each diode and I_z the current through the diodes without a load on the power supply. Generally, I_z can be set for half of the maximum allowable current. For example, $I_{z_{max}}$ for a 1-W 15-V zener diode can be found by the formula

$$I_{z_{max}} = \frac{P_{max}}{V_z} = \frac{1\text{ W}}{15\text{ V}} = 66.6\text{ mA}$$

Then

$$I_z = \frac{I_{z_{max}}}{2} = 33.3\text{ mA}$$

Of course, I_z can be set higher, depending on the load placed on the power supply, but it should never exceed $I_{z_{max}}$ or there is a definite possibility of destroying the zener diodes when there is no load current present.

For precision op-amp applications requiring very stable power supplies, an IC dual-polarity-tracking voltage regulator may be used, as shown in Figure 1-13b. These types of voltage regulators provide different voltages (LM325 = ±15 V, LM326 = ±12 V) and should be connected to the power supply according to the manufacturer's specification sheet.

In critical applications of low-level inputs and high gain, the power-supply requirements are extremely stringent. Voltage changes or noise on the power-supply lines can be coupled into an op amp and appear as equivalent input signals. Manufacturers take this into account in the design of op amps by minimizing these effects. The ability of an op amp to reject power-supply-induced noise and drift is called the power-supply rejection ratio (PSRR). Stated another way, the PSRR is the ratio of the change in input offset voltage to the change in power-supply voltage producing it. Ratings may be given for each power supply separately or together, giving a typical value and a maximum limit. The unit of measure may be μV/V or dB. (See Experiment 7-20.)

Some special op amps are designed to operate from single-voltage (usually positive) power supplies. Standard op amps may in some limited applications use a single power supply. In the quiescent state (no input signals present) the output should measure about halfway between ground and the maximum voltage supply.

1-5 TYPES OF OP AMPS AND PACKAGE CONFIGURATIONS

Op amps have been around for quite a number of years. But it was not until 1963, when Fairchild Semiconductor introduced the first usable IC op amp, the μA702, that the present trend in op amp usage was set. Since that time, solid-state manufacturers have improved and developed a vast amount of diversified types of op amps into the broad spectrum of devices available today. Some op amps have become so specialized that they are fabricated into specific circuit configurations, such as voltage comparators and voltage followers. The characteristics of op amps are constantly being improved, but regardless of the various manufacturers, op amps having similar characteristics are placed into group and/or family types.

General-purpose op amps (group I). General-purpose op amps have a gain–bandwidth product of approximately 1 MHz, a fairly high gain, input impedance of several megohms, and operate from power supplies of approximately ±5 to ±22 V.

The 709 family. The Fairchild μ709 IC op amp had significant improvements over the μA702, so that it became the first recognized industry standard and is still in use today. The μ702 had very limited common-mode input range, relatively low voltage gain, and use odd supply voltages, such as +12 V and −6 V. The μA709 overcame many of these problems and operates from a ±15-V supply. It has an input resistance of about 250 kΩ, an output resistance of about 150 Ω, a voltage gain of about 45,000, but no output short-circuit protection. It also had the problem of latch-up, where certain values of common-mode input signals would drive the output voltage to some level where it would remain. Some members of the 709 family are the Fairchild μ709, Motorola MC 1709, National Semiconductor LM709, and Texas Instruments SN72709.

The 101 family. The next evolutionary step came in 1967, when National Semiconductor Corporation introduced the LM 101. The 101 design solved many of the 709 problems, and in addition has an increased gain to 160,000 and a useful power-supply range from ±5 V to ±20 V. Op amps 101A, 107, and 301A belong to the 101 family.

The 741 family. In 1968, Fairchild Semiconductor introduced the μA741, the first internally compensated IC op amp. This op amp offers many features, which make their application nearly foolproof: overload protection on the input and output, no latch-up when the common-mode range is exceeded, and freedom from oscillations in most standard circuit configurations. The 741 is probably the most widely used industry standard today. It is readily available at most electronics component stores. This family includes the 741A, the 747 dual op amp, the 748, the LM 148 (quad-741 op amp), and the 1558 dual op amp.

The other numerous general-purpose op amps available have improved characteristics or are specialized versions of these basic families mentioned. Other groups and families emphasize extremely high input impedances. Some offer very wide bandwidth with high slew rates, while others are designed to operate at high voltages and currents. A general outline of the various groups is given in Table 1-1.

TABLE 1-1 Uses and features of op-amp groups.

Group	*Use*	*Salient Features*
I	General purpose	DC up to 1-MHz bandwidth
II	DC and low-level performance	Extremely high input impedance, low bias current, etc.
III	AC and high-level performance	Wide bandwidth and high slewing rates
IV	High voltage and power	Capable of driving loads directly
V	Unique devices	Special op amps, such as programmable, digitally addressable, etc., types

IC op amp package configurations. Integrated-circuit op amps are available in four commonly used packages, as shown in Figure 1-14a. The metal TO-type package is approximately 0.300 to 0.450 in. in diameter and 0.130 to 0.185 in. high, with 8 to 10 leads coming out of the bottom. The flat package has a body about 0.250 to 0.270 in. square and 0.050 to 0.070 in. thick, usually with five leads coming out each side. This package may be of metal, glass, or plastic. The dual-in-line package (DIP) can be made of metal, glass, ceramic or plastic. It measures about 0.750 in. long, 0.270 in. wide, and 0.190 in. thick, with seven leads protruding downward from each side. The mini-DIP is about half the size of a standard 14-pin DIP and has four leads protruding downward from each side.

The lead identification shown in Figure 1-14b is usually self-explanatory. The positive supply voltage is connected to the $+V$ terminal, and the negative supply voltage is connected to the $-V$ terminal. Input and output terminals are clearly indicated. The balance terminals (sometimes designated Offset Null) are connected to a potentiometer for null adjusting. Terminals marked NC means no connection and are included for physical ruggedness of the package.

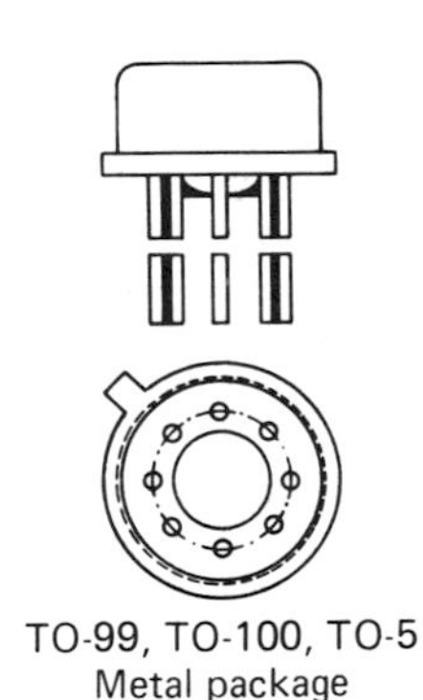

TO-99, TO-100, TO-5
Metal package

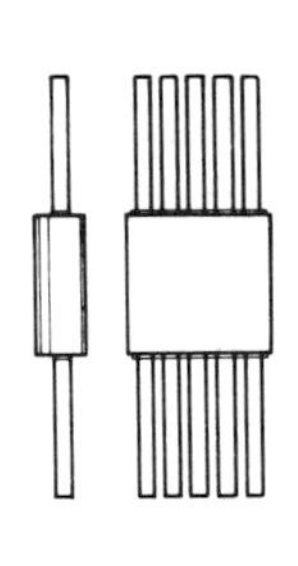

Plastic, ceramic
glass/metal
flat-pack

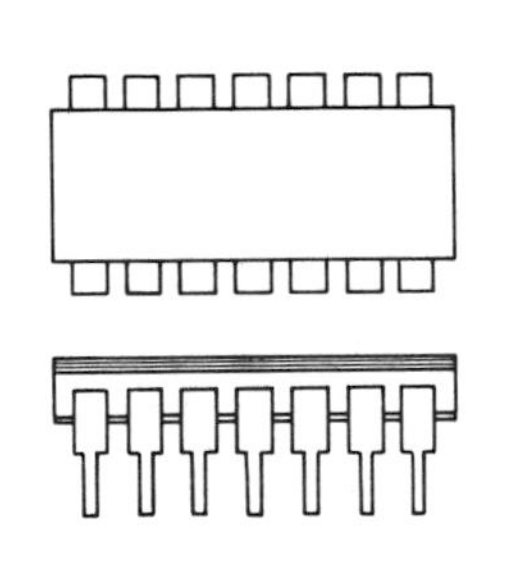

Plastic, ceramic
glass/metal DIP

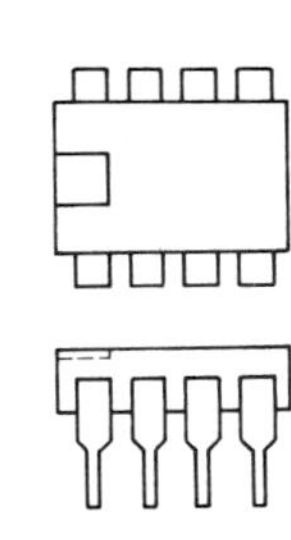

Mini-DIP

(a) IC packages — Top/bottom and side views

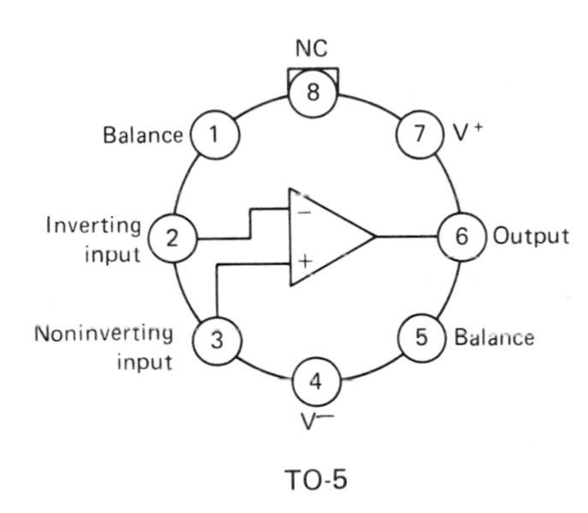

TO-5

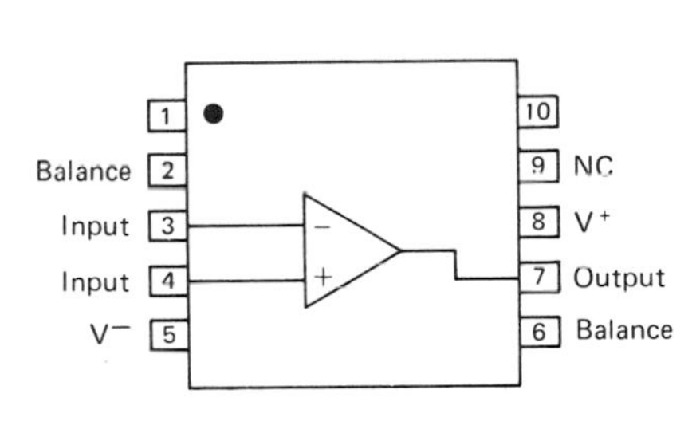

Flat package

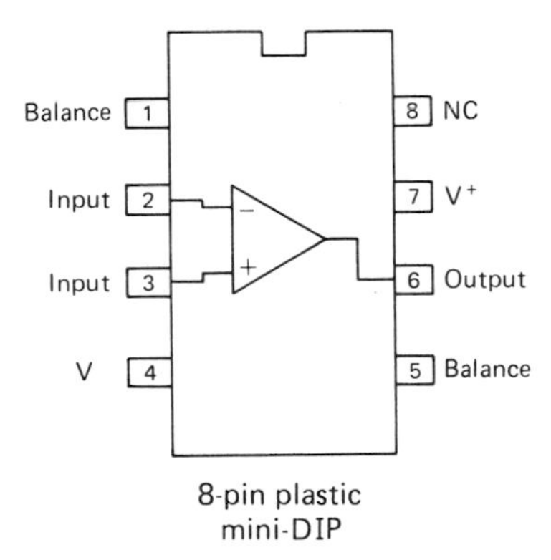

8-pin plastic
mini-DIP

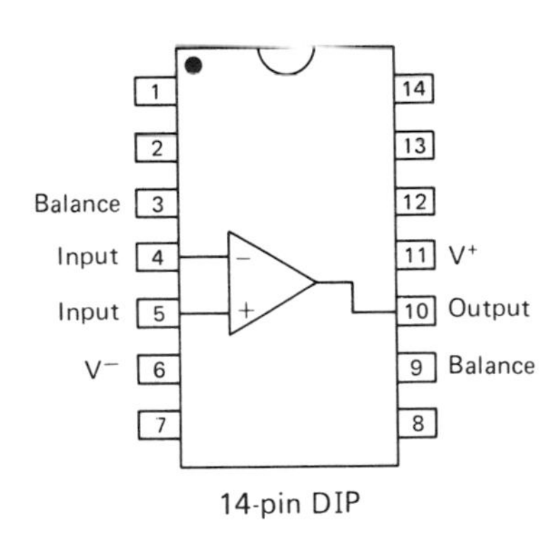

14-pin DIP

(b) Single op-amp IC packages

FIGURE 1-14 IC op-amp packages.

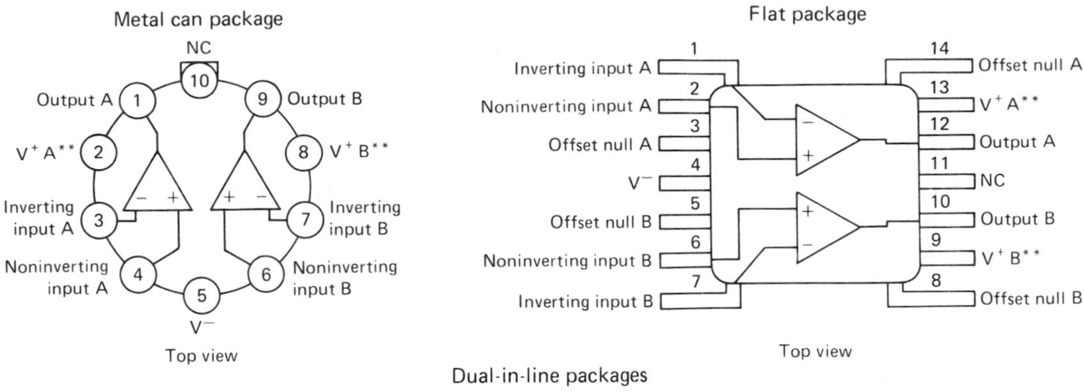

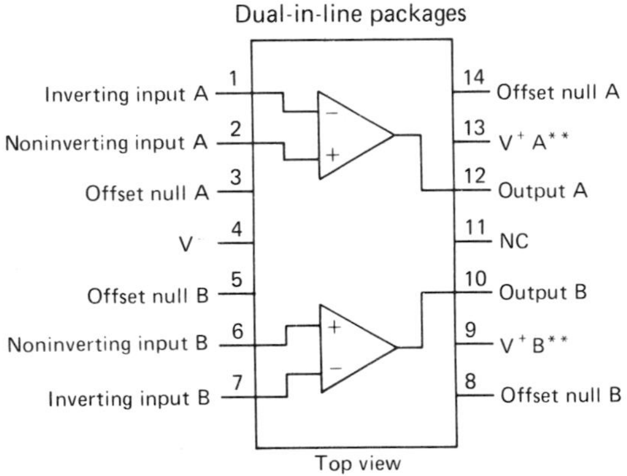

(c) Dual op-amp IC packages

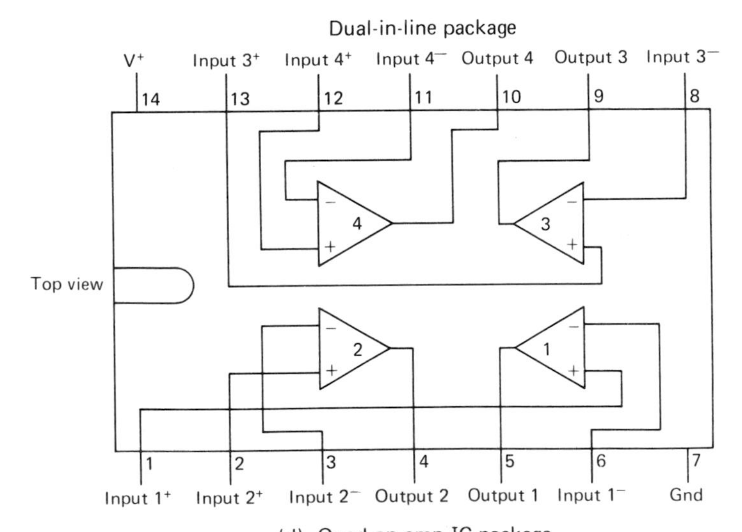

(d) Quad op-amp IC package

FIGURE 1-14 *(Continued)*

SUMMARY POINTS

1. Op amps have very high input impedance.
2. Op amps have very high open-loop gain.
3. Op amps have low output impedance.
4. Most op amps operate from dual (±) power supplies.
5. Controlled closed-loop gain is determined by the ratio of the feedback resistor to the input resistor at the inverting input and is expressed $A_v = R_F / R_{in}$.
6. When the inverting input is more positive than the noninverting input, the output will be negative, and vice versa.
7. Input-offset voltage can be compensated by null adjusting (the output to zero) with external circuitry.
8. Op amps may be externally or internally frequency-compensated to prevent self-oscillation.
9. The slew rate is the maximum change in output voltage to the change in time, expressed $SR = \Delta V_{out(max)} / \Delta t$.
10. Op amps with high slew rates have wider bandwidths.
11. In the open-loop mode, gain falls off rapidly with frequency increase, resulting in a very narrow bandwidth.
12. The greater the feedback in closed-loop mode, the wider the bandwidth.
13. The gain-bandwidth product is gain times bandwidth, which equals the unity-gain frequency: $GBP = G \times BW$.
14. Noise and temperature affect all electronic circuits and must also be taken into consideration when using op amps.
15. The common-mode rejection ratio (CMRR) is the ability of an op amp to amplify the differential input signal while rejecting the common-mode input signal, expressed $CMRR = A_D / A_{cm}$.
16. Op amps must be operated within their absolute maximum ratings to ensure circuit dependability.
17. There are five general groups of op amps: general purpose, DC and low-level performance, AC and high-level performance, high voltage and power, and unique or special devices.
18. Op amps come in four general packages: metal TO-group, flat pack, DIP, and mini-DIP.

19. An IC package may contain one, two, or four op amps.
20. The op amp is a versatile solid-state device, because it can be used in so many types of circuits.

TERMINOLOGY EXERCISE

Write a brief definition for each of the following terms:

1. Open-loop gain
2. Closed-loop gain
3. $A_V = -R_F/R_{in}$
4. $A_V = R_F/R_{in} + 1$
5. Output-offset voltage
6. Input-offset voltage
7. Input-offset current
8. Offset nulling
9. Op-amp frequency compensation
10. Slew rate
11. Frequency response
12. Unity gain
13. GBP
14. CMRR
15. Noise gain (show formula)

PROBLEMS AND EXERCISES

1. Draw the block diagram of an op amp.
2. Draw the schematic symbol of an op amp and identify all parts.
3. What is the controlled gain of an op amp when R_{in} = 2.2 kΩ and R_F = 68 kΩ?

4. What is the controlled gain of an op amp when R_{in} = 4.7 kΩ and R_F = 4.7 kΩ?
5. Draw a schematic symbol of an op amp, indicating offset nulling.
6. Draw two examples of slew-rate limiting on voltage waveforms.
7. What is the gain–bandwidth product of an op-amp circuit with a gain of 45 and a bandwidth of 50 kHz.
8. What is the upper frequency limit of an op-amp circuit with a gain of 25 and a unity-gain frequency of 1 MHz?
9. Draw a voltage gain versus frequency graph for an op amp with a unity-gain frequency of 1 MHz and an open-loop gain of 200,000. Indicate a closed-loop gain of 10,000 and 100.
10. Find the value of R_s for a zener diode regulator for a ± 12-V supply where I_z = 80 mA and V_{in} = 30 V.

SELF-CHECKING QUIZ

Match each expression in column A with its best associated meaning or related expression in column B. (Some items in column A may have two answers.)

Column A	*Column B*
1. Open-loop mode has	a. feedback
2. Closed-loop mode has	b. gain × bandwidth
3. Controlled gain accomplished by	c. narrow bandwidth
4. Input-offset voltage corrected by	d. high
5. Self-oscillation eliminated by	e. increases bandwidth
6. High slew rate	f. wider bandwidth
7. Input impedance	g. two inputs
8. Output impedance	h. internal compensation
9. Differential amplifier has	i. nulling
10. GBP	j. low
	k. controlled gain
	l. high gain

True-or-False Questions

11. A positive voltage on the noninverting input will cause the output to swing negative.
12. The op amp consists basically of a differential amplifier, a high gain voltage amplifier, and a low impedance output amplifier.
13. The closed-loop-gain formula is $A_v = R_{in}/R_F$.
14. Input-offset voltages can cause error in the output voltage.
15. An op amp has a unity-gain frequency of 4 MHz. For a gain of 50, its bandwidth is 80 kHz.
16. With a gain of 60 and a bandwidth of 100 kHz, the unity-gain frequency of a particular op amp is 6 MHz.
17. The formula for slew rate is $SR = \Delta V/\Delta t$.
18. Self-oscillations of an op amp can be reduced or stopped by offset nulling.
19. The higher the common-mode rejection ratio, the better the quality of an op amp.
20. Using a dual (±) power supply, the output of an op amp can swing positive and negative.

(Answers at back of book)

Chapter 2

Basic Op-Amp Circuits

This chapter concentrates on the principles involved with basic op-amp circuits. Understanding these principles will provide you with the necessary foundation to use and test more complex op-amp circuits.

We will begin with a basic voltage comparator circuit that uses the open-loop gain of the op amp. This circuit configuration will show you how both inputs are actively used. It will then be shown how comparators are used for AC sensing and voltage-level detection.

More-in-depth information is given about amplification using the op amp and how the external resistors determine the gain for a particular circuit.

You will be able to understand how the op amp amplifies in the inverting and noninverting configurations. Easy-to-use formulas allow you to construct circuits that will perform to your expectations.

Special circuits, such as the summing amplifier and the difference amplifier, are presented to show you some of the versatility of op amps.

The popular 741 op amp is shown in most of the illustrations to enable you to construct these circuits right from the book.

2-1 VOLTAGE COMPARATOR

A voltage comparator compares the voltage on one input to the voltage on the other input. Figure 2-1 shows the basic voltage comparator. In this

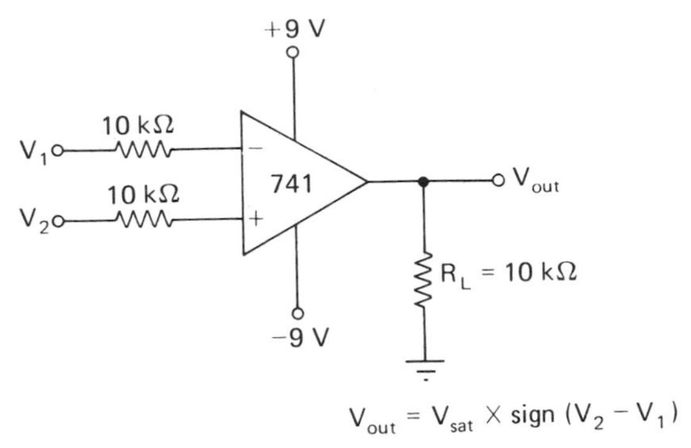

$V_{out} = V_{sat} \times \text{sign}\,(V_2 - V_1)$

(a) Schematic diagram

Input voltage		Output voltage
V_1	V_2	$\pm V_{sat}$
+1	+2	+8
+2	+1	−8
0	0	0
+1	−1	−8
−1	+1	+8
−1	−2	−8
−2	−1	+8

(b) Input/output voltage table

FIGURE 2-1 Voltage comparator: (a) schematic diagram; (b) input/output voltage table.

simplest circuit configuration, the open-loop mode, any minute difference between the two inputs will drive the op-amp output into saturation. The direction in which the output goes into saturation is determined by the polarity of the input signals. When the voltage on the inverting input is more positive than the voltage on the noninverting input, the output will swing to negative saturation ($-V_{sat}$). Similarly, when the voltage on the inverting input is more negative than the voltage on the noninverting input, the output will swing to positive saturation ($+V_{sat}$). Referring to the table in Figure 2-1, you can see that with +1 V on the inverting input and +2 V on the noninverting input, the former is 1 V negative with respect to the latter. Therefore, the output will go into positive saturation. When the input voltages are reversed (+2 V on −input and +1 V on +input), the inverting input is 1 V positive with respect to the noninverting input and the output goes into negative saturation. If the input voltages are the same amplitude and polarity, the output will be zero. Negative voltages applied to the inputs have the same effect on the output of the op amp as shown in the table.

Remember that the polarity relationship of the inverting input to the noninverting input will cause the output to be 180° out of phase.

2-1-1 Sensing a Sine Wave on the Inverting Input

A comparator can be used to detect a changing voltage on one input with a fixed reference voltage on the other input. In Figure 2-2, the inverting input

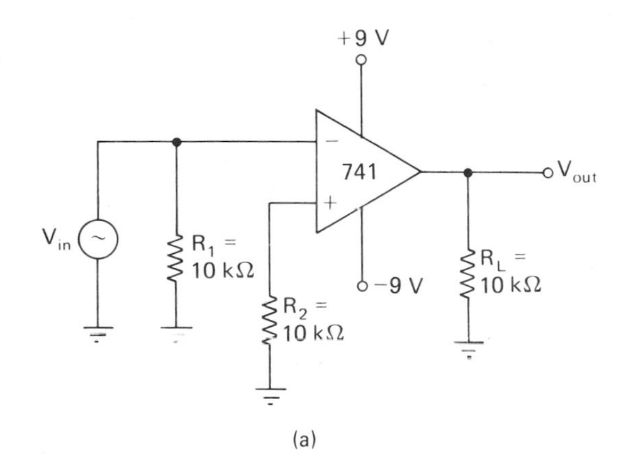

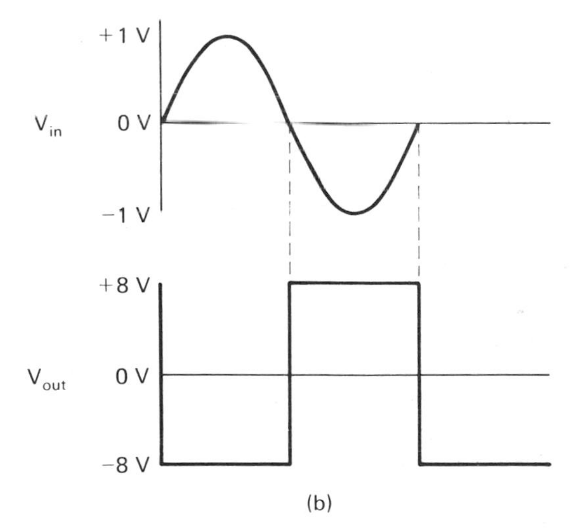

FIGURE 2-2 Comparator sensing a sine wave on the inverting input: (a) schematic diagram; (b) input/output voltage relationship.

is used to sense a sine-wave signal. A signal source is placed on the inverting input. Since the input impedance of the op amp could be considered infinite, resistor R_1 acts as the signal source load, which results in more effective operation of the circuit. The noninverting input is grounded through resistor R_2. This resistor is used to balance the inputs for any input offset current that might exist.

The noninverting input is at the reference voltage (0 V). During the positive alternation of the input signal, the output is at $-V_{sat}$. At the point where the signal goes through zero to the negative alternation, the output swings to $+V_{sat}$. Notice that the output is 180° out of phase with the input.

2-1-2 Sensing a Sine Wave on the Noninverting Input

We can place the signal source on the noninverting input as shown in Figure 2-3. The inverting input is now the 0 V reference point. When the positive alternation of the signal is present, the output is at $+V_{sat}$. Similarly, when the signal goes through zero to the negative alternation, the output swings to $-V_{sat}$. With this circuit configuration, the output is in phase with the input.

These two circuits are sometimes referred to as zero detectors. Each time the input signal crosses zero, the output voltage swings to the opposite polarity. It is also interesting to note how a square wave can be produced from a sine wave with these circuits.

2-1-3 Detecting Phase Difference

A comparator may be used to detect the phase difference of two signals with the same frequency as shown in Figure 2-4. Whenever the two signals are out of phase, there will be a differential voltage at the inputs and the output will be at $\pm V_{sat}$. When V_1 is more positive than V_2, the output will be at $-V_{sat}$, and vice versa. As the two signals become in phase, the output goes to zero (the effect of common-mode rejection).

2-1-4 Positive-Voltage-Level Detector

A specific voltage level other than zero can be detected by a comparator, as shown in Figure 2-5. This positive-voltage-level detector uses the inverting input to sense the changing voltage, while a resistive voltage divider network establishes the reference voltage (V_{ref}) at the noninverting input. The resistive divider is connected between the $+V$ supply and ground. The reference voltage can be determined by the ratio formula

$$V_{ref} = \frac{R_2}{R_2 + R_3}(+V)$$

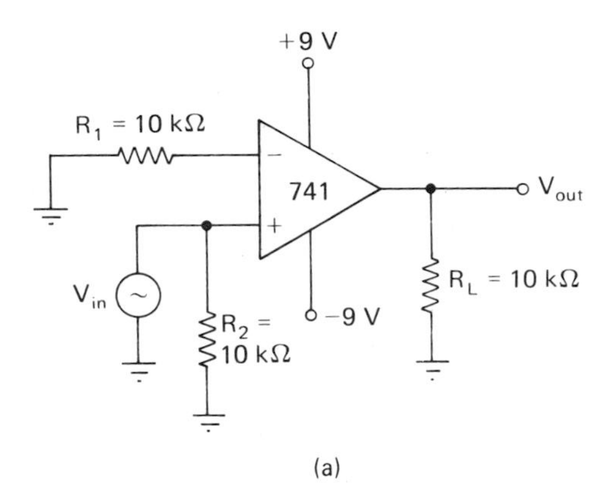

(a)

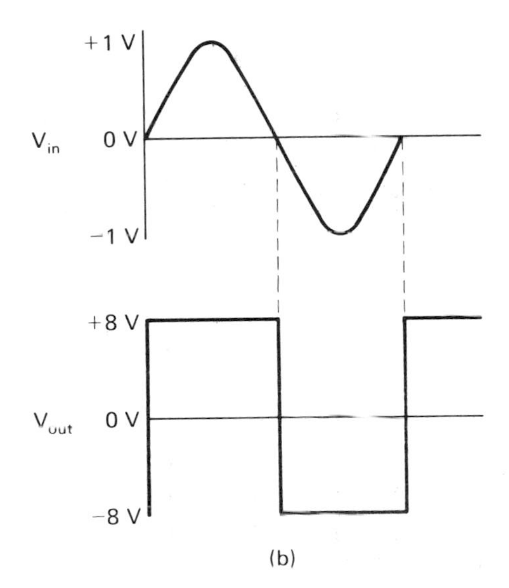

(b)

FIGURE 2-3 Comparator sensing a sine wave on the non-inverting input: (a) schematic diagram; (b) input/output voltage relationship.

By substituting the values of R_2, R_3, and $+V$ in the formula, we have

$$V_{ref} = \frac{10\ \text{k}\Omega}{10\ \text{k}\Omega + 22\ \text{k}\Omega}\ (+9\ \text{V}) = 0.3125\ (+9\ \text{V}) = +2.8\ \text{V}$$

The noninverting input is +2.8 V above ground. As long as the changing voltage on the inverting input is below (negative to) +2.8 V, the output will be at $+V_{sat}$ ($\approx$+8 V). The instant that the voltage on the inverting input becomes greater (more positive) than +2.8 V, the output will swing to

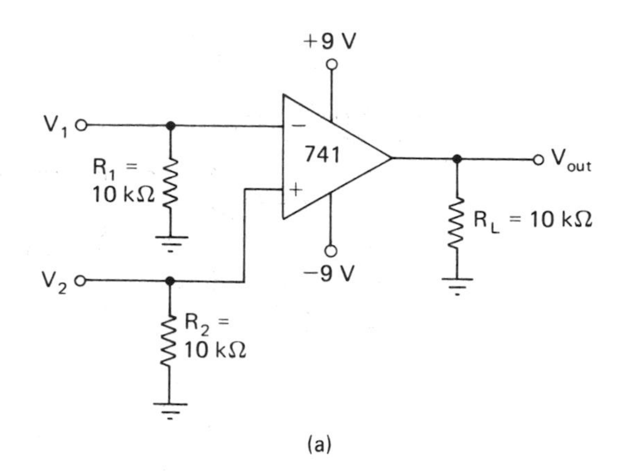

(a)

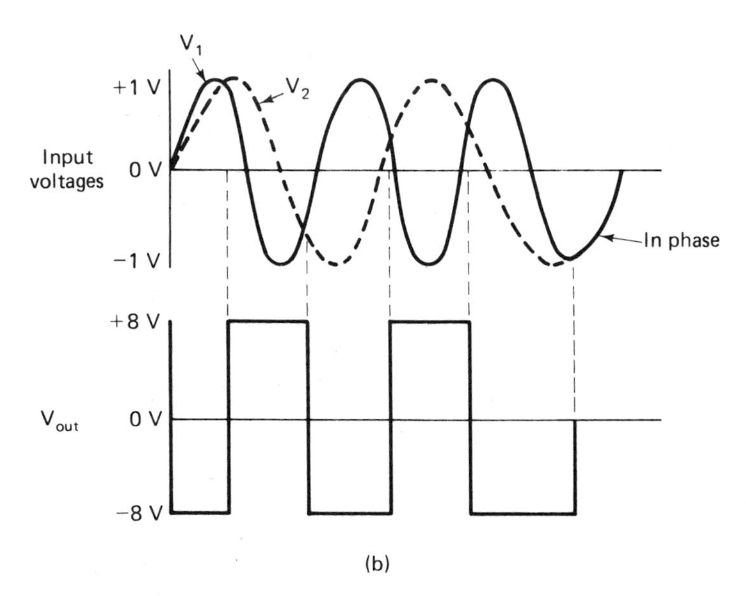

(b)

FIGURE 2-4 Comparator sensing an out-of-phase sine wave on both inputs: (a) schematic diagram; (b) input/output voltage relationship.

$-V_{sat}$ (≈ -8 V), indicating that the comparator has detected a +2.8-V level. If this voltage on the inverting input falls below +2.8 V, the output will again swing to $+V_{sat}$.

2-1-5 Negative-Voltage-Level Detector

A negative-voltage-level detector can be constructed by connecting the resistive divider between $-V$ supply and ground, as shown in Figure 2-6. In this case, the V_{ref} is -2.8 V with reference to ground. If the changing voltage

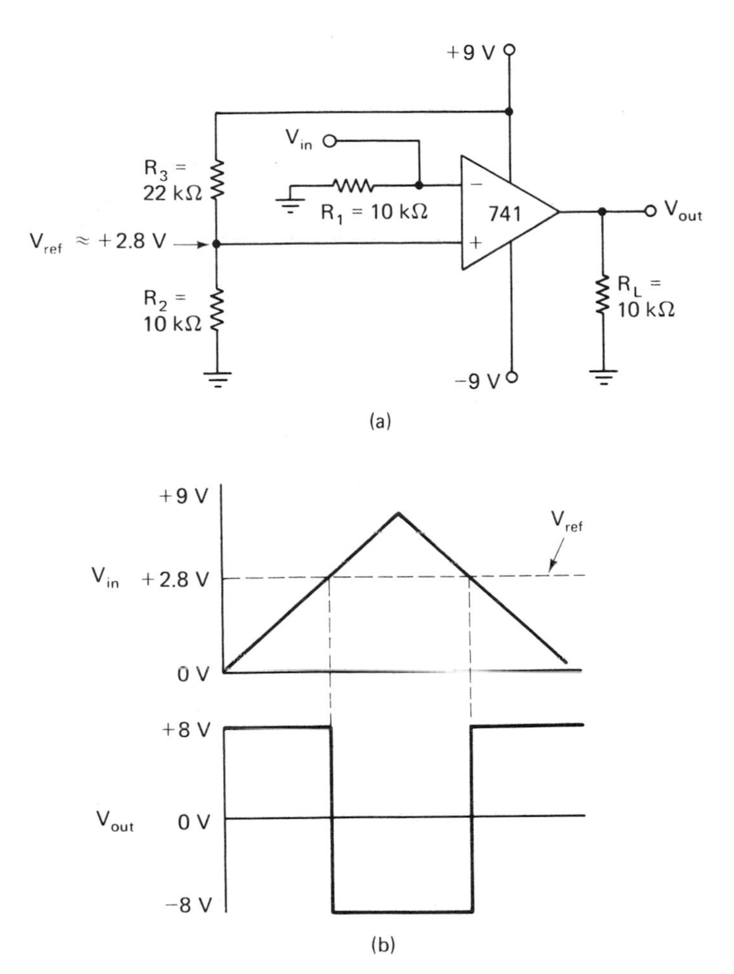

FIGURE 2-5 Positive voltage level detector: (a) schematic diagram; (b) input/output voltage relationship.

on the inverting input is more positive than the V_{ref}, the output will be at $-V_{sat}$. The instant detection occurs, the output will swing to $+V_{sat}$.

Voltage-level detectors can be designed with the noninverting input as the sensing input and the reference voltage applied to the inverting input. The output voltage will swing in opposite directions to the detectors that have been discussed.

2-1-6 Determining Resistive Voltage Divider

One method of determining the resistive voltage divider for the reference voltage is illustrated in Figure 2-7. First, you would select or determine

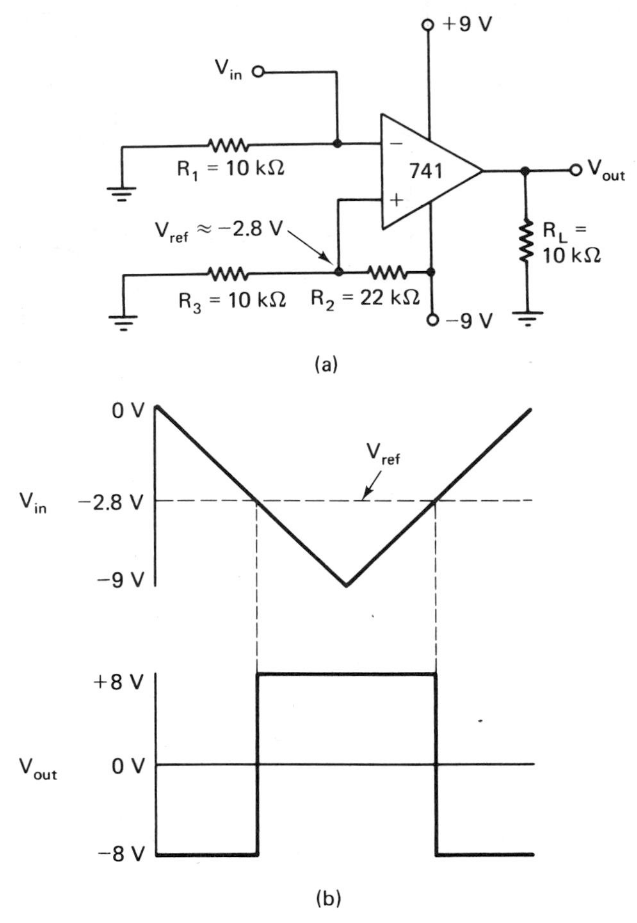

FIGURE 2-6 Negative voltage level detector: (a) schematic diagram; (b) input/output voltage relationship.

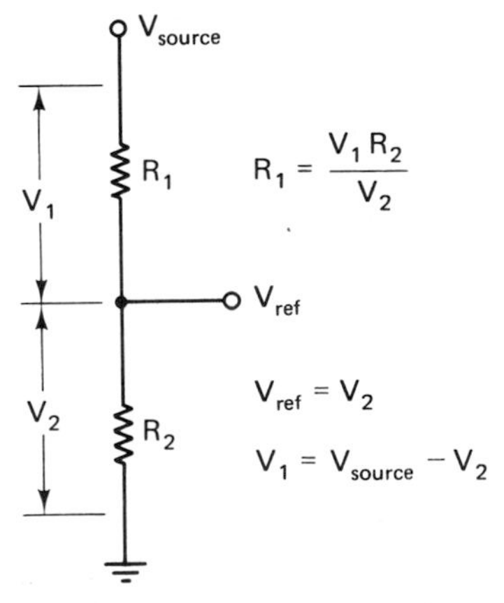

FIGURE 2-7 Determining resistor divider for V_{REF}.

the V_{ref}. This is the same as the voltage drop V_2 across resistor R_2. Voltage V_1 is found by subtracting V_2 from V_{source}. Resistor R_2 is one input resistor to the op amp and is generally the same value as the other input resistor. Since we select V_{ref}, assume the value of R_2, and are able to find V_1, we only need to solve for R_1.

If the voltage drops and resistance are directly proportional:

$$\frac{V_1}{V_2} = \frac{R_1}{R_2}$$

then, rearranging the formula, we find

$$R_1 = \frac{V_1 R_2}{V_2}$$

(See Experiment 7-21.)

2-2 INVERTING AMPLIFIER

An amplifier accepts a small voltage or current at its input and produces a larger voltage or current at its output. An op-amp amplifier has relatively linear gain and the output is controlled as a function of the input. The basic op-amp inverting amplifier is shown in Figure 2-8.

In Chapter 1 you were introduced to the concept of closed-loop mode of operation for the op amp. It was mentioned how the gain of the op amp can be controlled or determined by an external resistive ratio network in the closed-loop mode.

A closed-loop arrangement of this type is called negative (degenerative) feedback. The out-of-phase voltage at the output is fed back to the inverting input, where it tends to cancel the original input voltage. The feedback voltage will greatly reduce the effects of the input voltage and keep the inverting input at nearly 0 V. Of course, the feedback voltage cannot completely cancel the input voltage or there would be no feedback. In other words, nothing at all would happen in the circuit. However, there is a slight change of perhaps a few microvolts at the inverting input. This change is amplified by the extremely high gain of the op amp to produce the voltage change at the output.

The voltage gain of a circuit is found by the formula

$$A_v = \frac{V_{out}}{V_{in}}$$

and the gain factor in the closed-loop mode for an inverting amplifier was given as

$$A_v = -\frac{R_F}{R_{in}}$$

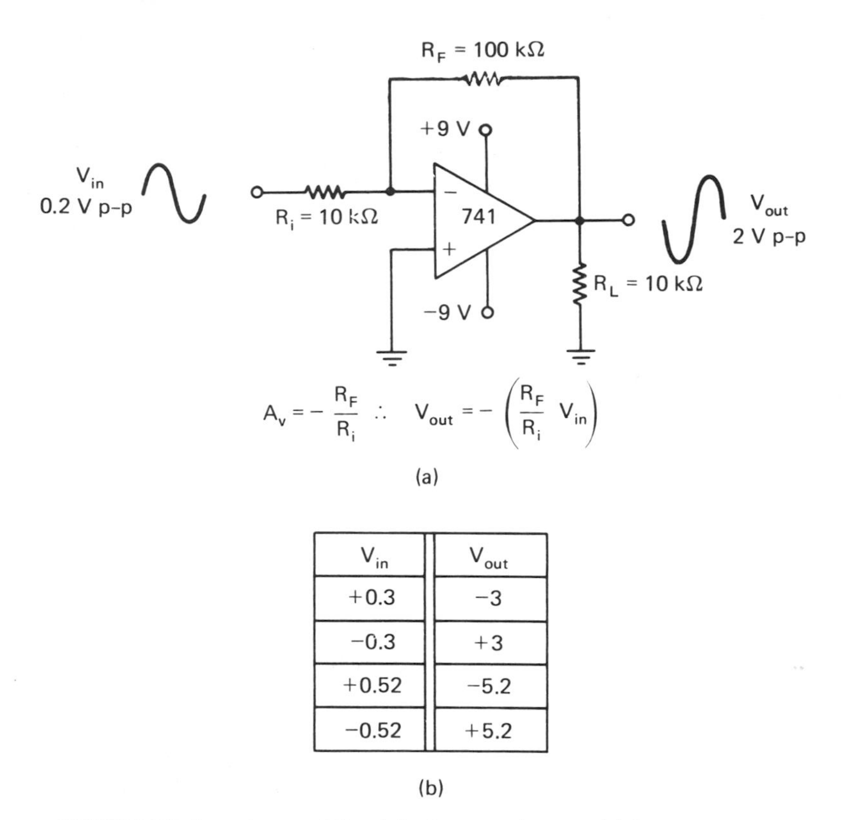

V_{in}	V_{out}
+0.3	−3
−0.3	+3
+0.52	−5.2
−0.52	+5.2

(b)

FIGURE 2-8 Inverting amplifier: (a) schematic diagram; (b) DC voltage table.

The output voltage can be determined if the known input voltage is multiplied by the gain factor:

$$V_{\text{out}} = -(A_v V_{\text{in}}) \quad \text{or} \quad -\left(\frac{R_F}{R_{\text{in}}} V_{\text{in}}\right)$$

The minus sign is disregarded in calculations and indicates only that the output is out of phase with the input.

As an example, referring to Figure 2-8, the gain is

$$-A_v = -\frac{R_F}{R_{\text{in}}} = -\frac{100\text{ k}\Omega}{10\text{ k}\Omega} = -10$$

and the output voltage is

$$V_{\text{out}} = -A_v V_{\text{in}} = -10 \times 0.2\text{ V p-p} = -2\text{ V p-p}$$

Also shown in Figure 2-8 is a sample DC voltage table. The input voltage is amplified by the gain factor of 10, and the output voltage indicates this gain together with the proper polarity for the inverting amplifier. This basic inverting amplifier circuit is the starting point for other more specific DC and AC amplifiers to be covered in subsequent chapters.

A closer analysis of this circuit will enable you to understand how the resistive ratio network (R_{in} and R_F) actually determine the closed-loop gain. Referring to Figure 2-9, you can see that a +1 V at the input will cause a current of 0.1 mA to flow by the formula

$$I_{in} = \frac{V_{in}}{R_{in}} = \frac{1\text{ V}}{10\text{ k}\Omega} = 0.1\text{ mA}$$

(Electron flow is indicated for the examples in this book; however, conventional current flow can also be used depending on the reader's preference.)

Now, since the input impedance of an op amp is considered very high or even infinite, no current can flow into or out of the input terminals for most practical purposes. Therefore, I_{in} must flow through R_F and is indicated by I_F (the feedback current). Since R_{in} and R_F are in series, then $I_{in} = I_F$. However, R_F is 10 times greater than R_{in}, and if the same current is supposed to flow through both resistors, then the voltage drop across R_F must be 10 times greater. The internal circuitry of the op amp adjusts accordingly and the output swings to −10 V (notice that the polarity ensures that current will flow in the same direction). Because no current can flow into or out of the inverting input and R_{in} and R_F drop the total voltage from V_{in} to V_{out}, this point is at 0 V, commonly referred to as virtual ground. The output voltage (V_{out}) is across R_F, and I_F can be proven by the formula

$$I_F = -\frac{V_{out}}{R_F}$$

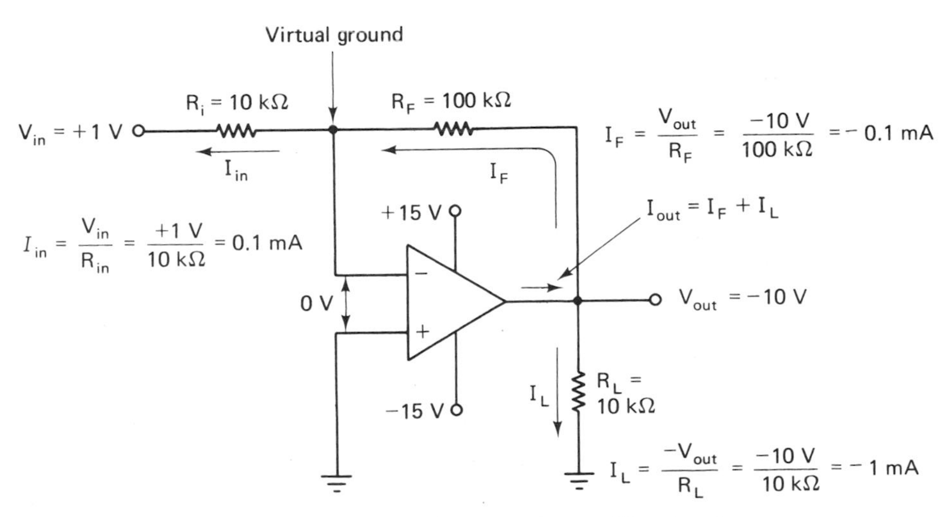

FIGURE 2-9 Currents in an inverting amplifier circuit.

If $I_{in} = I_F$, then

$$\frac{V_{in}}{R_{in}} = -\frac{V_{out}}{R_F}$$

Rearranging the equation, we obtain

$$-\frac{V_{out}}{V_{in}} = -\frac{R_F}{R_{in}}$$

You will recall that the voltage gain of any inverting amplifier stage can be expressed as

$$A_v = -\frac{V_{out}}{V_{in}}$$

Therefore,

$$A_v = -\frac{R_F}{R_{in}}$$

The op amp inverting-amplifier-stage gain factor is the ratio of the external resistors R_F and R_{in}.

The load current I_L flows out of the op amp toward ground and is determined by load resistor R_L, which can be found by the formula

$$I_L = -\frac{V_{out}}{R_L}$$

Therefore, the total current being supplied by the op amp can be expressed as

$$I_{out} = I_F + I_L$$

If the polarity of V_{in} is reversed (made negative), V_{out} would be positive and the current will flow into the op amp (see Experiments 7-2, 7-7, and 7-8).

2-2-1 Virtual Ground

The concept of virtual ground may be understood better from Figure 2-10a. The two series-aiding batteries equal 20 V and the total circuit resistance equals 20 kΩ. Therefore, the current in the circuit is 1 mA. According to Ohm's law ($E = IR$), each resistor drops 10 V. The sum of the voltage drops of both resistors equals the total voltage. Point A will be 0 V with reference

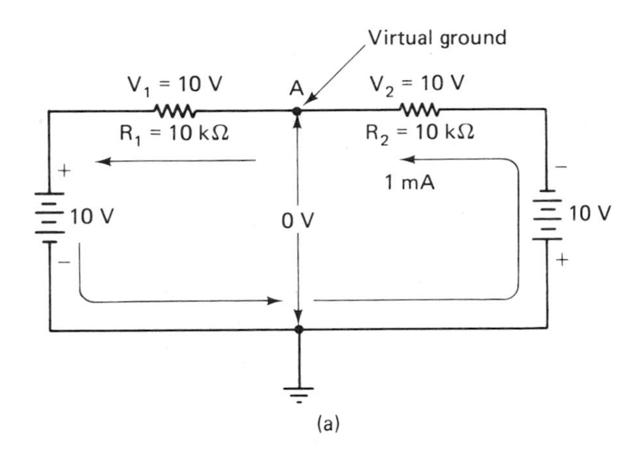

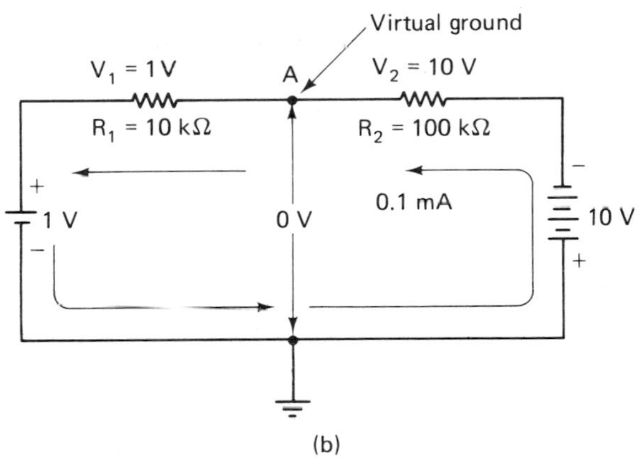

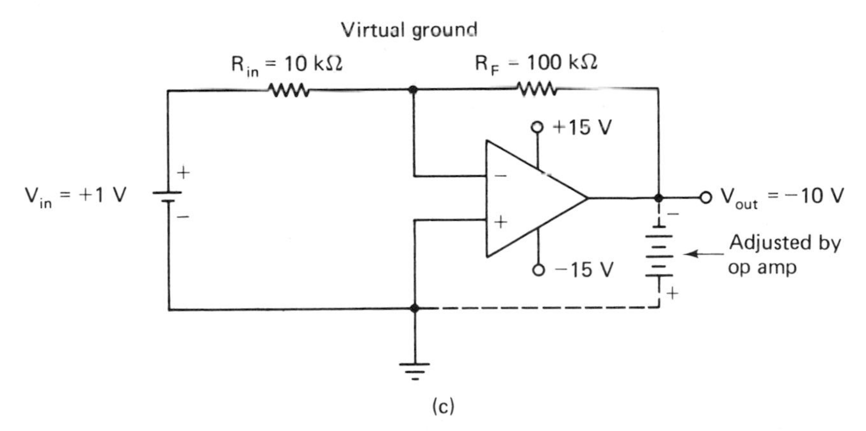

FIGURE 2-10 Examples of virtual ground: (a) equal resistors and voltages; (b) unequal resistors and voltages; (c) op-amp circuit.

to ground and yet be isolated from ground. The same situation will occur for unequal voltages and proportionately unequal resistors, as shown in Figure 2-10b.

This circuit approximates the analogy of the op-amp inverting amplifier shown in Figure 2-10c. The fixed V_{in} will cause the op amp to adjust to the proper V_{out} according to the ratio of R_F to R_{in}. Thus, the inverting input will tend to be driven toward the same potential as the grounded noninverting input. It must be remembered that a wire short placed from point A to ground of the resistor networks shown in Figure 2-10a and b has little or no effect, while a wire short placed across the op-amp inputs of Figure 2-10c will affect the circuit, since the op amp is producing V_{out}.

2-2-2 Input Impedance

As mentioned earlier, the input impedance of an op amp is very high. But the input impedance of an inverting amplifier is determined by R_{in}. Therefore, the input impedance in Figure 2-8 is equal to 10 kΩ.

2-3 NONINVERTING AMPLIFIER

The op amp can be used as a noninverting amplifier shown in Figure 2-11. In this circuit configuration the feedback to control gain is still applied to the inverting input, while V_{in} is applied to the noninverting input. The output voltage will be in phase with the input voltage for this circuit.

Resistors R_F and R_{in} form a resistive ratio network to produce the feedback voltage (V_A) needed at the inverting input. Feedback voltage (V_A) is developed across R_{in}. Since the potential at the inverting input tends to be the same as the noninverting input (as pointed out with the description of virtual ground), for most practical purposes

$$V_{in} = V_A$$

Since $V_A = V_{in}$, the gain of the stage can be expressed

$$A_v = \frac{V_{out}}{V_A}$$

However, V_A is determined by the resistance ratio of R_{in} and R_F; thus,

$$V_A = \frac{R_{in}}{R_F + R_{in}} (V_{out})$$

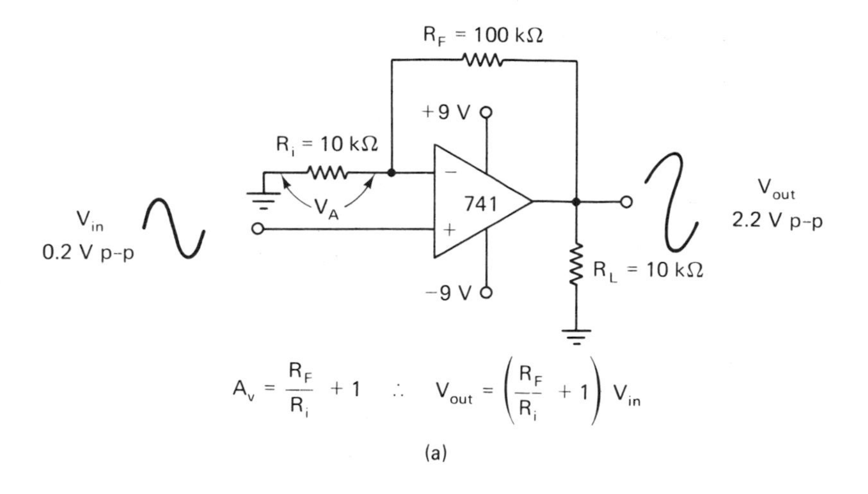

V_{in}	V_{out}
+0.3	+3.3
−0.3	−3.3
+0.52	+5.72
−0.52	−5.72

(b)

FIGURE 2-11 Noninverting amplifier: (a) schematic diagram; (b) DC voltage table.

If the equation is rearranged so that the voltages are on one side, the result is

$$\frac{V_A}{V_{\text{out}}} = \frac{R_{\text{in}}}{R_F + R_{\text{in}}}$$

Inverting the equation and simplifying, we obtain

$$\frac{V_{\text{out}}}{V_A} = \frac{R_F + R_{\text{in}}}{R_{\text{in}}}$$

$$\frac{V_{\text{out}}}{V_A} = \frac{R_F}{R_{\text{in}}} + \frac{R_{\text{in}}}{R_{\text{in}}}$$

$$\frac{V_{\text{out}}}{V_A} = \frac{R_F}{R_{\text{in}}} + 1$$

The stage-gain formula is

$$A_v = \frac{V_{out}}{V_A}$$

Therefore,

$$A_v = \frac{R_F}{R_{in}} + 1$$

Finally, the output voltage can be found by

$$V_{out} = \left(\frac{R_F}{R_{in}} + 1\right) V_{in}$$

The output voltage of the circuit shown in Figure 2-11 can now be found:

$$V_{out} = \left(\frac{R_F}{R_{in}} + 1\right) V_{in} = \left(\frac{100\ \text{k}\Omega}{10\ \text{k}\Omega} + 1\right) 0.2\ \text{V p-p} = 2.2\ \text{V p-p}$$

The DC voltage table in Figure 2-11b shows some sample input voltage times a gain of 11. Notice that the input and output polarities are in phase (see Experiment 7-3).

2-4 VOLTAGE FOLLOWERS

Voltage followers are usually defined as circuits with a gain of 1 or less with the output voltage following the input voltage. There exists an impedance isolation between input and output. The op amp is particularly useful as a voltage follower, which is shown in Figure 2-12.

With the basic noninverting voltage follower (Figure 2-12a), the output is connected directly to the inverting input with the input voltage at the noninverting input. The feedback resistance equals 0; therefore, according to the stage gain for a noninverting amplifier,

$$A = \frac{R_F}{R_{in}} + 1 = \frac{0}{R_{in}} + 1 = 1$$

or the gain of the follower is 1. In other words, with 100% feedback, the output voltage follows the input voltage. Notice that the inverting input will always be the same potential as the noninverting input. Thus, the voltage difference between the two inputs will always be approximately zero.

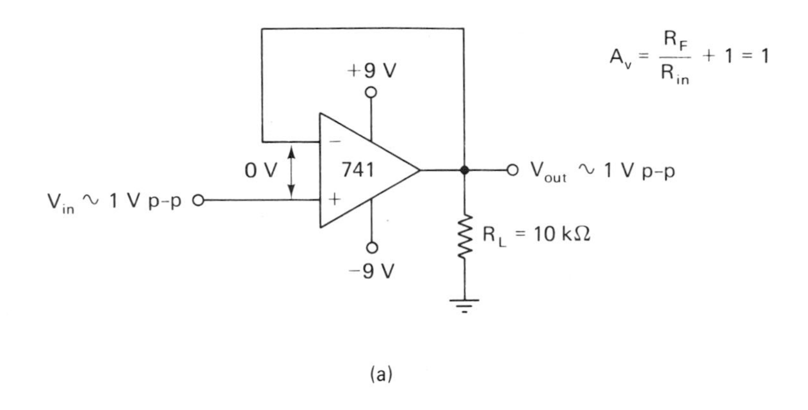

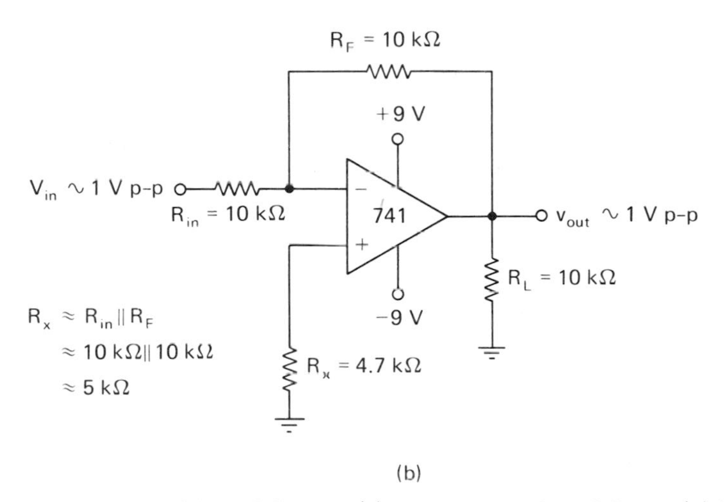

FIGURE 2-12 Voltage followers: (a) noninverting voltage follower; (b) inverting voltage follower.

The advantage of this circuit is an extremely high input impedance and a low output impedance, which is ideal for buffering or isolation between circuits.

If a particular application needed a voltage follower with phase inversion, the circuit in Figure 2-12b could be used. Since R_{in} equals R_F, the gain formula states

$$A_v = -\frac{R_F}{R_{in}} = \frac{10\ \text{k}\Omega}{10\ \text{k}\Omega} = -1$$

The limitations of this circuit would be a greatly reduced input impedance (since R_{in} equals the input impedance). Resistor R_x at the noninverting input is used to reduce input offset currents and is equal to the value of R_{in} in parallel with R_F: $R_x = R_{in} || R_F$.

2-5 VOLTAGE SUMMING AMPLIFIER

By using the basic inverting amplifier circuit and adding another input resistor, we can create an inverting summing amplifier or analog adder, as shown in Figure 2-13. The output voltage is inverted and equals the algebraic sum of each input voltage times the ratio of its appropriate input resistor to the feedback resistor, which can be expressed

$$V_{out} = -\left(\frac{R_F}{R_1} V_1 + \frac{R_F}{R_2} V_2 + \ldots + \frac{R_F}{R_N} V_N\right)$$

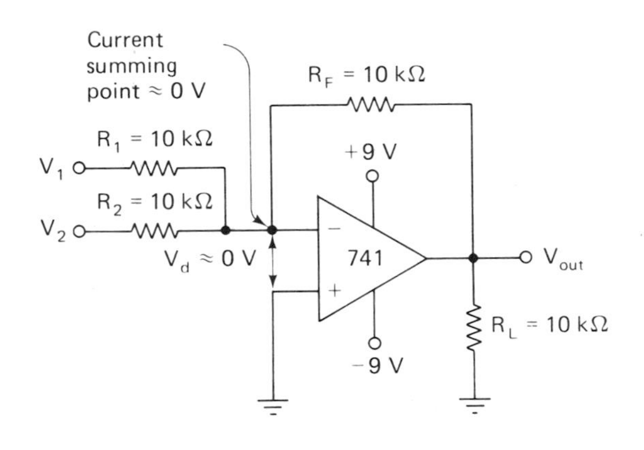

(a)

Input voltage		Output voltage
V_1	V_2	Algebraic sum
+1	+1	−2
+1	−1	0
+2	+1	−3
−1	+1	0
−1	+2	−1
−2	+1	+1

$$V_{out} = -\frac{R_F}{R_1} V_1 + \frac{R_F}{R_2} V_2 + \ldots + \frac{R_F}{R_n} V_n$$

When $R_1 = R_2 = R_F = \ldots R_n$

$$V_{out} = -(V_1 + V_2 + \ldots + V_n)$$

(b)

FIGURE 2-13 Voltage summing amplifier: (a) schematic diagram; (b) input/output voltage table.

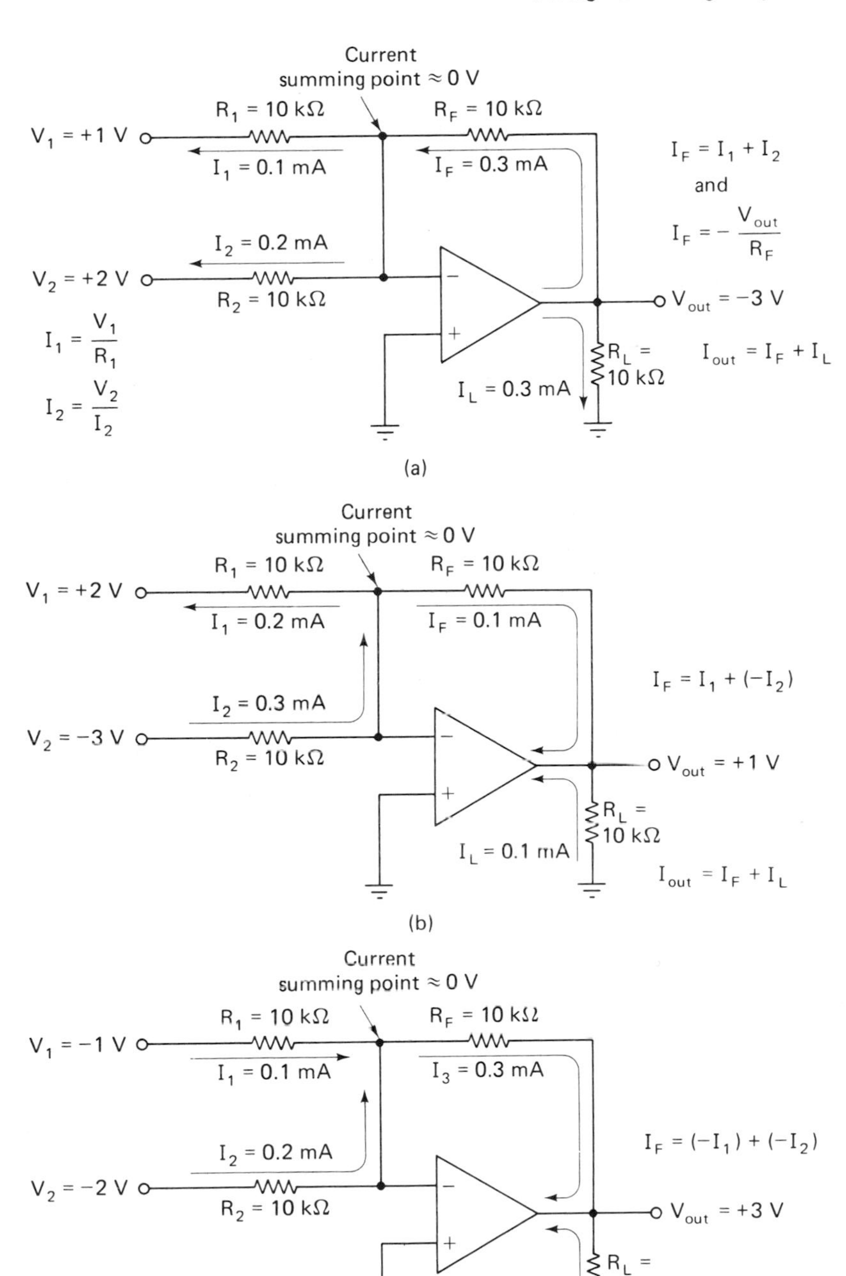

FIGURE 2-14 Currents in an inverting summing amplifier: (a) both inputs positive; (b) one input positive and one input negative; (c) both inputs negative.

The $R_F/R_N(V_N)$ in the formula means that there can be more than two inputs. If all of the external resistors equal one another ($R_F = R_1 = R_2 = \ldots = R_N$), the output voltage can be found simply by algebraically adding up the input voltages and is expressed as

$$V_{out} = -(V_1 + V_2 + \ldots + V_N)$$

The input/output voltage table shows the results of various input voltages. Remember that the output will be inverted from the resulting polarity of the algebraic summation.

The virtual ground, described earlier, is referred to as the current summing point for this type of circuit. Understanding the concept of the summing point can be achieved by analyzing the currents in the summing amplifier as illustrated in Figure 2-14. Since the summing point is the virtual ground, the voltage at this point will always be about the same as the noninverting input ($\approx$0 V).

With both input voltages positive, the current through each input resistor will flow the same direction. In the case shown in Figure 2-14a, $I_1 = 0.1$ mA and $I_2 = 0.2$ mA. Therefore, I_F must equal 0.3 mA and the output will go to −3 V to accomplish this.

If one input voltage is positive and the other input voltage is negative, as shown in Figure 2-14b, one input current (0.3 mA) will flow toward the summing point, while the other input current (0.2 mA) will flow away from the summing point. Because the amount of current entering a point must be the same amount leaving that point, 0.1 mA must flow away from the summing point through R_F. The output voltage will go to +1 V in order to develop this needed current.

When both input voltages are negative, as shown in Figure 2-14c, both input currents will flow toward the summing point (0.1 mA and 0.2 mA). The current flowing through R_F must equal the sum of these two currents (0.3 mA). Again, the output voltage will go to +3 V to accomplish this.

2-5-1 Summing Amplifier with Gain

A summing amplifier may be constructed to provide a gain greater than 1, as shown in Figure 2-15. To accomplish this, R_F must be greater than the input resistors. The gain of each input is found and then summed to get the resulting output. For example:

$$V_{out} = -\left(\frac{R_F}{R_1} V_1 + \frac{R_F}{R_2} V_2\right)$$

$$= -\left[\left(\frac{100\text{ k}\Omega}{10\text{ k}\Omega}\right)(0.1\text{ V}) + \left(\frac{100\text{ k}\Omega}{10\text{ k}\Omega}\right)(0.2\text{ V})\right]$$

$$= -[(10)(0.1\text{ V}) + (10)(0.2\text{ V})]$$
$$= -(1\text{ V} + 2\text{ V})$$
$$= -3\text{ V}$$

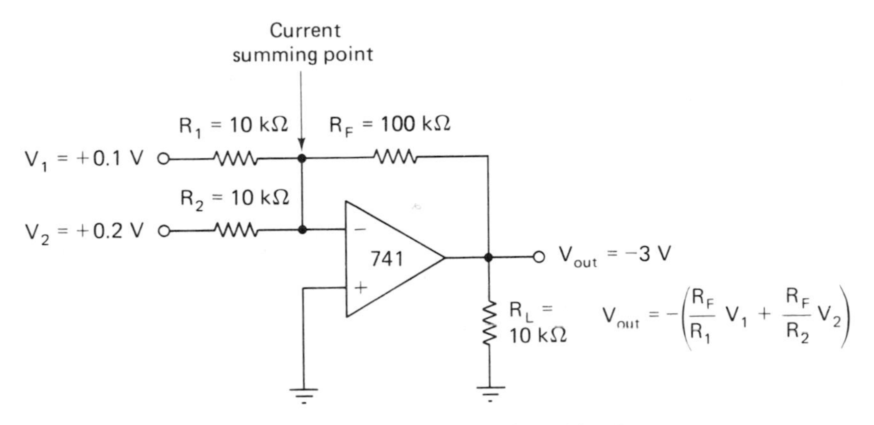

FIGURE 2-15 Summing amplifier with gain.

The V_{out} can also be expressed:

$$V_{out} = A_{v1}V_1 + A_{v2}V_2$$

where $$A_{v1} = -\frac{R_F}{R_1}$$

$$A_{v2} = -\frac{R_F}{R_2}$$

2-5-2 Scaling Adder Amplifier

Some applications of a summing amplifier may require one input to influence the output voltage more than another input. Different gains are then required for the various inputs.

Consequently, the input resistors will be of different values, as shown in Figure 2-16. This circuit is referred to as a scaling adder amplifier.

The same output voltage formula is used with this circuit as with the other summing amplifier circuits.

For example:

$$V_{out} = -\left(\frac{R_F}{R_1}V_1 + \frac{R_F}{R_2}V_2 + \frac{R_F}{R_3}V_3\right)$$

$$= -\left[\left(\frac{10\ \text{k}\Omega}{10\ \text{k}\Omega}\right)(3\ \text{V}) + \left(\frac{10\ \text{k}\Omega}{4.7\ \text{k}\Omega}\right)(2\ \text{V}) + \left(\frac{10\ \text{k}\Omega}{2.2\ \text{k}\Omega}\right)(1\ \text{V})\right]$$

$$= -[(1)(3\ \text{V}) + (2.13)(2\ \text{V}) + (4.5)(1\ \text{V})]$$

$$= -(3\ \text{V} + 4.26\ \text{V} + 4.5\ \text{V})$$

$$= -11.76\ \text{V}$$

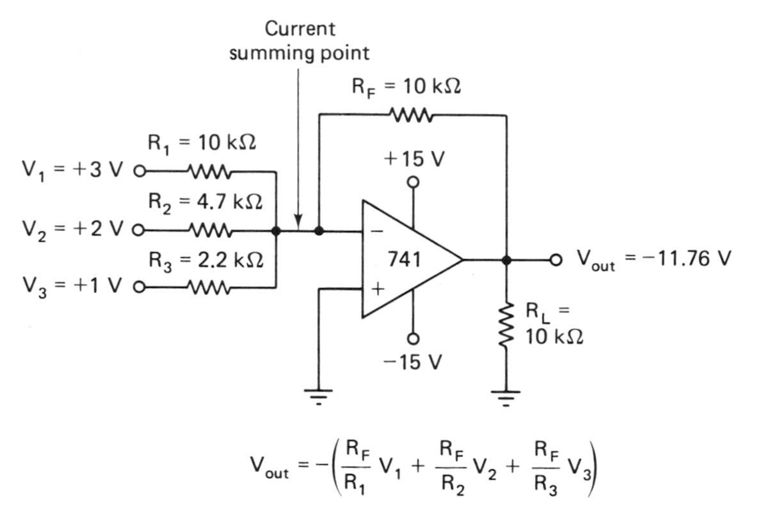

FIGURE 2-16 Scaling adder amplifier.

2-6 VOLTAGE DIFFERENCE AMPLIFIER

The voltage difference amplifier is similar to the comparator discussed in Section 2-1. Both inputs are used to sense a difference of potential between them, but the circuit utilizes the closed-loop mode, which results in a controlled and predictable output voltage. If all the external resistors are equal, the voltage difference amplifier functions as an analog mathematical circuit and is often called a voltage subtractor, as shown in Figure 2-17a. The output voltage is the inverted algebraic difference between the two input voltages and can be found by using the formula

$$V_{\text{out}} = -\frac{R_F}{R_1} V_1 + \left(\frac{R_g}{R_2 + R_g}\right)\left(\frac{R_1 + R_F}{R_1}\right) V_2$$

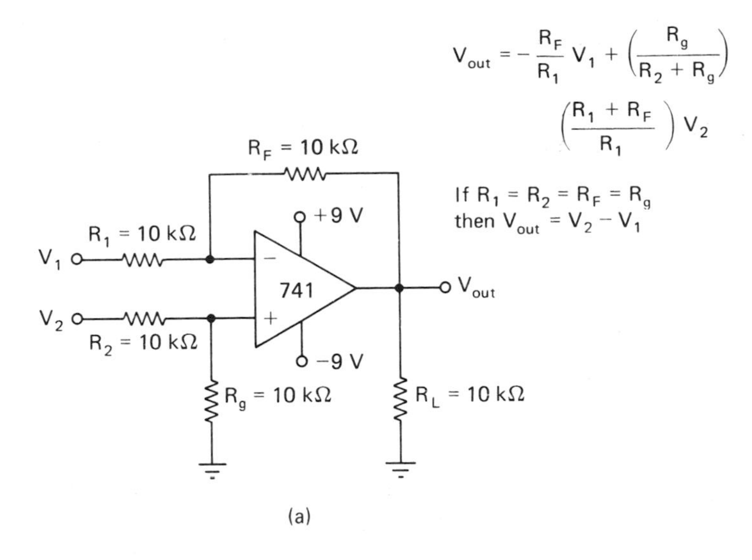

(a)

Input/voltage		Output/voltage
V_1	V_2	Algebraic difference
+2	+4	+2
+4	+2	−2
+4	−2	−6
−2	+4	+6
−4	+2	+6
+2	−4	−6
−4	−2	+2
−2	−4	−2

(b)

FIGURE 2-17 Voltage difference amplifier: (a) schematic diagram; (b) input/output voltage table.

For example, referring to Figure 2-17, if $V_1 = +2$ V and $V_2 = +4$ V, then

$$V_{out} = -\frac{10\ \text{k}\Omega}{10\ \text{k}\Omega}(+2\ \text{V}) + \left(\frac{10\ \text{k}\Omega}{10\ \text{k}\Omega + 10\ \text{k}\Omega}\right)\left(\frac{10\ \text{k}\Omega + 10\ \text{k}\Omega}{10\ \text{k}\Omega}\right)(+4\ \text{V})$$

$$= -1(+2\ \text{V}) + \left(\frac{10}{20}\right)\left(\frac{20}{10}\right)(+4\ \text{V})$$

$$= -(+2\ \text{V}) + (0.5)(2)(+4\ \text{V})$$

$$= -(+2) + 1(+4\ \text{V})$$

$$= -2 + 4$$

$$= +2\ \text{V}$$

As with the comparator, the polarity of the output voltage will be positive if the voltage at the inverting input is more negative than the voltage at the noninverting input (as just proven with the formula), and vice versa.

The input/output voltage table of Figure 2-17b shows the proper polarity and algebraic difference output voltage for various input voltages.

2-6-1 Voltage Difference Amplifier with Gain

If the ratio of the resistors is changed (as shown in Figure 2-18), the voltage difference circuit can provide amplification.

The formula previously given can be used to find the output voltage.

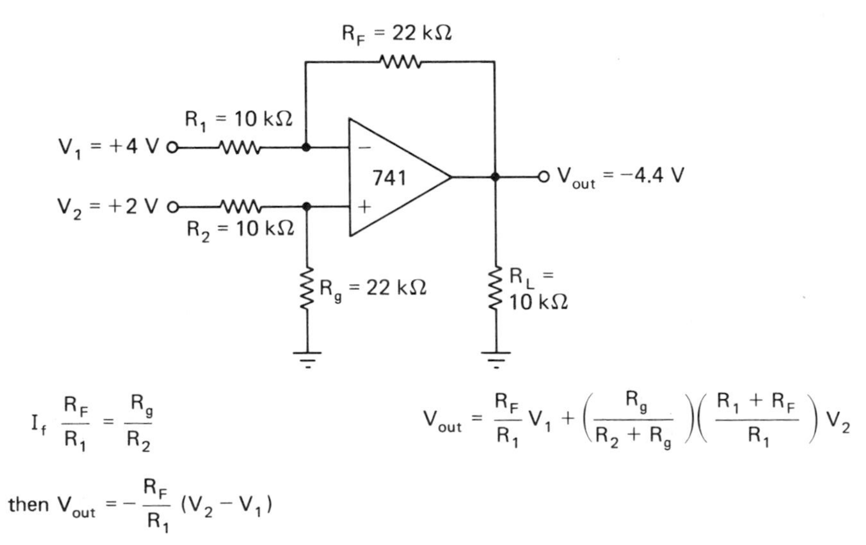

$$\text{If } \frac{R_F}{R_1} = \frac{R_g}{R_2}$$

$$\text{then } V_{out} = -\frac{R_F}{R_1}(V_2 - V_1)$$

$$V_{out} = \frac{R_F}{R_1}V_1 + \left(\frac{R_g}{R_2 + R_g}\right)\left(\frac{R_1 + R_F}{R_1}\right)V_2$$

FIGURE 2-18 Voltage difference amplifier with gain.

However, if the ratio of R_F to R_1 equals the ratio of R_g to R_2, which is normally the case, then the output voltage can easily be found by

$$V_{out} = \frac{R_F}{R_1}(V_2 - V_1)$$

Although this description deals primarily with an algebraic subtractor, a major advantage of a voltage difference amplifier is the ability to sense a small differential voltage buried within larger signal voltages. Unfortunately, the input impedance is very low and this type of circuit may require voltage followers for buffering or isolation.

SUMMARY POINTS

1. A voltage comparator senses which input voltage is more positive (or negative) than the other, depending on which input is used as the reference voltage.
2. A basic comparator output will be 0 V for 0 V between the inputs, or $+V_{sat}$ (or $-V_{sat}$) for an input differential voltage.
3. The comparator output voltage polarity will be 180° out of phase with the inverting input voltage polarity in relation to the non-inverting input voltage.
4. Comparators can be used as voltage-level detectors.
5. Degenerative (negative) feedback is used in a closed-loop mode to increase stability and bandwidth of an op amp.
6. An inverting amplifier inverts the signal 180°.
7. A virtual short exists between the two inputs of the op amp.
8. The inverting input of an op amp will tend to seek the same voltage potential as the noninverting input.
9. Because of the high input impedance, for most practical purposes, no current will flow into or out of the op-amp input terminals.
10. Current through the inverting input resistor (R_{in}) is the same as the current through the feedback resistor (R_F) for amplifier circuits.
11. A noninverting amplifier does not invert the signal.
12. Voltage followers have a gain of 1 and are used as isolating circuits.

13. An inverting summing amplifier will give an inverted algebraic summation output of the voltage at its inputs.
14. The summing amplifier may have gain if the feedback resistor is larger than the input resistors.
15. A scaling adder amplifies some inputs more than others, which depends on the various values of input resistors.
16. The virtual ground at the inverting input becomes the current summing point for a summing amplifier.
17. A voltage difference amplifier output provides the difference voltage between the inputs. The output may be algebraic proportional, include gain, or reflect a different gain for each input.

TERMINOLOGY EXERCISE

Write a brief definition for each of the following terms:

1. Voltage comparator
2. Voltage-level detector
3. Degenerative feedback
4. Feedback resistor
5. Inverting amplifier
6. Noninverting amplifier
7. Virtual ground
8. Voltage follower
9. Voltage summing amplifier
10. Scaling adder amplifier
11. Current summing point
12. Voltage difference amplifier

PROBLEMS AND EXERCISES

1. Design two voltage-level detectors using ±15-V power supplies. Draw your circuits and show all values for the V_{ref} given.

a. $V_{ref} = +5$ V

b. $V_{ref} = -3$ V

2. Draw the output voltage waveform for the circuit shown in Figure 2-19. Make sure that the output follows the same time reference as the input.

3. Referring to Figure 2-20, find R_F, V_{out}, I_{in}, I_F, and I_L for an inverting amplifier with the various values of R_{in} gain and input voltage given. (Let $R_L = 10$ kΩ.)

a. $R_{in} = 4.7$ kΩ
$A_v = 10$
$V_{in} = 1$ V
$R_F = ?$
$V_{out} = ?$
$I_{in} = ?$
$I_F = ?$
$I_L = ?$

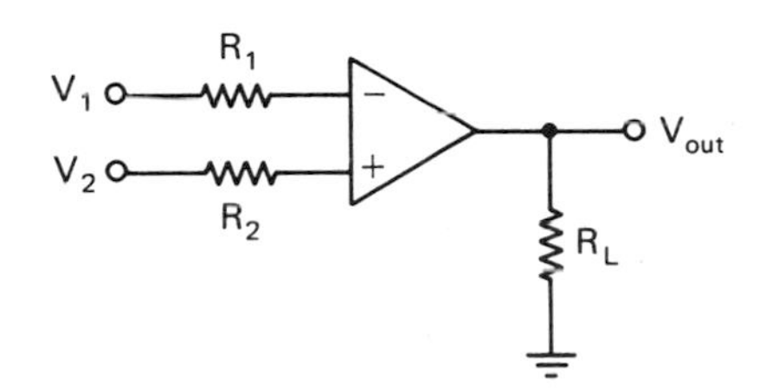

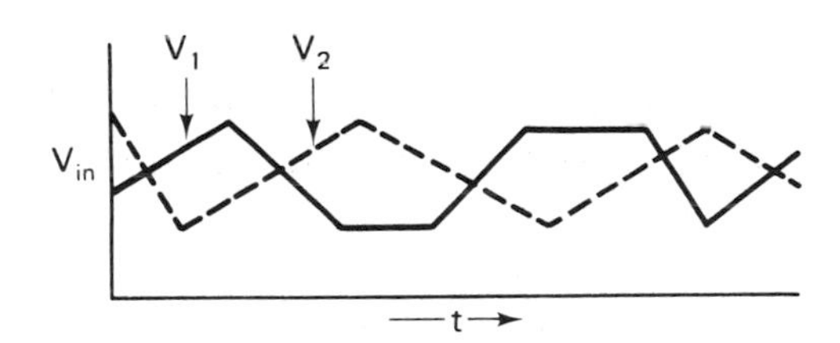

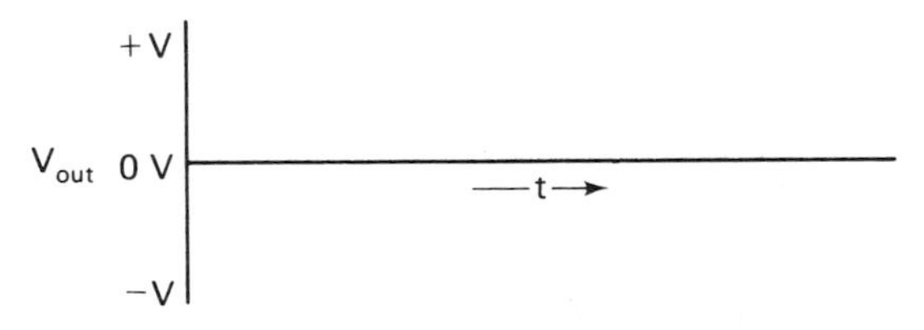

FIGURE 2-19

b. $R_{in} = 1\ \text{k}\Omega$
$A_v = 22$
$V_{in} = -0.3\ \text{V}$
$R_F = ?$
$V_{out} = ?$
$I_{in} = ?$
$I_F = ?$
$I_L = ?$

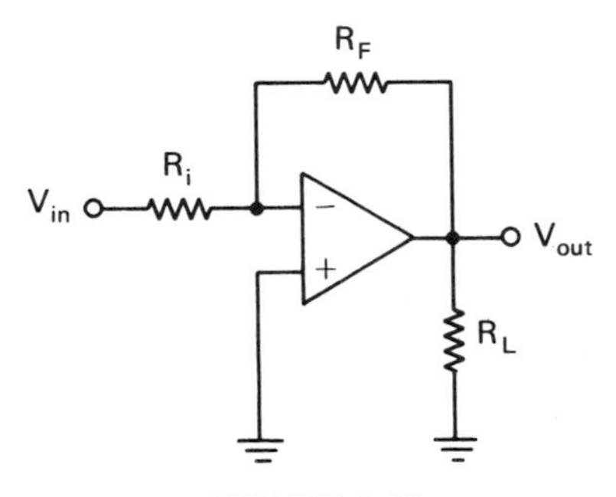

FIGURE 2-20

c. $R_{in} = 10\ \text{k}\Omega$
$A_v = 100$
$V_{in} = +0.05\ \text{V}$
$R_F = ?$
$V_{out} = ?$
$I_{in} = ?$
$I_F = ?$
$I_L = ?$

4. Referring to Figure 2-20, find R_{in}, V_{out}, I_{in}, I_F, and I_L for an inverting amplifier with the various values of R_F, gain, and input voltage given. (Let $R_L = 47\ \text{k}\Omega$.)

a. $R_F = 100\ \text{k}\Omega$
$A_v = 20$
$V_{in} = -0.6\ \text{V}$
$R_{in} = ?$
$V_{out} = ?$
$I_{in} = ?$
$I_F = ?$
$I_L = ?$

b. $R_F = 68\ \text{k}\Omega$
$A_v = 10$

V_{in} = +0.58 V
R_{in} = ?
V_{out} = ?
I_{in} = ?
I_F = ?
I_L = ?

c. R_F = 10 kΩ
A_v = 1
V_{in} = -3 V
R_{in} = ?
V_{out} = ?
I_{in} = ?
I_F = ?
I_L = ?

5. Calculate the gain and output voltage for the noninverting amplifier shown in Figure 2-21 for the various resistor values and input voltages given.

a. R_{in} = 10 kΩ
R_F = 22 kΩ
R_L = 10 kΩ
V_{in} = +0.2 V
A_v = ?
V_{out} = ?

b. R_{in} = 4.7 kΩ
R_F = 22 kΩ
R_L = 100 kΩ
V_{in} = -0.3 V
A_v = ?
V_{out} = ?

c. R_{in} = 100 kΩ
R_F = 220 kΩ

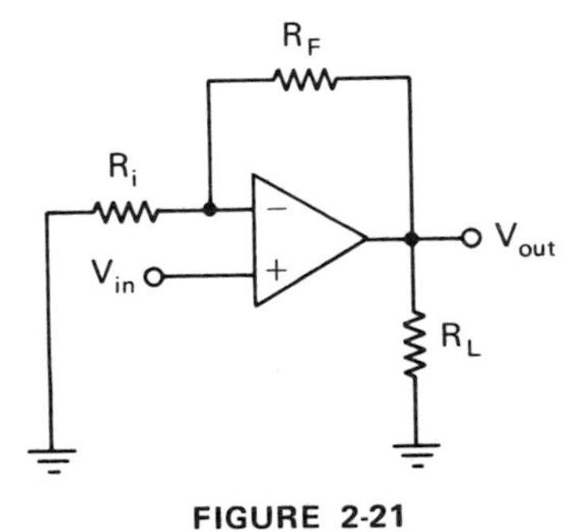

FIGURE 2-21

R_L = 47 kΩ
V_{in} = +0.5 V
A_v = ?
V_{out} = ?

6. Draw the schematic diagram for a basic noninverting voltage follower.

7. For a summing amplifier, where all resistors are equal, calculate the output voltage for the input voltages given. (See Figure 2-13.)

   a. V_1 = +3 V
      V_2 = +2 V

   b. V_1 = +5 V
      V_2 = −2 V

   c. V_1 = −4 V
      V_2 = +3 V

   d. V_1 = −5 V
      V_2 = +2 V

   e. V_1 = +2 V
      V_2 = +5 V
      V_3 = −4 V

   f. V_1 = −5 V
      V_2 = +2 V
      V_3 = −3 V
      V_4 = +1 V

   g. V_1 = +1.2 V
      V_2 = −0.5 V
      V_3 = +2.3 V
      V_4 = −0.7 V

8. Calculate the output voltage for the scaling adder shown in Figure 2-22 for the values given.

   a. R_1 = 10 kΩ
      R_2 = 4.7 kΩ
      R_3 = 2.2 kΩ
      R_F = 10 kΩ
      V_1 = +1 V
      V_2 = +2 V
      V_3 = +0.5 V

   b. R_1 = 22 kΩ
      R_2 = 33 kΩ
      R_3 = 47 kΩ

$R_F = 100\ \text{k}\Omega$
$V_1 = +0.3\ \text{V}$
$V_2 = -0.7\ \text{V}$
$V_3 = +1.5\ \text{V}$

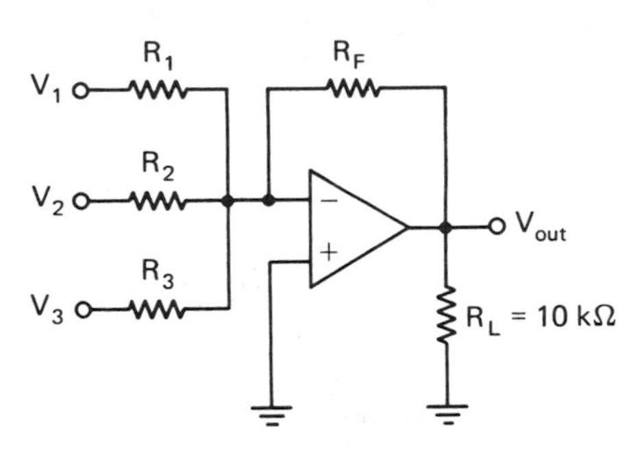

FIGURE 2-22

9. Find the output voltage of a difference amplifier where all resistors are equal, for the input voltages given. (See Figure 2-17.)

   a. $V_1 = +5\ \text{V}$
      $V_2 = -3\ \text{V}$

   b. $V_1 = -5\ \text{V}$
      $V_2 = +3\ \text{V}$

   c. $V_1 = -5\ \text{V}$
      $V_2 = -3\ \text{V}$

   d. $V_1 = +5\ \text{V}$
      $V_2 = +3\ \text{V}$

10. If each input equals +1.0 V and each op amp is considered ideal, what is the output voltage of each circuit shown in Figure 2-23? (*Hint:* Remember all formulas given in Chapter 2.)

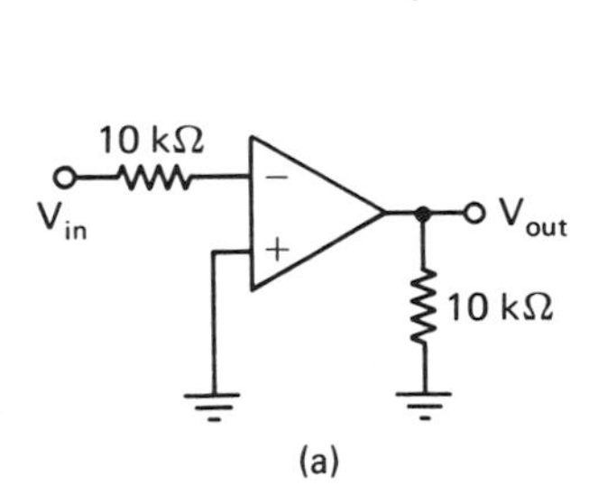

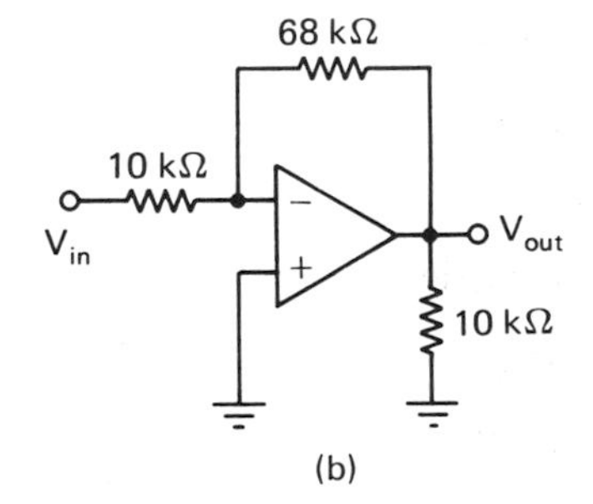

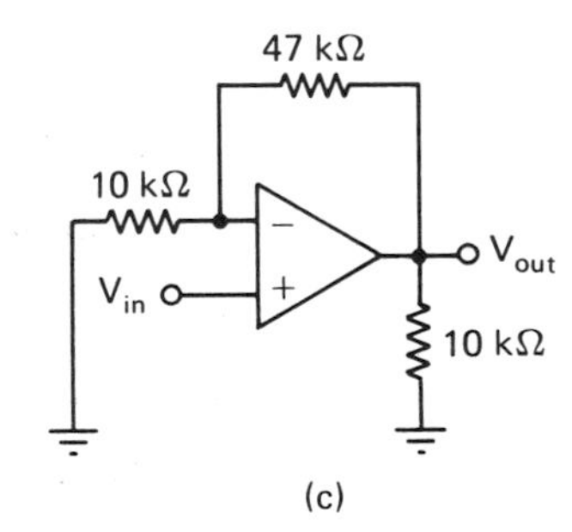

FIGURE 2-23

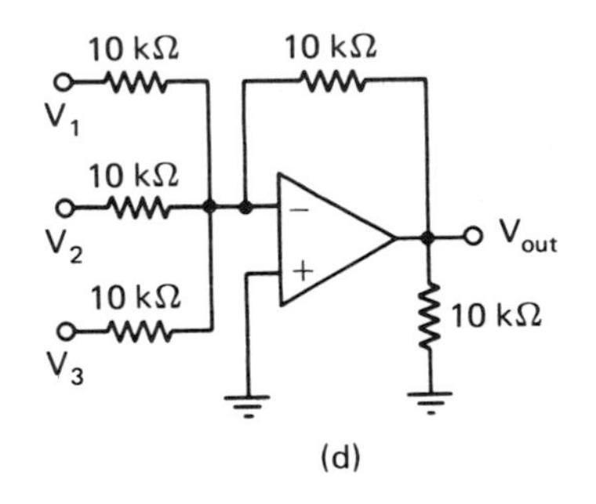

(d)

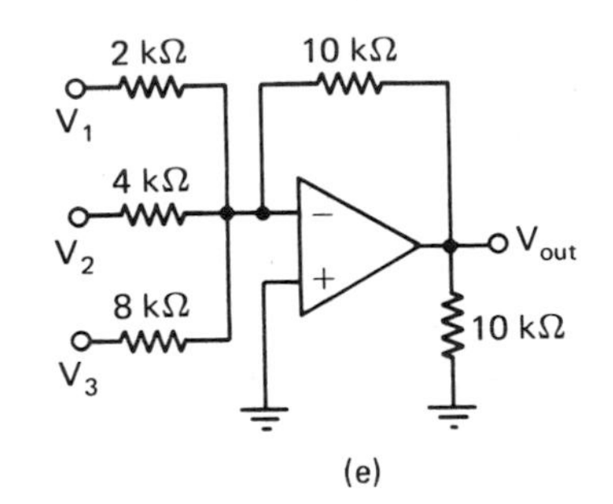

(e)

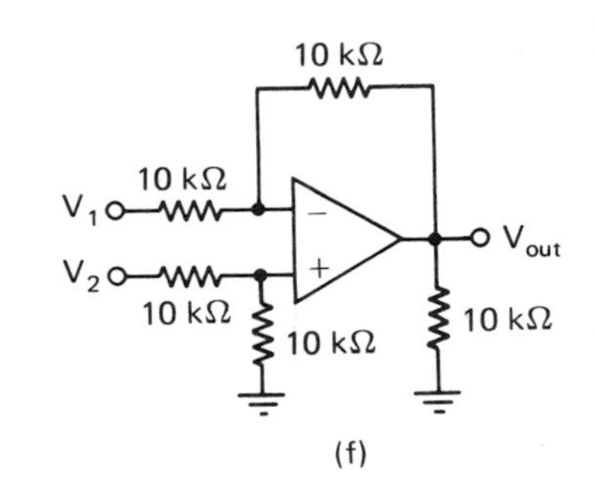

(f)

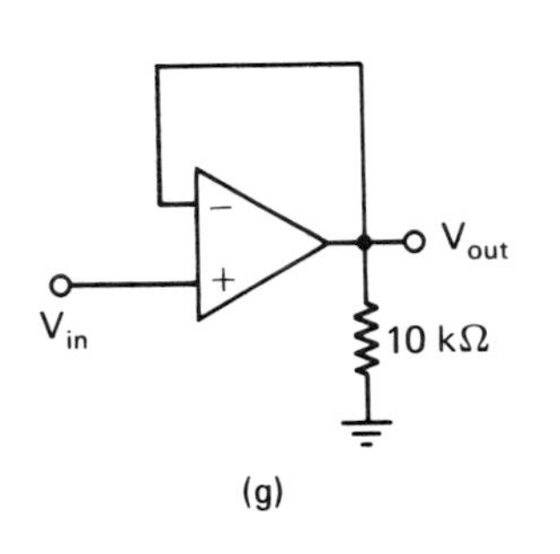

(g)

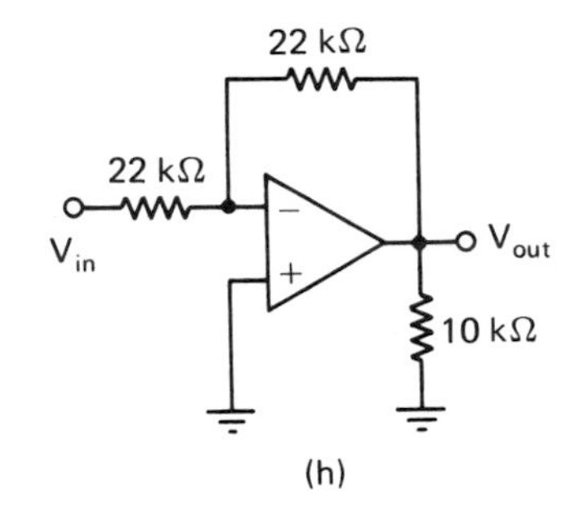

(h)

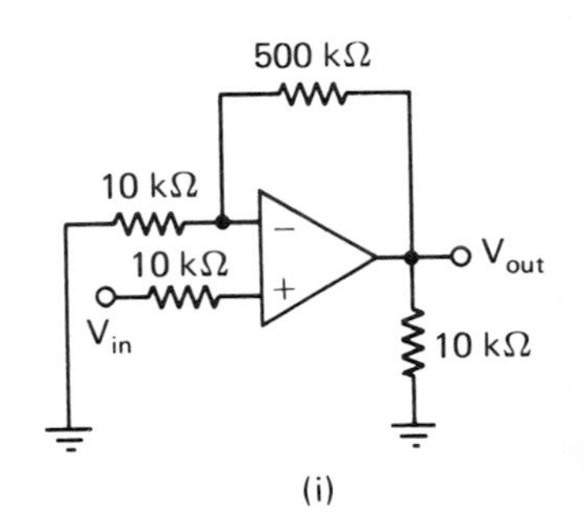

(i)

Power supply for all circuits = ±15 V.

FIGURE 2-23 (*Continued*)

SELF-CHECKING QUIZ

Match the names in column A with their proper schematic diagram in Figure 2-24.

Column A

1. Inverting amplifier
2. Summing amplifier
3. Comparator
4. Difference amplifier
5. Voltage follower
6. Noninverting amplifier

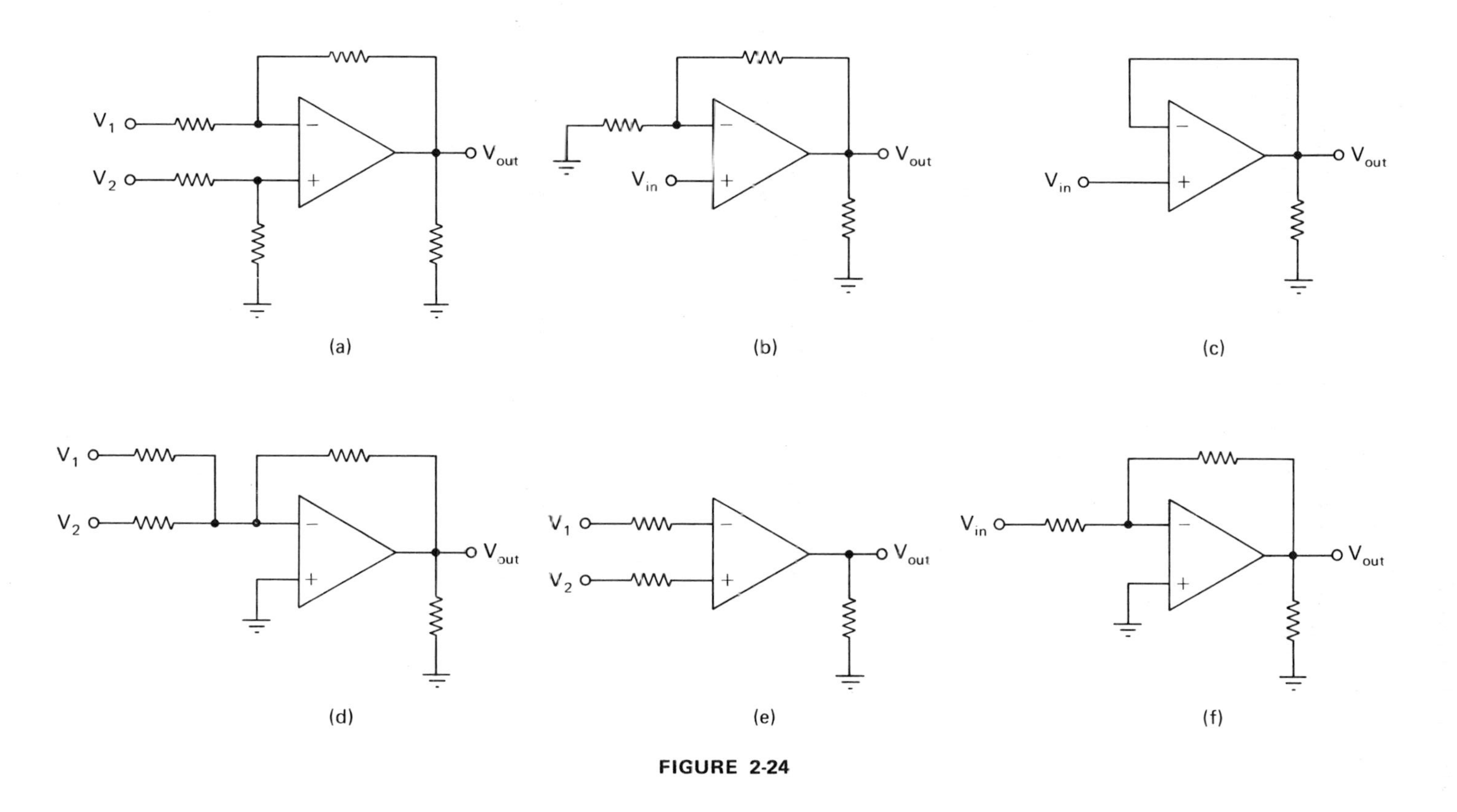

V1
V2
Vout
(a)
Vin
Vout
(b)
Vin
Vout
(c)
V1
V2
Vout
(d)
V1
V2
Vout
(e)
Vin
Vout
(f)

FIGURE 2-24

Multiple-Choice Questions

7. The output of a voltage comparator with +2.5 V on the inverting input and +2.7 V on the noninverting input will be:

   a. $+V_{sat}$

   b. $-V_{sat}$

   c. +0.2 V

   d. -0.2 V

8. For a particular inverting amplifier, R_{in} = 22 kΩ, R_F = 68 kΩ and V_{in} = +0.5 V p-p. The output voltage will equal about:

   a. -0.5 V p-p

   b. +15 V p-p

   c. -1.5 V p-p

   d. 0 V

9. For a particular noninverting amplifier, R_{in} = 10 kΩ, R_F = 120 kΩ, and V_{in} = +0.6 V p-p. The output voltage will equal about:

   a. +7.8 V p-p

   b. -7.2 V p-p

   c. -8.2 V p-p

   d. +8.8 V p-p

10. A noninverting voltage follower has an input voltage of +5.5 V p-p. Its output will be about:

    a. 0 V

    b. $+V_{sat}$

    c. $-V_{sat}$

    d. +5.5 V p-p

11. Referring to Figure 2-25, if V_1 = +2 V, V_2 = +3 V, and V_3 = -1 V, V_{out} will be:

    a. +4 V

    b. -4 V

    c. +6 V

    d. -6 V

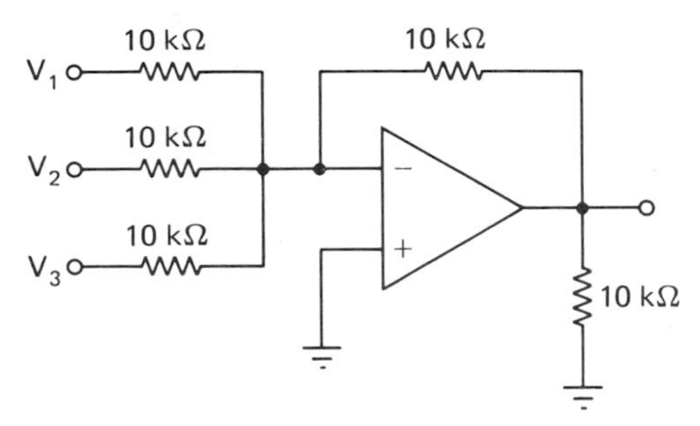

FIGURE 2-25

12. Referring to Figure 2-25, if V_1 = -3 V, V_2 = -2 V, and V_3 = +4 V, V_{out} will be:

    a. +1 V

    b. -1 V

    c. +9 V

    d. -9 V

13. Which of the following is *not* true for a voltage follower?

    a. Low output impedance

    b. Gain of 1

    c. Low input impedance

    d. Used as a buffer amplifier

14. The type of feedback used in the closed-loop mode to stabilize and increase bandwidth of an op-amp circuit is:

    a. Positive

    b. Negative

    c. Regenerative

    d. Degenerative

    e. (b) and (c) are correct

    f. (b) and (d) are correct.

15. The value of V_{ref} in Figure 2-26 is about:

    a. +1.5 V

    b. +4.7 V

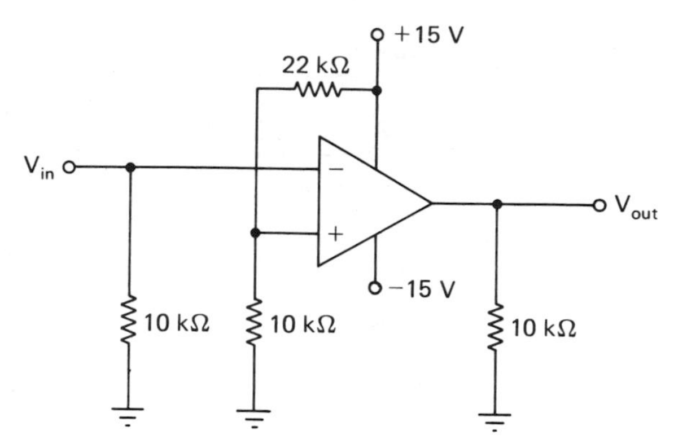

FIGURE 2-26

c. +10.0 V

d. -4.7 V

16. Referring to Figure 2-26, when V_{in} = +3 V, the output will be about:

 a. +13.5 V

 b. -13.5 V

 c. +4.7 V

 d. -4.7 V

17. Referring to Figure 2-27, when V_1 = +4 V and V_2 = +1 V, V_{out} will equal:

 a. +3 V

 b. -3 V

 c. +5 V

 d. -5 V

18. Referring to Figure 2-27, when V_1 = -5 V and V_2 = +2 V, V_{out} will equal:

 a. +3 V

 b. -3 V

 c. +7 V

 d. -7 V

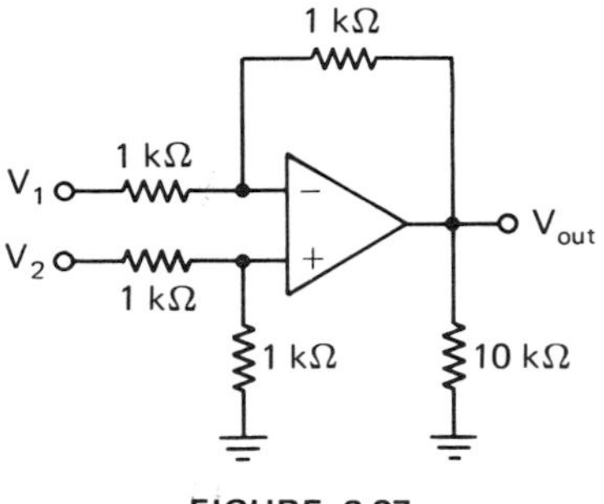

FIGURE 2-27

19. Virtual ground is also the:
    a. Common ground
    b. Summing point
    c. Earth ground

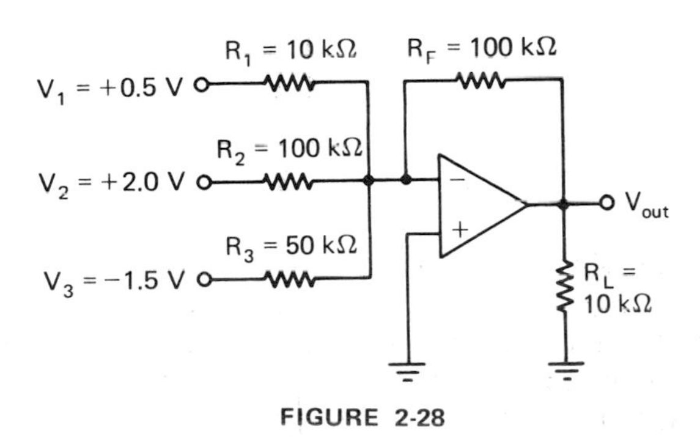

FIGURE 2-28

20. Referring to Figure 2-28, the output voltage will be about:
    a. +1 V
    b. -1 V
    c. +4 V
    d. -4 V

(Answers at back of book)

Chapter 3

Signal Processing with Op Amps

Signal processing involves the use of special circuits that change or modify given input signals. The output signals of these circuits can then be used to perform different functions. Wave-shaping circuits will actually change the signal's overall form, such as integration and differentiation. Other special circuits will attenuate the input signal at certain frequencies and are referred to as filters.

These signal-processing circuits can use passive components, such as resistors, capacitors, and inductors. With a passive circuit, there is a certain amount of signal lost or consumed, and they tend to be more bulky in size. Active circuits usually combine minimal passive components: resistors and capacitors, together with active devices, such as transistors and other discrete solid-state amplifiers. These active circuits amplify or at least maintain the amplitude of the input signal.

In this chapter you will see how op amps are particularly suited for signal-processing circuits. In many areas, op-amp circuits are less expensive, easier to use, more efficient, require less external components, take less space, and are much lighter in weight, compared to circuits using discrete components.

3-1 INTEGRATOR CIRCUIT

An integrator circuit continuously adds up a quantity being measured over a period of time. The output waveform is proportional to the time interval of the input signal.

In a basic passive *RC* integrator circuit as seen in Figure 3-1, the output is taken across the capacitor. The voltage across the capacitor takes time to build up, depending on the *RC* time constant. Because the current decreases as the capacitor is charging, the voltage across the capacitor rises at an exponential rate, as indicated by the bending (or rounding-off) waveform. When the input pulse falls to 0 V, the capacitor discharges at an exponential rate also. Notice that the input voltage is divided across the resistor and capacitor. Hence, a loss is incurred and the output voltage amplitude is less than the input voltage. Integration occurs when the input pulse duration is much less than one time constant ($TC = RC$).

When an op amp is used in an integrator circuit, as shown in Figure 3-2, the capacitor becomes the feedback element. From the previous discussion in Chapter 2, remember that the feedback current must equal the input current. Therefore, the op amp provides a linear rising output voltage to keep the charging current constant. When the input pulse goes to 0 V, the falling output voltage is also linear. The resulting output voltage waveform is triangular. Since the inverting input is used, the output will be a negative-going waveform. This basic op-amp integrator is in the open-loop mode, and even when V_{in} is 0 V, the input bias current will cause the capacitor to charge. The capacitor will continue to charge until the output is at saturation and the circuit is unusable.

A large-value feedback resistor placed in parallel with the feedback capacitor, as shown in Figure 3-3, prevents output saturation and provides a practical integrator with reduced noise, less offset drift, and better stability. The gain can be set from 10 to 100, depending on the amount of output

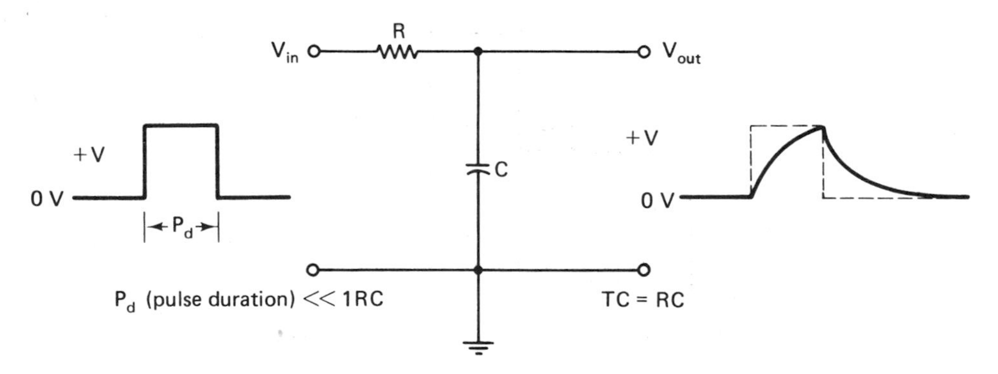

FIGURE 3-1 Simple passive integrator.

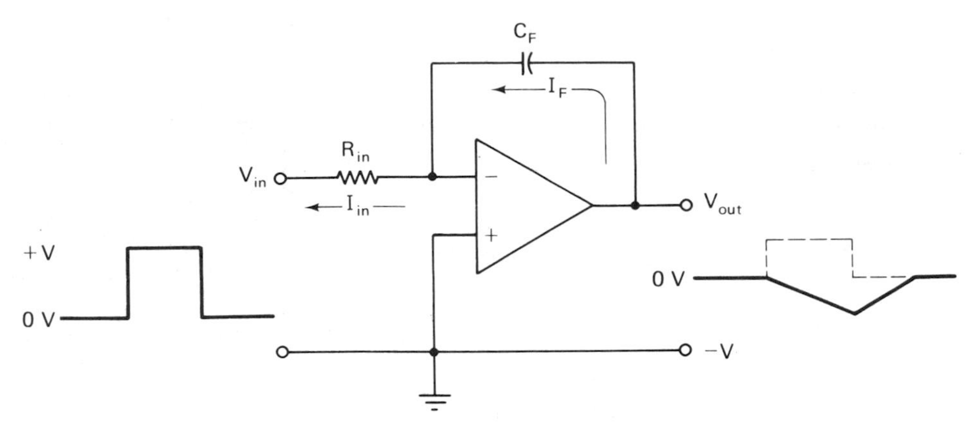

FIGURE 3-2 Basic op amp integrator.

voltage needed. When V_{in} equals 0 V, V_{out} will equal 0 V. Remember that the capacitive reactance (X_c) varies with frequency. For lower frequencies, X_c increases, less signal is fed back, and the output voltage will increase. When the frequency is increased, X_c decreases, more signal is fed back, and the output voltage decreases. Because the X_c changes, the integrator circuit behaves like a low-pass filter.

As long as V_{in} is constant, the output voltage can be found with the expression

$$V_{out} = -\frac{1}{R_{in}\,C_F}\int_0^t dV_{in}\;dt$$

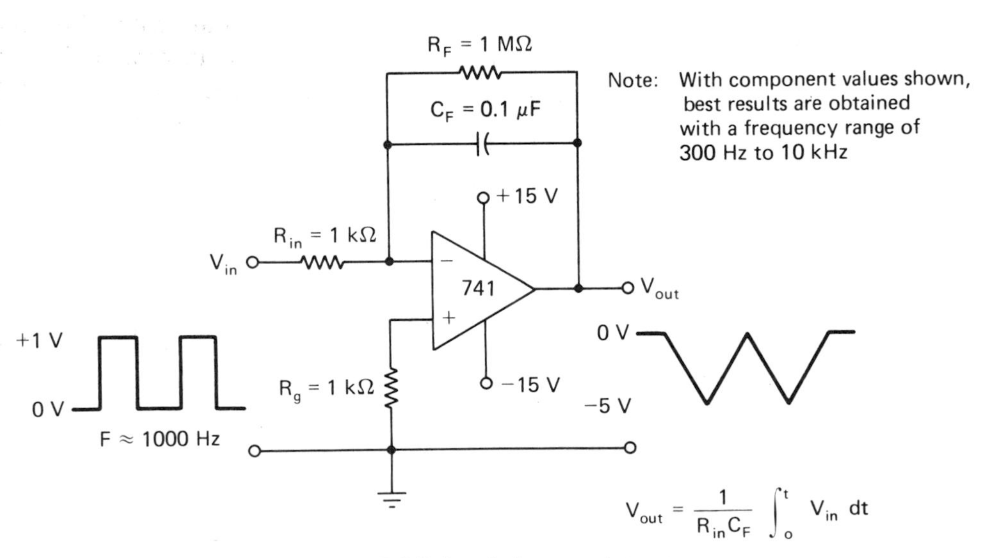

FIGURE 3-3 Practical op amp integrator.

The integral sign $\int_0^t$ indicates the period or limits of integration to be calculated, where V_{in} is constant and dt is the time or period of integration. For example, if the input signal of Figure 3-3 is symmetrical, the period of one cycle is 0.001 s. [Period (one cycle) = $1/F$, where F is frequency.] The pulse is at +1 V for half of this time, which is 0.0005 s. Now, substituting the values into the formula, we obtain

$$V_{out} = -\frac{1}{1\ \text{k}\Omega \times 0.1\ \mu\text{F}} \int_0^{0.0005\ \text{s}} (+1\ \text{V} \times 0.0005\ \text{s})$$

$$= -10{,}000\ (0.0005)$$

$$= -5\ \text{V}$$

The minus sign only indicates that the output is 180° out of phase with the input.

3-2 DIFFERENTIATOR CIRCUIT

The reverse concept of the integrator is the differentiator. The output of a differentiator circuit is proportional to the rate of change of the input signal. The output is taken across the resistor, as shown in the simple passive differentiator of Figure 3-4. Initially, when the leading edge of the pulse is applied to the input, maximum current flows. The I_R drop across the resistor (V_{out}) is also maximum at this time. As the voltage on the capacitor begins to build up, less current is required for the charging process. The V_{out} then drops off at an exponential rate. When the input pulse falls to 0 V, the capacitor discharges in the opposite direction and the same process occurs, but in the negative direction. Here again, the input voltage is divided across the capacitor and resistor, resulting in a loss at the output.

By using an op amp in the differentiator circuit shown in Figure 3-5, the output can be made equal to or larger than the input. The output voltage

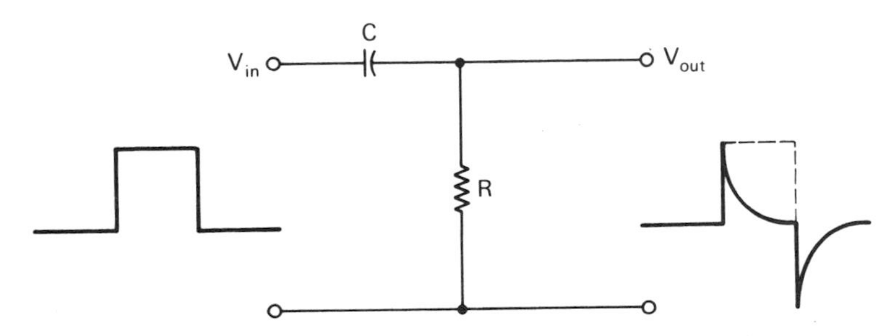

FIGURE 3-4 Simple passive differentiator.

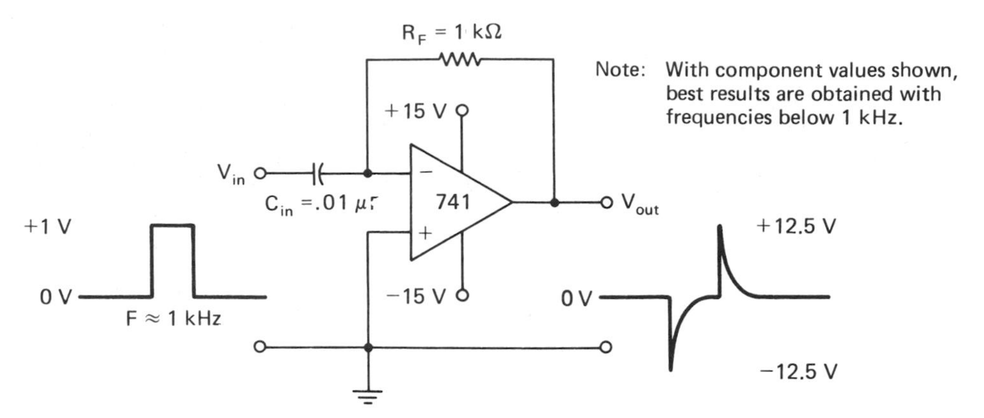

FIGURE 3-5 Basic op amp differentiator.

can be found by the expression

$$V_{out} = -2R_F C_{in} \frac{dV_{in}}{dt}$$

where dV_{in} is the change in input voltage and dt the change in time that it occurs. The minus sign only indicates phase inversion.

However, this input waveform is difficult to calculate because of the steep rise and fall times of the input pulse. The triangular input voltage shown in Figure 3-6 is easier to compute and shows the reverse effect of differentiation to that of integration. For example, with an input frequency of 1 kHz, the period of one cycle is 0.001 s. The change in time when the input voltage is either rising or falling is 0.0005 s. Substituting the values

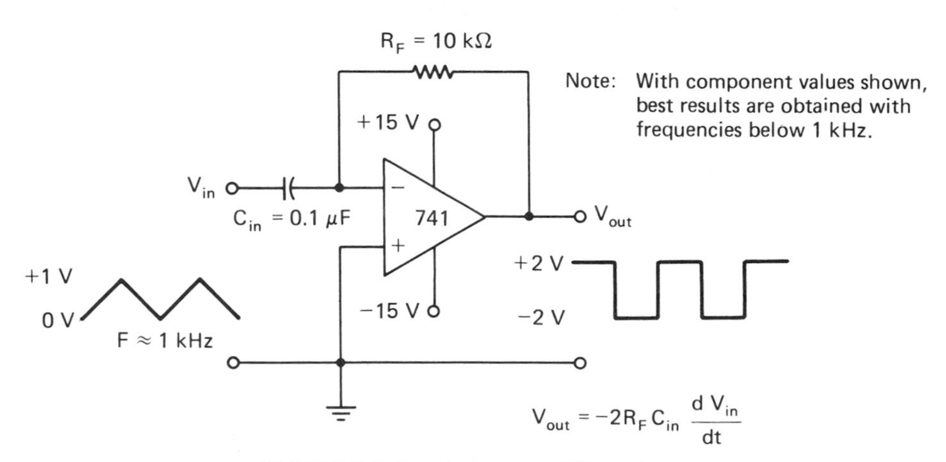

FIGURE 3-6 Practical op amp differentiator.

into the formula, we have

$$V_{out} = -2 \times 10\ k\Omega \times 0.1\ \mu F \frac{1\ V}{0.0005\ s}$$

$$= -0.002 \times 2000$$

$$= -4\ V$$

The differentiator circuit is useful in producing a sharp trigger pulse to drive other circuits. When the input signal increases in frequency, the X_c of the input capacitor decreases and the output signal increases. Therefore, the differentiator circuit serves as a basic high-pass filter.

3-3 ACTIVE LOW-PASS FILTER

A low-pass filter has a constant output voltage from DC up to a specific cutoff frequency (f_c). This cutoff frequency, f_c, is also called the 0.707 frequency, the -3-dB frequency, the corner frequency, or the "breakpoint" frequency. Frequencies above the f_c are attenuated (decreased). The range of frequencies below f_c are called the pass band, while the frequencies above f_c are known as the stop band. A low-pass-filter frequency-response curve is shown in Figure 3-7. The dashed line indicates an ideal cutoff. However,

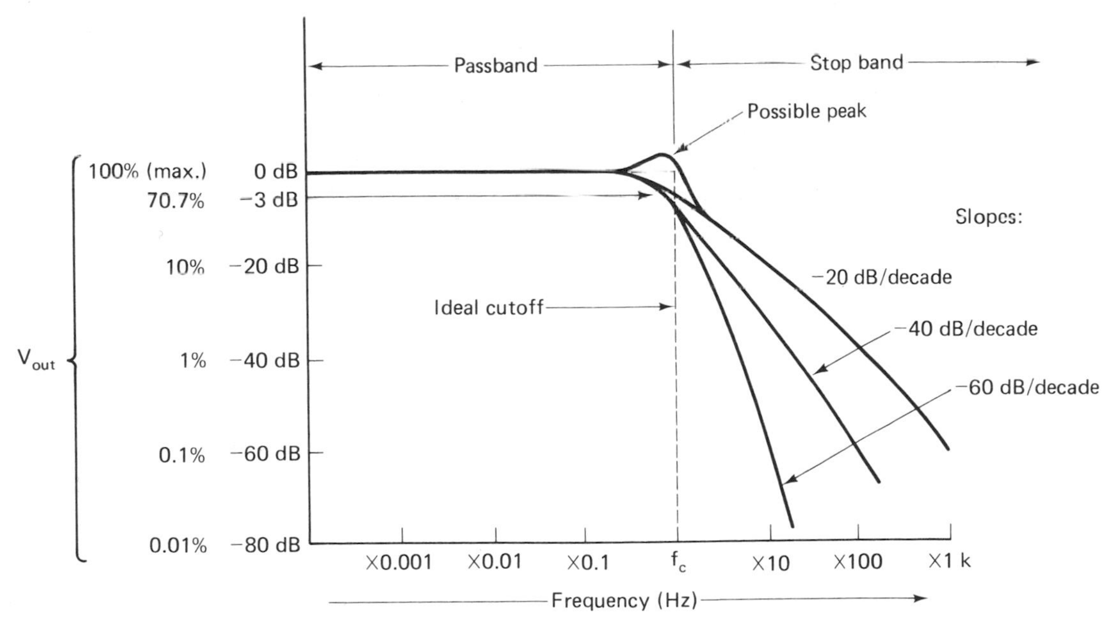

FIGURE 3-7 Low pass filter frequency response curve.

filters are usually not this efficient and tend to roll off or even peak and then roll off. The practical f_c occurs at the half-power point, or 70.7% of the maximum output voltage. This is also expressed in decibels (dB) and can be derived from the formula

$$\text{dB} = -20 \log \frac{V_{\text{out}}}{V_{\text{in}}}$$

$$= -20 \log \frac{0.707 \text{ V}}{1 \text{ V}}$$

$$= -20(0.15059)$$

$$= -3 \text{ dB}$$

Op-amp filters can be designed to have different roll-off characteristics, resulting in various slopes. A slope of -20 dB/decade means that as the frequency increases by ten (×10) from f_c, the output voltage will decrease 20 dB. The more decibel loss per decade results in a steeper slope. Remember that it is desirable to have the largest dB loss per decade, since this represents a sharper cutoff filter.

A simple low-pass filter is shown in Figure 3-8. The circuit configuration is a voltage follower. Resistor R and capacitor C at the noninverting input form a voltage divider. For frequencies of V_{in} below f_c, the capacitor's X_c is large, and nearly all of V_{in} is dropped across C. With V_{in} being large, V_{out} is also large. The gain of the stage is maximum for these lower frequencies. When the frequencies of V_{in} increase above f_c, the capacitor's X_c decreases and most of V_{in} is dropped across the resistor. In effect, capacitor C shunts much of V_{in} to ground. With V_{in} small, V_{out} is also small; hence, the gain of the stage is less than maximum for higher frequencies.

The f_c for this circuit can be approximated by the formula

$$f_c \approx \frac{1}{2\pi RC}$$

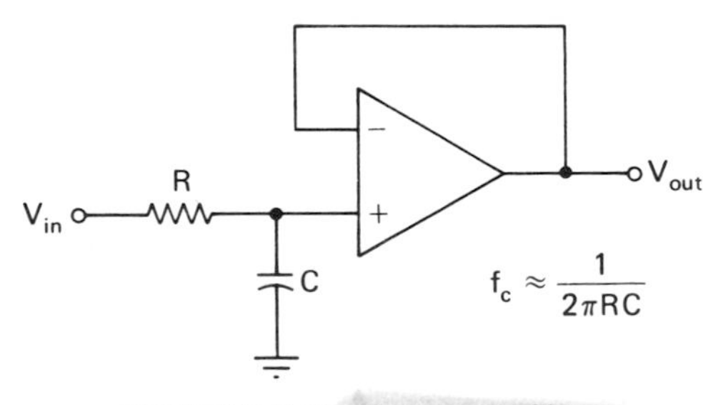

FIGURE 3-8 Simple low pass filter.

For example, if $R = 10\ \text{k}\Omega$ and $C = 0.1\ \mu\text{F}$, then

$$f_c \approx \frac{1}{6.28(10 \times 10^3)(0.1 \times 10^{-6})}$$
$$\approx \frac{1}{6.28(0.001)}$$
$$\approx \frac{1}{0.00628}$$
$$\approx 159\ \text{Hz}$$

This simple low-pass filter has a slope of approximately -20 dB/decade.

Because of the capacitors, filters do not have a constant phase angle (input-to-output phase) at f_c. A basic low-pass filter of -20 dB/decade has a phase angle of about -45° at f_c. An increase of -20 dB/decade will cause the phase angle to increase by -45°. For example, a -40-dB/decade filter will have a phase angle of -90°.

A sharper cutoff low-pass filter with a slope of about -40 dB/decade is shown in Figure 3-9. Capacitor C_2 shunts current away from the input for frequencies above f_c. The X_c of capacitor C_1 is low for frequencies in this stop-band range and more negative feedback is applied at the input. Therefore, the gain is reduced drastically.

The corner frequency for this circuit can be approximated with the formula

$$f_c \approx \frac{1}{2\pi\sqrt{R_1 R_2 C_1 C_2}}$$

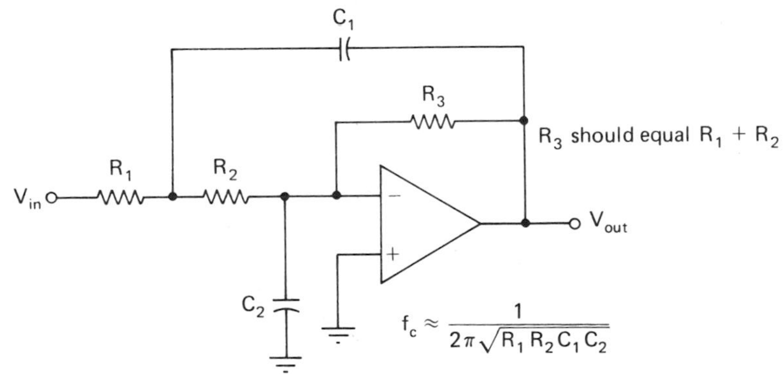

FIGURE 3-9 -40 dB/decade low pass filter.

For example, let $R_1 = 10\ \text{k}\Omega$, $R_2 = 10\ \text{k}\Omega$, $C_1 = 0.1\ \mu\text{F}$, and $C_2 = 0.01\ \mu\text{F}$. Then

$$f_c \approx \frac{1}{6.28\sqrt{(10 \times 10^3)(10 \times 10^3)(0.1 \times 10^{-6})(0.01 \times 10^{-6})}}$$

$$\approx \frac{1}{6.28\sqrt{(100 \times 10^6)(0.001 \times 10^{-12})}}$$

$$\approx \frac{1}{6.28(0.316 \times 10^{-3})}$$

$$\approx \frac{1}{1.99 \times 10^{-3}}$$

$$\approx 500\ \text{Hz}$$

This formula will be fairly accurate if $R_1 = R_2$ and C_1 is greater than C_2. Resistor R_3 is used for DC offset and should be equal to $R_1 + R_2$.

There are numerous types of filters for specific applications involving complex calculations. These basic filters, with a few calculations, are presented in the sections of this chapter to acquaint you with the applications of op amps to these circuits (see Experiment 7-22).

3-4 ACTIVE HIGH-PASS FILTER

A high-pass filter performs the opposite function from that of a low-pass filter. The high-pass filter attenuates all frequencies below a specific cutoff frequency f_c and passes all the frequencies above the f_c. Figure 3-10 shows a high-pass filter frequency-response curve. Similar to the low-pass filter, the practical f_c for a high-pass filter also occurs at 70.7% of the maximum output voltage. A high-pass filter can have different slopes, depending on the design of the circuit.

By interchanging the R and C components of a low-pass filter, a high-pass filter can be created. A simple high-pass filter is shown in Figure 3-11. With V_{in} to the noninverting input, C and R form a voltage divider. When V_{in} is below f_c, the X_c of C is large and drops most of V_{in}. The voltage drop across R is low and since the circuit is a follower, V_{out} is also low. When V_{in} increases above f_c, the X_c of C is low, allowing more V_{in} to be dropped across R; hence V_{out} is larger. This circuit has a slope of about -20 dB/decade, and the same formula used to find f_c for a simple low-pass filter can also be used.

A more dependable high-pass filter with a slope of about -40 dB/decade is shown in Figure 3-12. In this circuit, C_1 should equal C_2 and R_2 should be twice as large as R_1. Resistor R_3 should equal R_2 and is used for DC

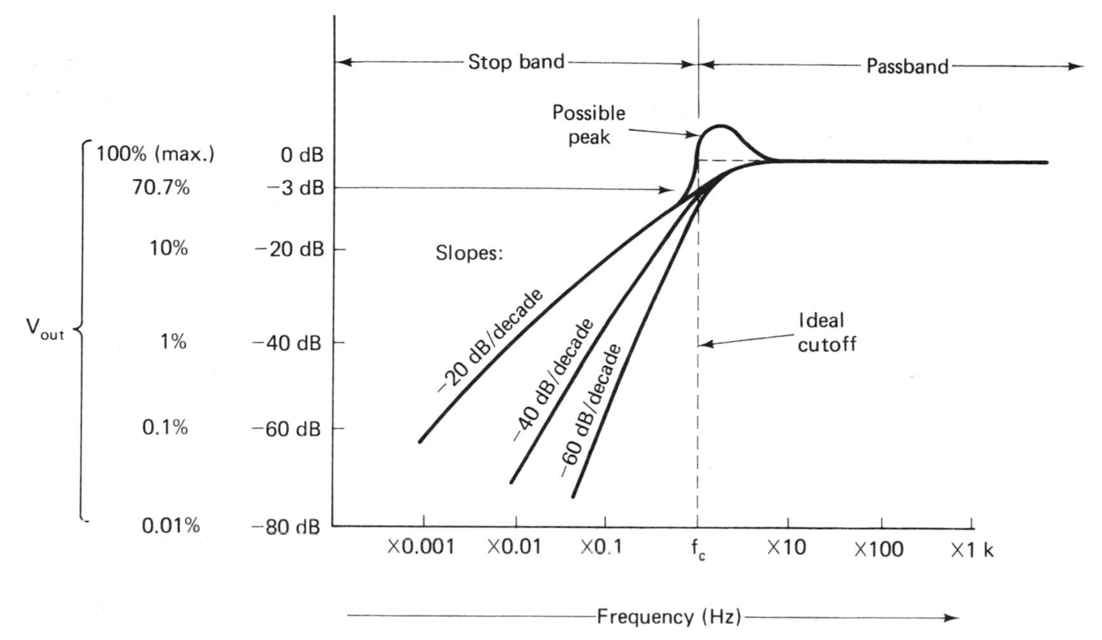

FIGURE 3-10 High pass filter frequency response curve.

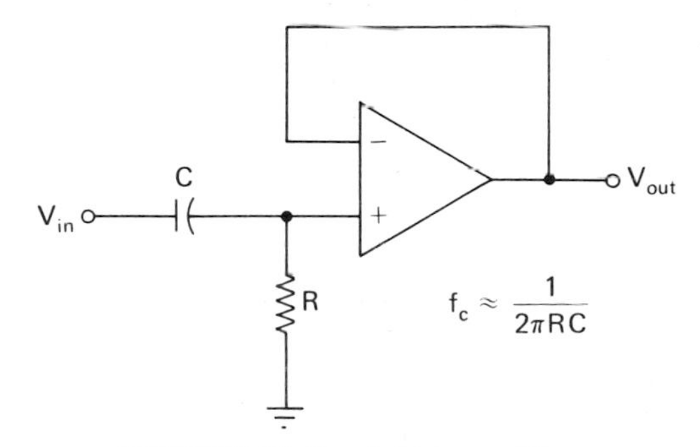

FIGURE 3-11 Simple high pass filter.

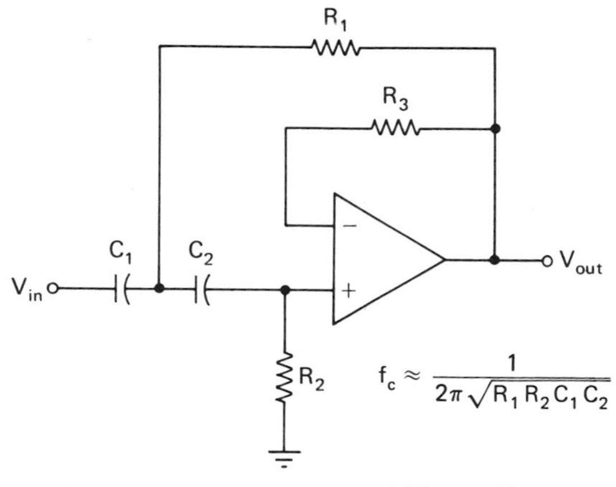

FIGURE 3-12 −40 dB/decade high pass filter.

offset. The X_c of C_1 and C_2 perform the same function as C with the simple high-pass filter of Figure 3-11. Feedback resistor R_1 connected to the junction of C_1 and C_2 provides a sort of double-filtering action.

The same formula can be used for this circuit to find f_c as that used for the low-pass filter of Figure 3-9 (see Experiment 7-23).

3-5 ACTIVE BANDPASS FILTER

A bandpass filter will pass a certain group of frequencies while rejecting all others. A typical frequency-response curve for a bandpass filter is shown in Figure 3-13. The maximum output voltage of this type of filter will peak at one specific frequency known as the resonant frequency f_r. When the frequency varies from resonance, the output voltage decreases. The point above and below f_r that V_{out} falls to 70.7% determines the bandwidth of the filter. The upper frequency that this point occurs can be designated f_H, while the lower frequency point can be f_L. The frequencies between f_L and f_H establish the bandwidth of the circuit (BW = $f_H - f_L$). If the bandwidth is less than 10% of f_r, the filter is considered as a narrow-bandpass filter, whereas greater than 10% would classify it as a wide-bandpass filter.

The narrower the bandwidth of a filter, the more selective it is said to

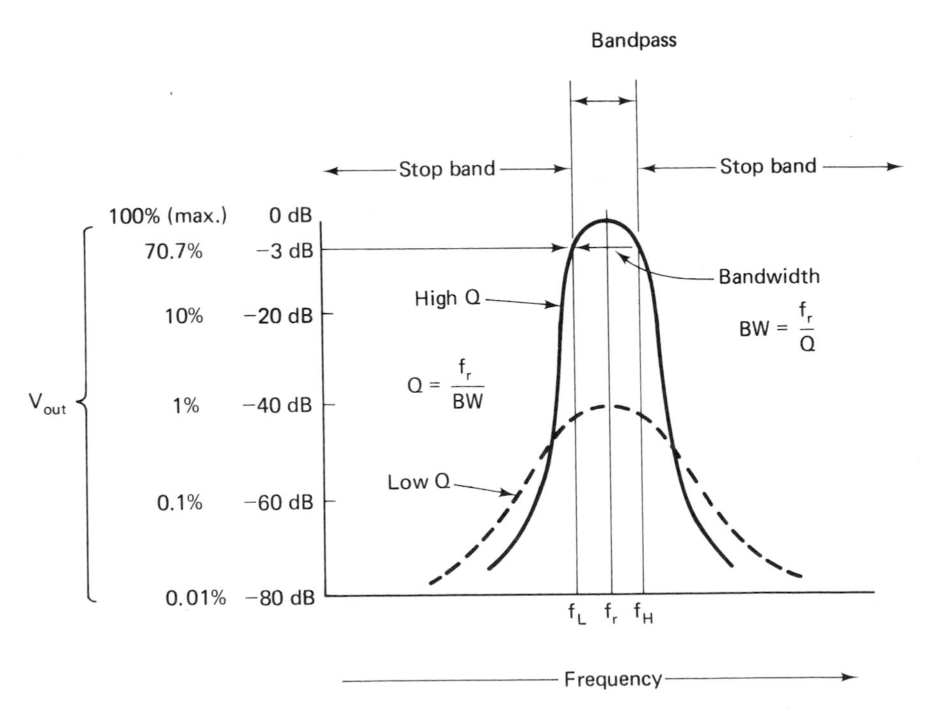

FIGURE 3-13 Bandpass filter frequency response curve.

be. The amount of selectivity is expressed as the quality factor, Q, of the circuit. The Q of a circuit can be determined by dividing the resonant frequency by its bandwidth,

$$Q = \frac{f_r}{\text{BW}}$$

The bandwidth of a circuit can be determined similarly:

$$\text{BW} = \frac{f_r}{Q}$$

A high-Q filter has a narrow bandwidth and tends to have a large V_{out}, while a low-Q filter has a wider bandwidth and tends to have less V_{out}. The dashed line in Figure 3-13 indicates a low-Q filter.

By combining low-pass and high-pass filtering circuits and selecting specific frequencies, a bandpass active filter can be constructed as shown in Figure 3-14. Components R_1 and C_2 provide low-pass filtering, while components C_1 and R_2 provide high-pass filtering. The f_r for this circuit is found with the formula

$$f_r \approx \frac{1}{2\pi\sqrt{R_p R_3 C_1 C_2}}$$

where R_p is the equivalent parallel resistance of R_1 and R_2, which is found thus:

$$R_p = \frac{R_1 R_2}{R_1 + R_2}$$

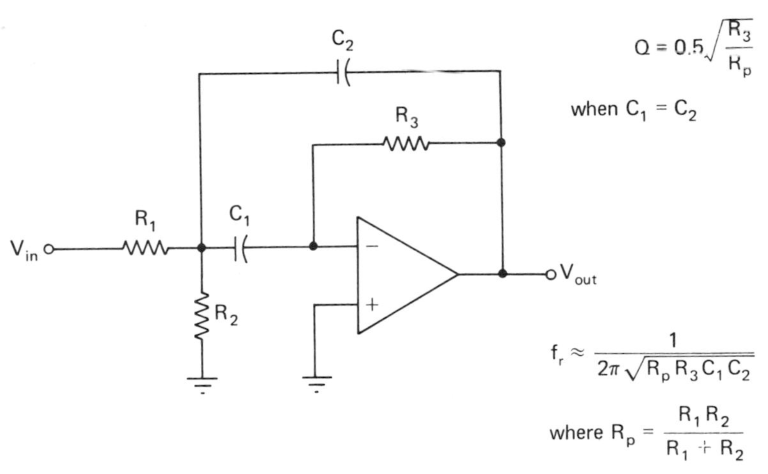

FIGURE 3-14 Active bandpass filter.

For simplification, the Q of this filter is found with the formula

$$Q = 0.5\sqrt{\frac{R_3}{R_p}}$$

when $C_1 = C_2$.

Feedback resistor R_3 plays an important part in this circuit. It not only establishes the gain of the circuit, but it affects the Q and f_r as well. When R_3 is relatively small, the f_r will be high and the Q low. But when R_3 is large, f_r is low and Q is high. For example, find the f_r, Q, and BW for the bandpass filter of Figure 3-14 when

$$R_1 = 10\ \text{k}\Omega \qquad C_1 = 0.01\ \mu\text{F}$$
$$R_2 = 10\ \text{k}\Omega \qquad C_2 = 0.01\ \mu\text{F}$$
$$R_3 = 100\ \text{k}\Omega$$

$$R_p = \frac{R_1 R_2}{R_1 + R_2} = \frac{(10\ \text{k}\Omega)(10\ \text{k}\Omega)}{10\ \text{k}\Omega + 10\ \text{k}\Omega} = \frac{100\ \text{k}\Omega}{20\ \text{k}\Omega} = 5\ \text{k}\Omega$$

$$f_r = \frac{1}{2\pi\sqrt{R_p R_3 C_1 C_2}}$$
$$\approx \frac{1}{6.28\sqrt{(5 \times 10^3)(100 \times 10^3)(0.01 \times 10^{-6})(0.01 \times 10^{-6})}}$$
$$\approx \frac{1}{6.28\sqrt{0.05 \times 10^{-6}}}$$
$$\approx \frac{1}{6.28(0.224 \times 10^{-3})}$$
$$\approx \frac{1}{1.4 \times 10^{-3}}$$
$$\approx 714\ \text{Hz}$$

$$Q \approx 0.5\sqrt{\frac{R_3}{R_p}} \approx 0.5\sqrt{\frac{100 \times 10^3}{5 \times 10^3}}$$
$$\approx 0.5\sqrt{20}$$
$$\approx 0.5(4.5)$$
$$\approx 2.25$$

$$\text{BW} = \frac{f_r}{Q} = \frac{714}{2.25} = 317\ \text{Hz}$$

Therefore,

$$f_H = f_r + \frac{\text{BW}}{2} = 714 + \frac{317}{2} \approx 873 \text{ Hz}$$

$$f_L = f_r - \frac{\text{BW}}{2} = 714 - \frac{317}{2} \approx 556 \text{ Hz}$$

If R_3 is increased to 1 MΩ, the f_r is

$$f_r \approx \frac{1}{6.28\sqrt{(5 \times 10^3)(1 \times 10^6)(0.01 \times 10^{-6})(0.01 \times 10^{-6})}}$$

$$\approx \frac{1}{6.28\sqrt{0.5 \times 10^{-6}}}$$

$$\approx 225 \text{ Hz}$$

$$Q = 0.5 \quad \frac{1 \times 10^6}{5 \times 10^3} = 7.07$$

$$\text{BW} = \frac{225}{7.07} \approx 32 \text{ Hz}$$

$$f_H = f_r + \frac{\text{BW}}{2} = 225 + \frac{32}{2} \approx 241 \text{ Hz}$$

$$f_L = f_r - \frac{\text{BW}}{2} = 225 - 16 \approx 209 \text{ Hz}$$

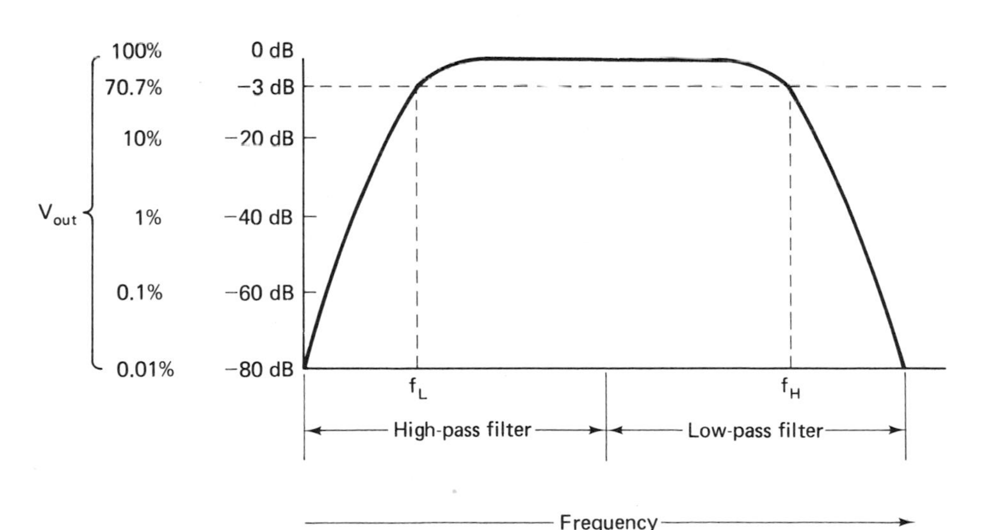

FIGURE 3-15 Wide-band pass frequency response curve using two filters.

If it is desired to have a wide-bandpass filter with a fairly constant output during the bandwidth, a low-pass filter and a high-pass filter can be connected together. The frequency-response curve would resemble that of Figure 3-15 on page 77. It makes no difference which filter comes first. The high-pass filter would have to be designed for an f_c at the low end of the bandwidth (f_L) and the low-pass filter would have to be designed for an f_c at the high end (f_H) (see Experiment 7-24).

3-6 ACTIVE NOTCH (BAND-REJECT) FILTER

A notch filter (formerly referred to as a band-reject filter) functions opposite to that of a bandpass filter. Figure 3-16 shows that this type of circuit passes all frequencies except a specific group. The output voltage will remain maximum until the applied frequency approaches f_r, where it is attenuated. The bandwidth occurs for amplitudes below 70.7% of V_{out}, and the same Q and BW formulas associated with the bandpass filter also apply to the notch filter. The notch filter is more highly selective than the older band-reject type and is used very often to reduce unwanted effects of specific frequencies, such as 60-Hz hum.

A basic active notch filter is shown in Figure 3-17. In this configuration, V_{in} is applied to both inputs. Components R_1, R_4, C_1, and C_2 form a frequency-selective feedback network. The ratio of the resistances to the

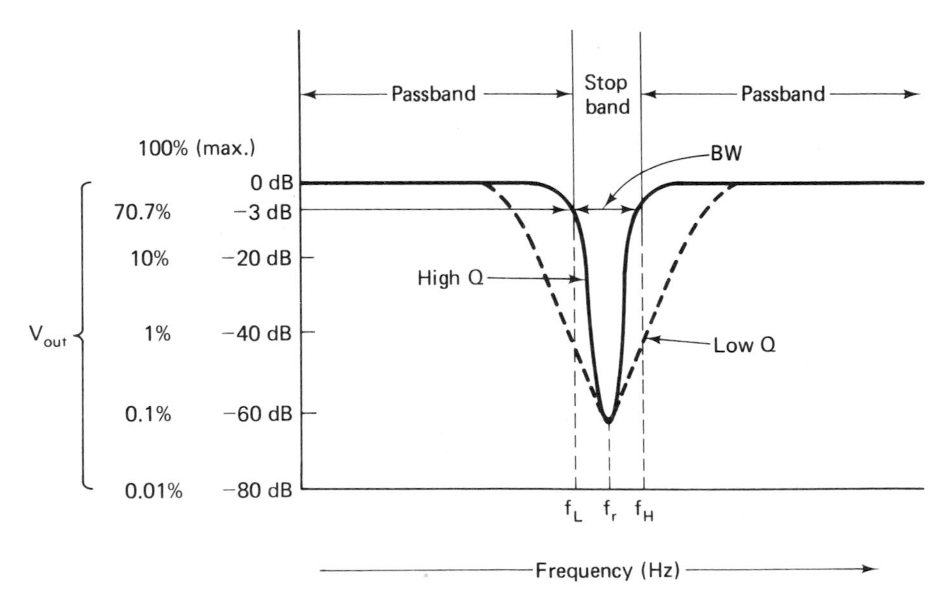

FIGURE 3-16 Notch filter frequency response curve.

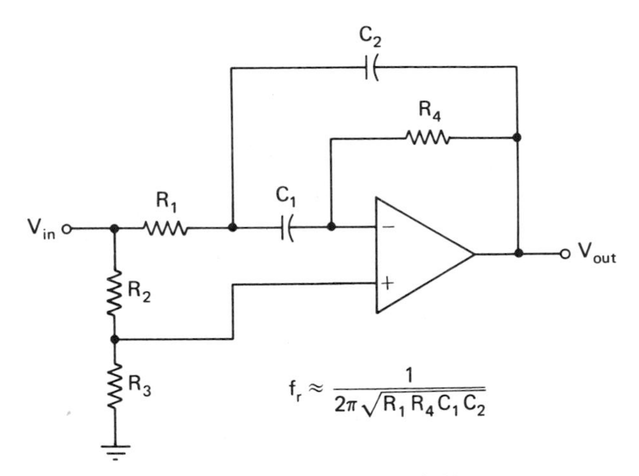

FIGURE 3-17 Active notch filter.

reactances of the capacitors determine the f_r of the circuit and is expressed

$$f_r = \frac{1}{2\pi\sqrt{R_1 R_4 C_1 C_2}}$$

The Q of the circuit is determined by the ratio of R_1 to R_4 when C_1 equals C_2 and is given as

$$Q = 0.5\sqrt{\frac{R_4}{R_1}}$$

The slope of the notch is determined by the Q, whereas the depth of the notch is dependant on the Q and the percentage of R_3 to $R_2 + R_3$, and can be expressed

$$N \text{ (slope of notch)} = \frac{\frac{1}{Q} - \frac{1 - R_p^*}{R_p}(2Q)}{1/Q} \qquad {}^*R_p = \frac{R_3}{R_2 + R_3}$$

Resistor R_3 is typically 50 times greater than R_2.

The R_2 and R_3 voltage divider produces a differential voltage at the inputs of the op amp. For frequencies below f_r, the X_c of the capacitors is very high and there is little feedback; thus the output is maximum. When the frequency of V_{in} approaches f_r, the reactances form the appropriate relationship and phase angle with the resistances to produce feedback, which decreases the output. As the frequency of V_{in} increases above f_r, the X_c of the capacitors decrease and the feedback factor approaches 1, or the gain of a voltage follower. As an example, let

$R_1 = 10\ \text{k}\Omega \quad C_1 = 0.0266\ \mu\text{F}$

$R_2 = 1\ \text{k}\Omega \quad C_2 = 0.0266\ \mu\text{F}$

$R_3 = 47\ \text{k}\Omega$ (*Hint:* Parallel one 0.02 μF with two 0.0033-μF capacitors.)

$R_4 = 1\ \text{M}\Omega$

Then

$$f_r = \frac{1}{6.28\sqrt{(10 \times 10^3)(1 \times 10^6)(0.0266 \times 10^{-6})(0.0266 \times 10^{-6})}}$$
$$= \frac{1}{6.28\sqrt{7.075 \times 10^{-6}}}$$
$$= \frac{1}{6.28(2.66 \times 10^{-3})}$$
$$\approx 60\ \text{Hz}$$

and

$$Q = 0.5\sqrt{\frac{1 \times 10^6}{10 \times 10^3}}$$
$$= 0.5\sqrt{100}$$
$$= 5$$

(See Experiment 7-25.)

Another popular notch filter is the twin-"T" notch filter shown in Figure 3-18. A basic twin-"T" notch filter (Figure 3-18a) simply consists of a double- or twin-"T" passive filter fed into an op-amp follower. This circuit usually has a very low Q (less than 1) and the op amp serves mainly as a buffer. The data notations indicate that R_3 is one-half the value of R_1 or R_2 and C_1 is twice the value of C_2 or C_3.

An identical circuit, with a slight modification, is shown in Figure 3-18b. In this case, C_3 and R_3 are not returned to ground but are connected to the low-impedance output of the op amp. The Q of this circuit can reach up to 50, creating an extremely sharp notch. It also permits the use of high resistances and very low values of capacitances for f_r at low frequencies. The same formulas apply to both of the twin-"T" notch filters.

To minimize frequency shift with temperature, many op amp filters use silver mica or polycarbonate capacitors with precision resistors.

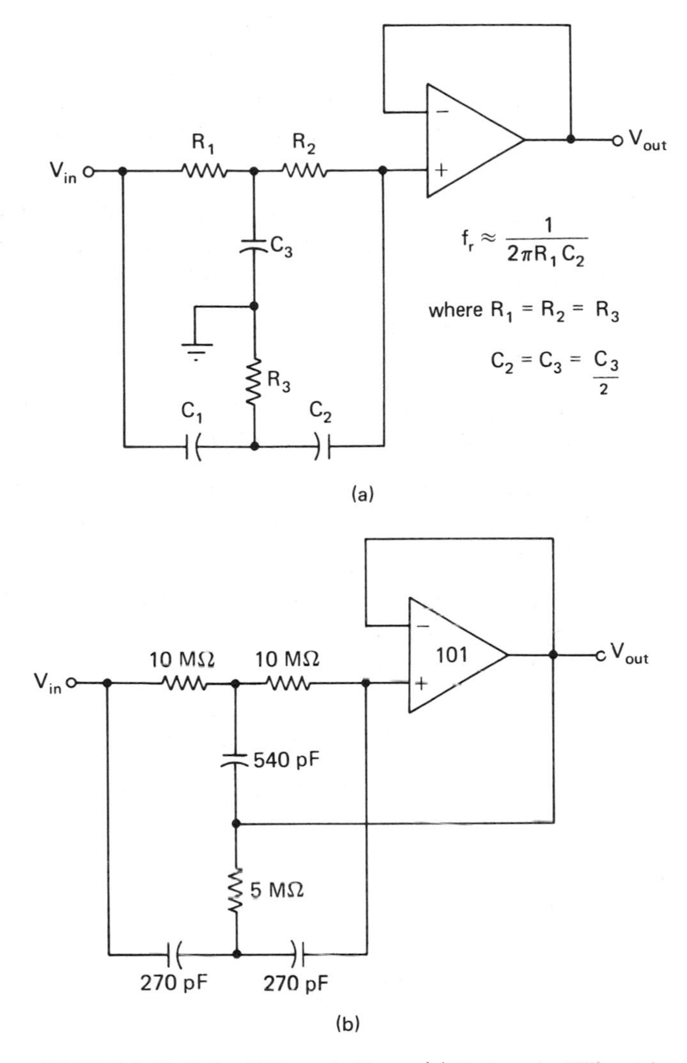

FIGURE 3-18 Twin "T" notch filters: (a) Basic twin "T" notch filter; (b) High Q notch filter.

SUMMARY POINTS

1. The capacitor is the major feedback element in an op-amp integrator.
2. An op-amp integrator produces a ramp output voltage for DC input voltage (a square wave produces a triangular wave).

3. An op-amp differentiator has a capacitor in series with the input.
4. An op-amp differentiator produces a sharp leading edge output for a ramp input voltage (a triangular input voltage produces a square-wave output).
5. A practical op-amp low-pass filter shunts high frequencies to ground through a capacitor at the input and feeds back high frequencies through a capacitor to reduce gain.
6. A practical op-amp high-pass filter blocks low frequencies with series capacitors at its input.
7. An op-amp bandpass filter combines both low-pass and high-pass filtering elements so that the output voltage only peaks at f_r.
8. An op-amp notch filter uses both inputs, where a frequency-selective feedback network causes the output voltage to attenuate sharply at f_r.
9. The cutoff frequency, also called the corner frequency or break frequency, occurs at 70.7% of the output voltage (or −3 dB down from maximum output).
10. The bandwidth of a filter is the upper (f_H) and lower (f_L) frequencies, where 70.7% of the maximum output voltage occurs.
11. The Q of a filter determines how selective the circuit is to the resonant frequency, f_r.
12. Decibel loss per decade means the loss in decibels of a filter as the frequency increases by a factor of 10.

TERMINOLOGY EXERCISE

Write a brief definition for each of the following terms:

1. Integrator circuit
2. Differentiator circuit
3. Low-pass filter
4. High-pass filter
5. Bandpass filter
6. Band-reject filter
7. Notch filter

8. Cutoff frequency
9. Corner frequency
10. Resonant frequency
11. Frequency response curve
12. Slope (pertaining to a response curve)
13. Stop band
14. Bandpass
15. Bandwidth
16. Q of a circuit

PROBLEMS AND EXERCISES

1. Referring to Figure 3-3, what is V_{out} when V_{in} is +2 V at 500 Hz?
2. Referring to Figure 3-6, what is V_{out} when V_{in} is +2 V at 200 Hz?
3. What is the approximate f_c in Figure 3-8 when R = 22 kΩ and C = 0.01 μF?
4. What is the approximate f_c in Figure 3-9 when R_1 = 22 kΩ, R_2 = 10 kΩ, C_1 = 0.01 μF, and C_2 = 0.01 μF?
5. What is the approximate value of f_c in Figure 3-11 when C = 0.016 μF and R = 10 kΩ?
6. Find the approximate f_c in Figure 3-12 when R_1 = 10 kΩ, R_2 = 100 kΩ, C_1 = 0.01 μF, and C_2 = 0.02 μF.
7. What is the Q of a filter if its f_r = 1 kHz and its BW = 125 Hz? Find f_H and f_L for this circuit.
8. What is the BW of a filter if its Q = 10 and its f_r = 400 Hz? Find f_H and f_L for this circuit.
9. Find the f_r, Q, BW, f_H, and f_L of the filter shown in Figure 3-14 when R_1 = 10 kΩ, R_2 = 22 kΩ, R_3 = 470 kΩ, C_1 = 0.1 μF, and C_2 = 0.1 μF. If the input voltage is 2.5 V p-p, what is the voltage at the corner frequency (f_H or f_L)?
10. Find the f_r, Q, BW, f_H, and f_L, of the filter shown in Figure 3-17 when R_1 = 4.7 kΩ, R_2 = 1 kΩ, R_3 = 50 kΩ, R_4 = 2.2 MΩ, C_1 = 0.01 μF, and C_2 = 0.01 μF.

SELF-CHECKING QUIZ

Match each name in column A with its proper output frequency-response curve shown in Figure 3-19.

Column A

1. Integrator
2. Differentiator
3. Low-pass filter
4. High-pass filter
5. Bandpass filter
6. Notch filter

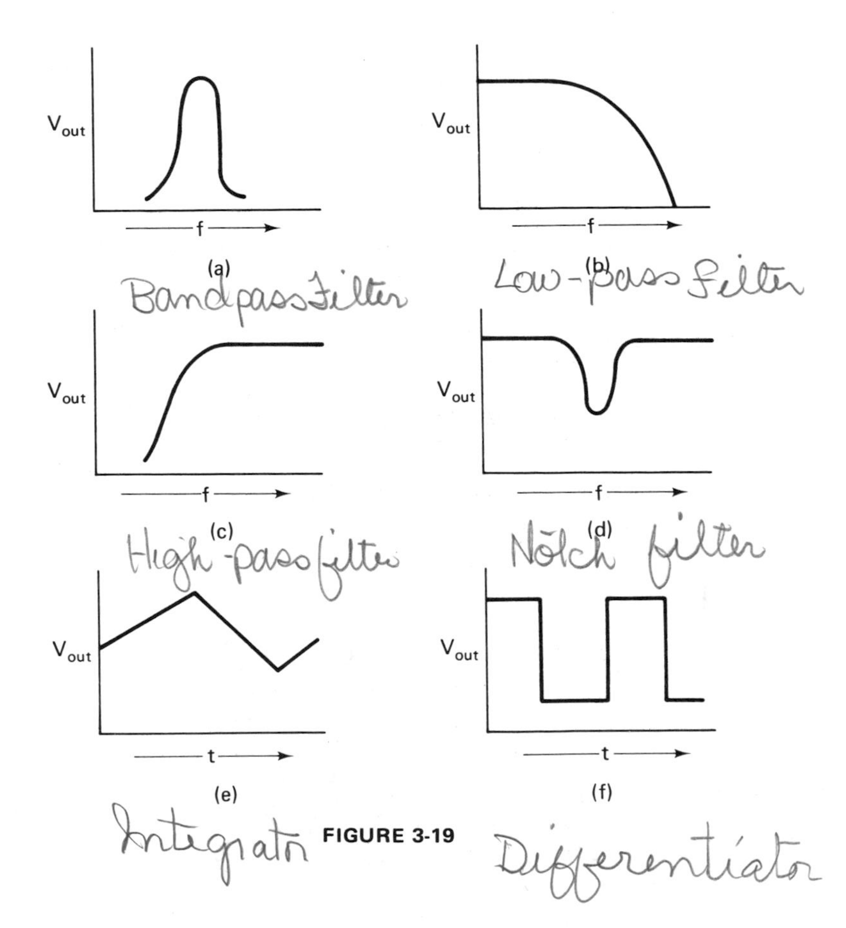

FIGURE 3-19

Match each name in column B with its proper schematic diagram shown in Figure 3-20 below and on page 86.

Column B

7. Integrator
8. Differentiator
9. Low-pass filter
10. High-pass filter
11. Bandpass filter
12. Notch filter

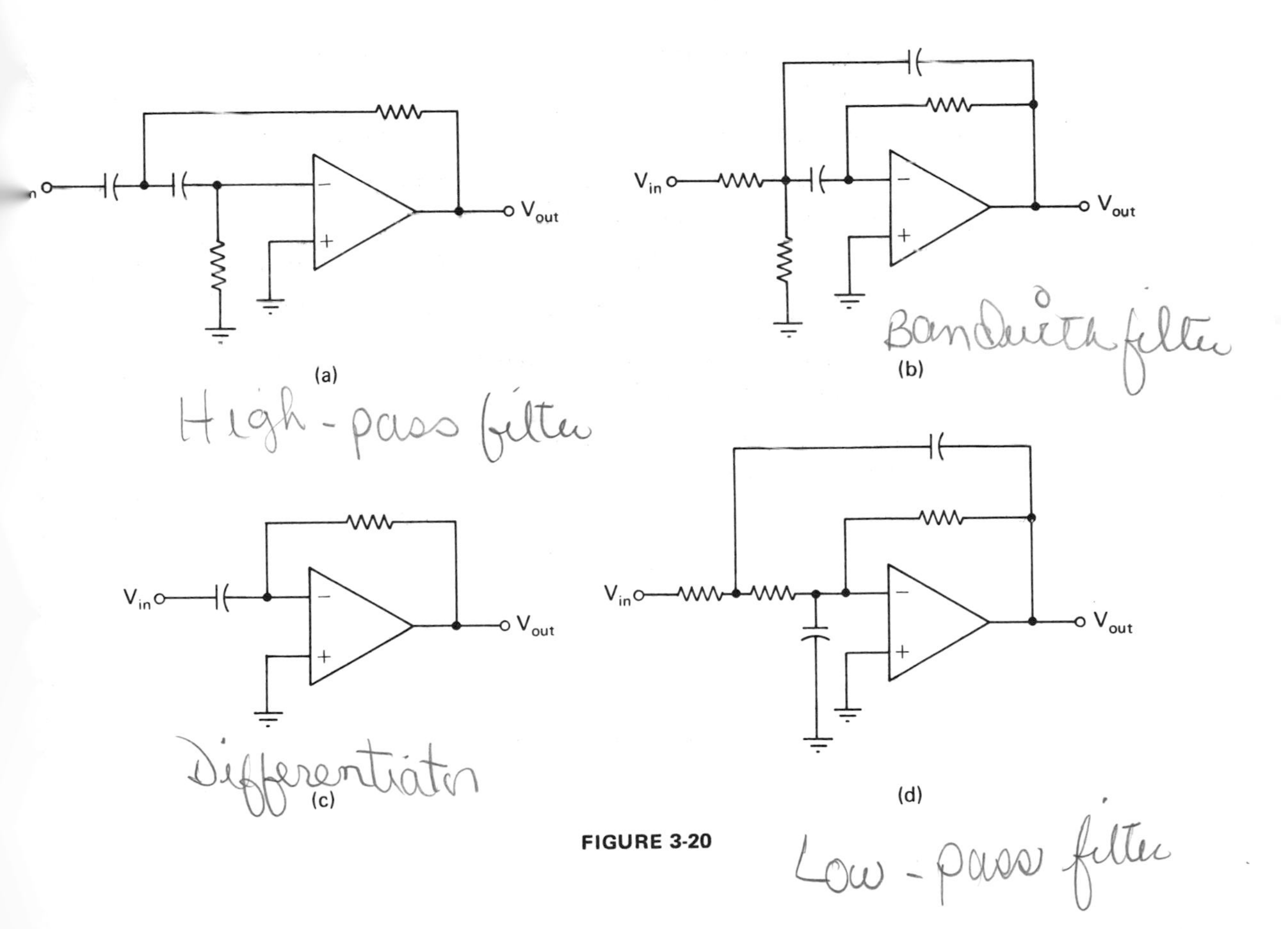

FIGURE 3-20

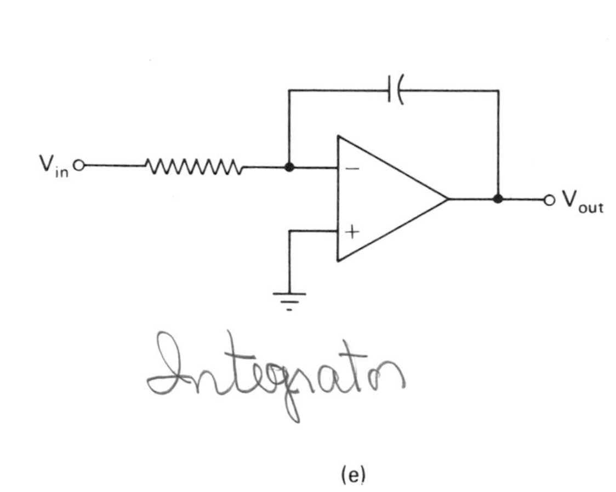

(e)

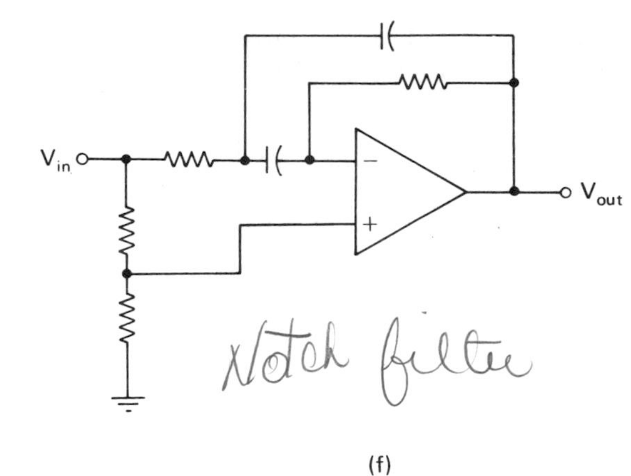

(f)

FIGURE 3-20 (*Continued*)

True-or-False Questions

T 13. A low-pass filter and a high-pass filter can be connected to produce a bandpass filter.

F 14. A notch filter passes the frequencies in its bandwidth.

F 15. A differentiator will produce a triangular output wave when a square wave is applied to the input.

T 16. The bandwidth of a filter can be found by dividing f_r by Q.

T 17. The depth of the notch in a notch filter is determined by Q.

T 18. An op-amp integrator will transform a square-wave input to a triangular-wave output.

F 19. A high-Q bandpass filter will have a wider bandwidth than a low-Q bandpass filter.

T 20. A voltage decrease from 8 V to 5.7 V is the same as a −3-dB loss.

(Answers at back of book)

Chapter 4

Op-Amp Oscillators

Oscillators convert DC voltage into AC voltage or pulsating DC voltage. The cycles per second at which the voltage changes is called the oscillator frequency. There are four basic oscillator waveforms: square wave, triangle wave, sawtooth wave, and sine wave. The term "signal generator" is often used interchangeably with oscillator. A signal generator is classified by the type of waveform it generates. The stability of an oscillator indicates how well it will maintain its output amplitude and remain on or close to the desired frequency for which it was designed. Signal generators are used to provide the signal source for other electronic circuits.

This chapter will show how op amps are applied to signal generators. Using op amps and a few passive components can produce signal generators that are fairly stable and easy to build.

4-1 SQUARE-WAVE GENERATOR

The square-wave generator belongs to that family of oscillators known as multivibrators. It is referred to as a free-running or astable multivibrator, since its output is constantly changing states (high and low) without any input signal. A basic square-wave generator is shown in Figure 4-1. There are two feedback paths for this circuit. One goes from the output to the

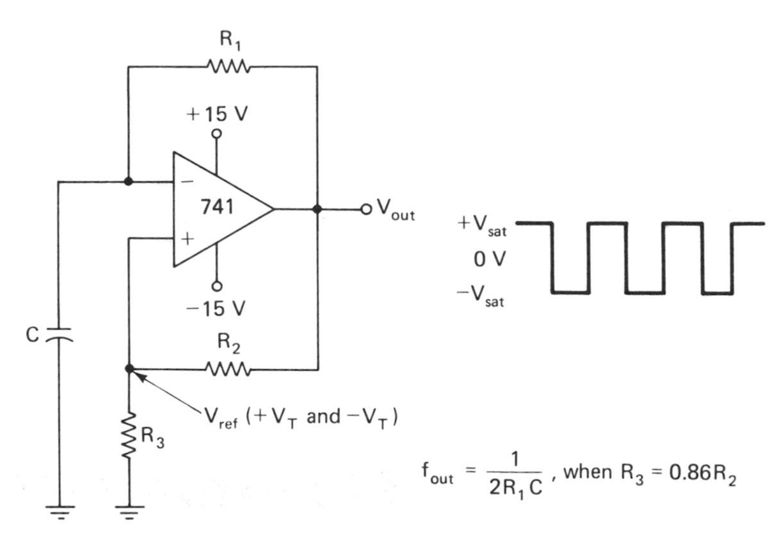

FIGURE 4-1 Basic square wave generator.

inverting input and contains a feedback resistor and a capacitor connected to ground. This *RC* combination determines the fundamental operating frequency of the generator. The other feedback path goes to the noninverting input and contains two resistors. These resistors form a voltage divider that produces a reference voltage (V_{ref}) at the noninverting input. If these resistors are selected so that R_3 is 86% of the value of R_2, the approximate frequency of the generator can be found by the simple expression

$$f_{out} = \frac{1}{2R_1C}$$

With the R_2 and R_3 voltage divider providing the V_{ref} at the noninverting input, the circuit behaves like a voltage-level detector.

For instance, when power is applied to the circuit, the capacitor begins to charge up through R_1 to the value of V_{out}. The op-amp output will be at $+V_{sat}$, and V_{ref} at the noninverting input will be at the positive threshold voltage, $+V_T$. When the voltage across the capacitor increases above $+V_T$, the op amp switches states and V_{out} goes in the negative direction to $-V_{sat}$. The V_{ref} at the noninverting input is now at the negative threshold voltage, $-V_T$. The capacitor now changes its charging direction and begins to charge toward $-V_{sat}$. The instant the voltage on the capacitor increases below $-V_T$, the op amp switches back to the original state and V_{out} goes to $+V_{sat}$. A cycle has now been completed and the process is repeated. Figure 4-2 illustrates the action of the capacitor voltage (V_c) and the op-amp output voltage, V_{out}.

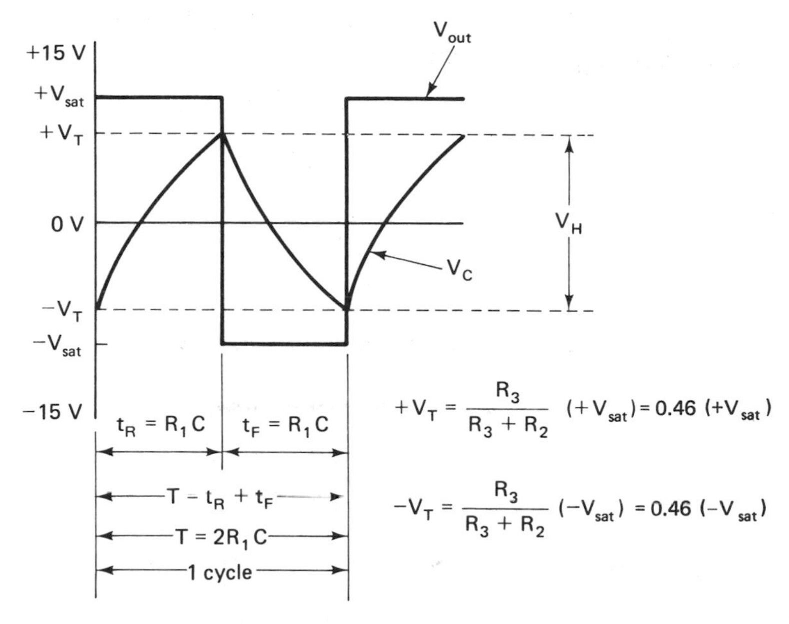

FIGURE 4-2 Capacitor voltage versus output voltage.

Threshold voltages $+V_T$ and $-V_T$ are determined by the resistor voltage divider of R_2 and R_3 and is expressed as

$$+V_T = \frac{R_3}{R_3 + R_2}(+V_{\text{sat}}) = 0.46(+V_{\text{sat}})$$

and

$$-V_T = \frac{R_3}{R_3 + R_2}(-V_{\text{sat}}) = 0.46\,(-V_{\text{sat}})$$

To build a signal generator with a 1-kHz test signal, let $R_1 = 10\ \text{k}\Omega$, $C = 0.05\ \mu\text{F}$, $R_2 = 100\ \text{k}\Omega$, and $R_3 = 86\ \text{k}\Omega$.

To verify the output frequency,

$$f_{\text{out}} = \frac{1}{2(10 \times 10^3)(0.05 \times 10^{-6})}$$

$$= \frac{1}{2(0.5 \times 10^{-3})}$$

$$= \frac{1}{1 \times 10^{-3}}$$

$$= 1\ \text{kHz}$$

If $+V_{sat}$ and $-V_{sat}$ equal +13.5 V and −13.5 V, respectively, the threshold voltage amplitudes can be found:

$$+V_T = 0.46(+13.5\text{ V})$$
$$= +6.21\text{ V}$$

and

$$-V_T = 0.46(-13.5\text{ V})$$
$$= -6.21\text{ V}$$

Therefore, the peak-to-peak threshold voltage V_H is

$$V_{H(\text{p-p})} = (+V_T) - (-V_T)$$
$$= +6.21 - (-6.21)$$
$$= 12.42\text{ V}$$

or, in other words, the V_H is twice $+V_T$ or twice $-V_T$:

$$V_{H(\text{p-p})} = 2(+V_T) \quad \text{or} \quad 2(-V_T)$$

(See Experiment 7-26.)

4-2 SAWTOOTH-WAVE GENERATOR

A sawtooth-wave generator, or a ramp-voltage generator as it is sometimes called, is shown in Figure 4-3a. Notice that it resembles an op-amp integrator circuit. If a −1 V is placed at the active (inverting) input, the capacitor will begin to charge up at a linear rate in a positive direction. It will continue to charge until $+V_{sat}$ is reached. If the switch is momentarily closed before $+V_{sat}$ is reached, the capacitor will discharge rapidly. When the switch is again open, the process will be repeated as shown in Figure 4-3b. The output voltage is determined by

$$V_{out} = V_{in}\left(\frac{1}{R_{in}C_f}\right) \times t$$

where t is the time in seconds that the switch is open. The slope is determined by V_{in}, R_{in}, and C_f. If a negative ramp voltage is needed, V_{in} should be positive.

Manual operation of this circuit, of course, results in an extremely low output frequency. Therefore, an electronic switch is used in place of the manual switch in order to produce a useful sawtooth-wave generator, as shown in Figure 4-4.

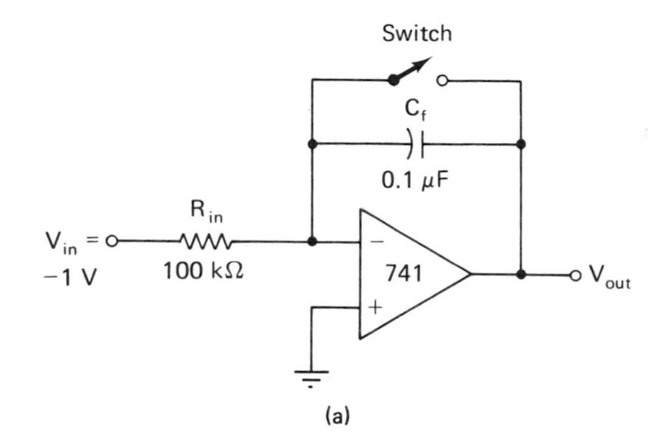

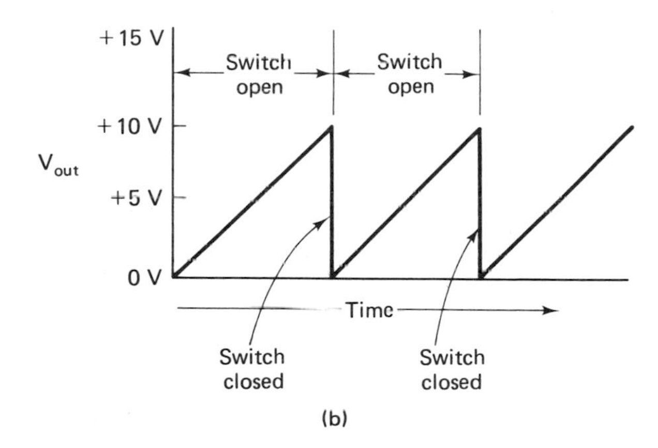

FIGURE 4-3 Basic sawtooth wave generator with manual control: (a) schematic diagram; (b) sawtooth waveform voltage output.

A programmable unijunction transistor (PUT) is used as the active switch. The PUT belongs to the thyristor family of electronic devices and resembles a silicon-controlled rectifier (SCR) in operation, the difference being that it is triggered on by a negative-going pulse. However, as shown, if the gate voltage is set at a determined positive voltage V_P (peak voltage) by R_4 and R_5 resistor voltage dividers and the anode (A) to cathode (K) voltage (V_{AK}) goes more positive than V_P, the PUT will fire (turn on). This is the same effect as a negative pulse to the gate. The PUT will remain on until the current through it falls below its minimum holding current value. It then turns off and becomes like an open switch.

Component values and variable resistors are given in Figure 4-4 so that you can change the frequency of the generator and verify the formula that will be given. Voltage dividers R_1 and R_2 are used to develop V_{ref}. Diodes D_1 and D_2 help to stabilize the voltage across R_2 when it is adjusted to vary

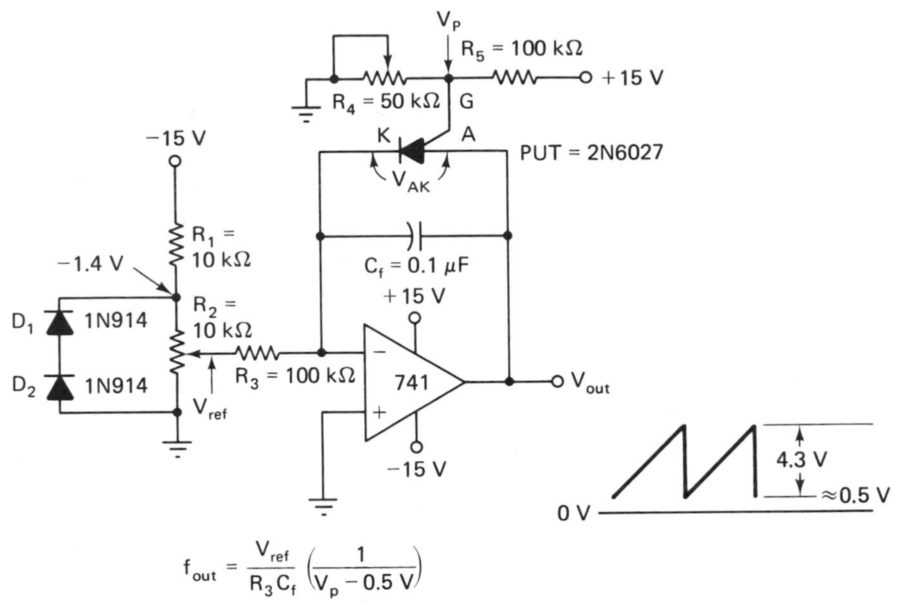

FIGURE 4-4 Self-generating sawtooth wave generator.

the frequency. The V_{out} amplitude is determined by R_4, which in turn affects the frequency.

Let us say that we set V_{ref} to -1 V and V_P to +4 V. The capacitor begins charging up linearly toward $+V_{sat}$. At some point just greater than V_P, the PUT fires, the capacitor discharges, and V_{out} goes to minimum. The current through the PUT falls below the minimum holding current, the PUT turns off, and the process begins again. When the PUT is on, it does not act like a normal closed switch because it has a forward voltage drop (V_F) of a few tenths of a volt up to 1 V, depending on the particular PUT used. We can however, approximate the output frequency with the expression

$$f_{out} = \frac{V_{ref}}{R_3 C_f} \left(\frac{1}{V_P - 0.5 \text{ V}} \right)$$

For instance, with the voltages given above and $R_3 = 100 \text{ k}\Omega$ and $C_f = 0.1\ \mu\text{F}$,

$$f_{out} = \frac{1}{(100 \times 10^3)(0.1 \times 10^{-6})} \left(\frac{1}{4 - 0.5} \right)$$

$$= \frac{1}{0.01} \left(\frac{1}{3.5} \right)$$

$$= (100)(0.286)$$

$$= 29 \text{ Hz}$$

The peak amplitude of f_{out} will be a few tenths of a volt greater than V_P (in this case about +0.3 V).

This equation shows how the amplitude of V_{ref} and V_P along with R_3 and C_f affect the f_{out} of the generator. How fast V_{out} rises is determined by V_{ref} and R_3 and C_f, as that part of the equation V_{ref}/R_3C_f, while $(V_P - 0.5)$ sets the value to which V_{out} can rise before the capacitor discharges. Since the voltage plays a part in f_{out}, a circuit like this is sometimes referred to as a voltage-to-frequency converter or a voltage-controlled oscillator.

4-3 TRIANGLE-WAVE GENERATOR

The triangle-wave generator usually requires at least two op amps to operate. A basic circuit is a square-wave generator connected to an integrator, as shown in Figure 4-5a. From the discussions on the op-amp integrator (Section 3-1) and the sawtooth generator (Section 4-2), it was shown that the V_{out} of the integrator can ramp up or ramp down. For this reason it is often called a ramp generator.

When the output of the square-wave generator goes positive, the output of the ramp generator ramps negative. Similarly, when the output of the square-wave generator goes negative, the output of the ramp generator ramps positive. This action is illustrated in Figure 4-5b, which produces the triangle-wave output (V_{tri}). There is also a square-wave output (V_{squ}). A signal generator such as this, that produces two or more different waveforms, is referred to as a function generator.

The frequency of the triangle-wave output is the same as the frequency of the square-wave generator, which can be determined from Section 4-1. It is desireable to have the RC time constant of R_4 and C_2 be twice as large as the time constant of R_1 and C_1 to prevent distortion of the triangle waveform. The amplitude of the square will be nearly $\pm V_{sat}$, while the amplitude of the triangle wave can be determined from the integrator circuit discussed in Section 3-1.

4-3-1 Positive Feedback

Another very popular triangle-wave generator uses a ramp generator combined with a voltage-sensing comparator. To understand the action of this type of triangle-wave generator, it will be of value to analyze an op amp with positive feedback. An op-amp circuit with positive feedback is shown in Figure 4-6. Notice that the inverting input is connected to ground while the noninverting input is placed above it. It is advisable to scrutinize a schematic diagram using op amps, since this arrangement is used very often. Realizing that a circuit contains this type configuration will aid you in analyzing the operation of a particular circuit.

When power is initially applied to the circuit, a slight differential voltage

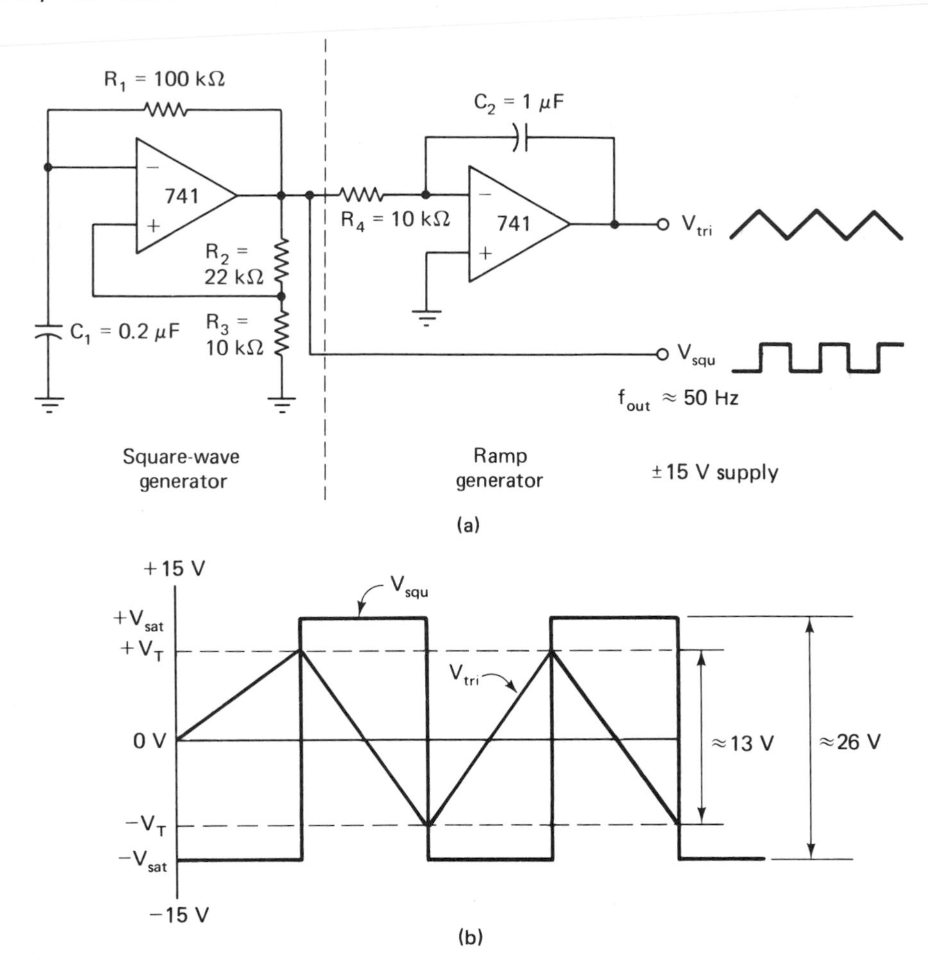

FIGURE 4-5 Simple triangle wave generator: (a) combination of two basic circuits; (b) output waveforms.

at the inputs or the offset voltage will cause V_{out} to saturate in either the positive or negative direction. This is because the regenerative action of positive feedback increases the voltage at the input, therefore driving the op amp harder in the direction in which its output is going. In Figure 4-6a, V_{out} is at $+V_{sat}$ and will remain in this state until V_{in} drops to the negative threshold voltage $(-V_T)$, at which time V_{out} will be driven to $-V_{sat}$. Remember from Section 2-1 on voltage comparators that each time the voltage on the active input crossed the zero reference point, V_{out} would swing to the opposite saturation voltage. At this time of changing states, the differential voltage V_d at the inputs is nearly 0 V. We can understand this better by analyzing the current path through the resistors. If V_{in} = 0 V and V_{out} = +13.5 V, then by Ohm's law ($I_{in} = I_F = V_{out}/R_{in} + R_F$), the current through R_{in} and R_F =

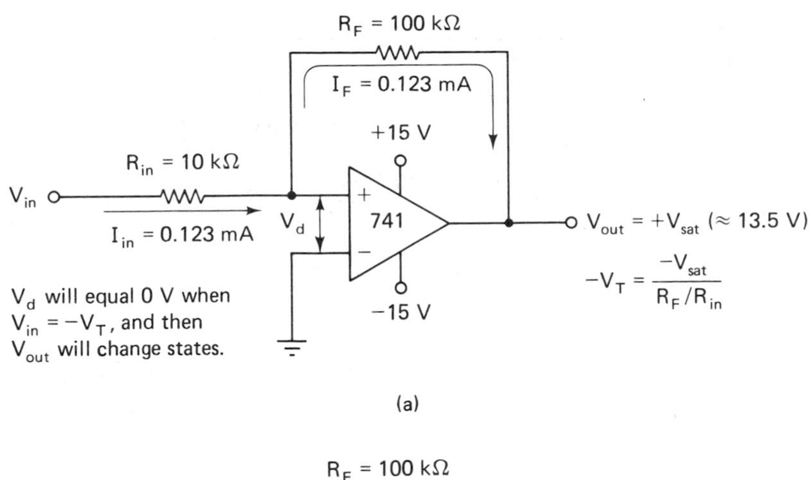

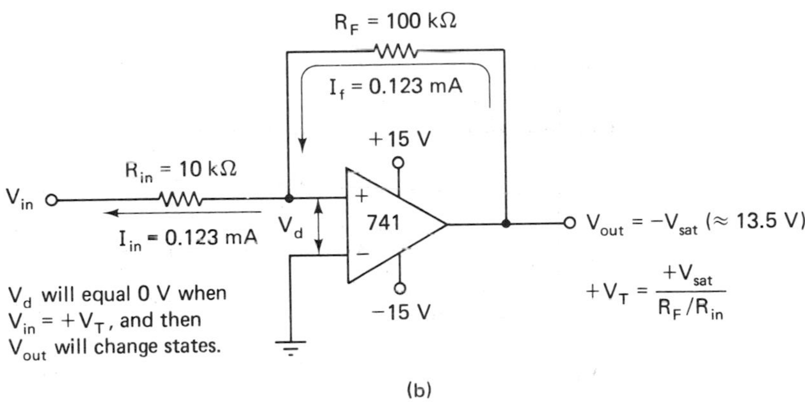

FIGURE 4-6 Op amp positive feedback: (a) in the +Vsat state; (b) in the −Vsat state.

0.123 mA (the current is the same since both resistors are in series). Therefore, $V_d = R_{in} \times I_{in} = +1.23$ V. As V_{in} is made more negative, V_d decreases, because of the algebraic summation of voltages of unlike polarities. When V_d reaches 0 V, V_{out} swings to $-V_{sat}$.

When V_{out} is at $-V_{sat}$ (Figure 4-6b), V_d will be -1.23 V. The same process given above must occur to force V_{out} to $+V_{sat}$, except that V_{in} must equal $+V_T$.

The threshold voltages, $+V_T$ and $-V_T$, are dependent on the ratio of R_{in} and R_f, as shown by the expressions

$$+V_T = \frac{+V_{sat}}{R_F/R_{in}} \quad \text{and} \quad -V_T = \frac{-V_{sat}}{R_F/R_{in}}$$

This type of comparator is used with a ramp generator to produce the triangle-wave generator shown in Figure 4-7a. The output of the ramp

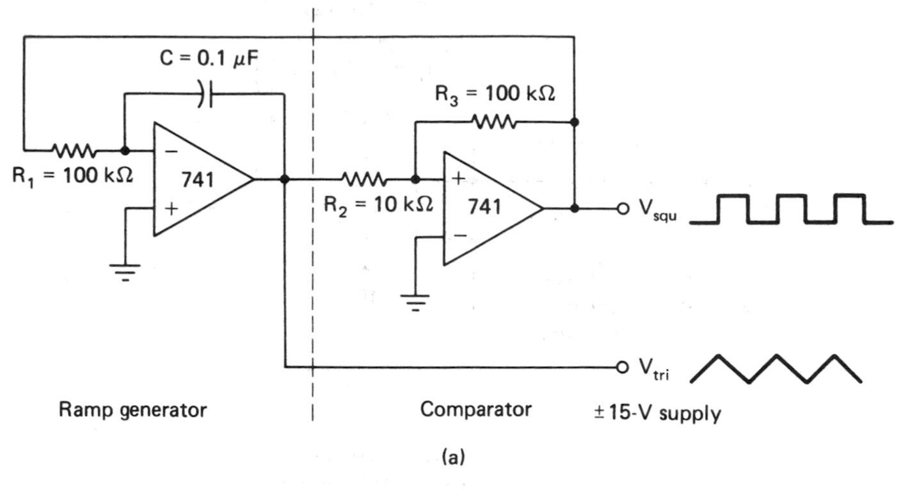

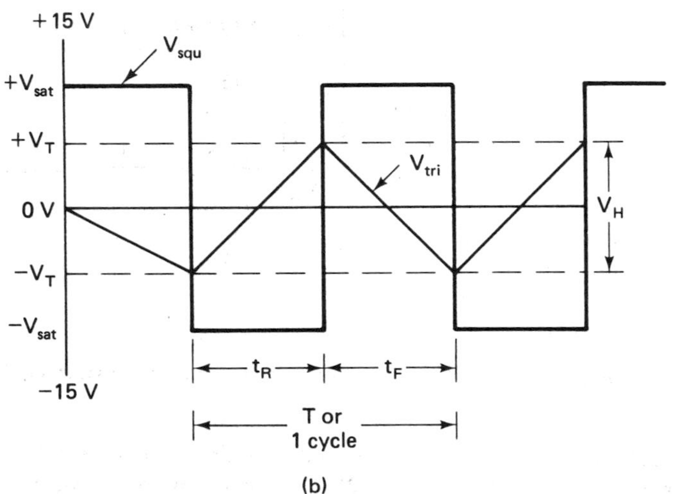

FIGURE 4-7 Basic triangle wave generator (with less components): (a) combination of basic circuits; (b) output waveforms.

generator is connected to the input of the comparator, while its output is fed back to the input of the ramp generator. Each time the ramp voltage reaches the threshold voltage, the comparator changes states, as shown in Figure 4-7b. Therefore, oscillation is sustained by the circuit.

The output frequency for this circuit can be determined by finding the rise time t_R and the fall time t_F of the triangle wave, which constitutes one cycle T, and then solve for the reciprocal. The rise and fall times can be found by the expression

$$t_R = \frac{V_H}{-V_{\text{sat}}}(R_1 C) \quad \text{and} \quad t_F = \frac{V_H}{+V_{\text{sat}}}(R_1 C)$$

where V_H, the hysteris voltage as it is called, is the difference between $+V_T$ and $-V_T$; thus,

$$V_H = +V_T - (-V_T)$$

or twice the value of either threshold voltage.

One cycle or T, then, is

$$T = t_R + t_F$$

and the output frequency f_{out} is

$$f_{out} = \frac{1}{T}$$

For example, using the values shown in Figure 4-7a,

$$+V_T = \frac{+13.5}{10} = +1.35 \text{ V} \qquad \text{and} \qquad -V_T = \frac{-13.5}{10} = -1.35 \text{ V}$$

$$V_H = +1.35 \text{ V} - (-1.35 \text{ V}) = 2.7 \text{ V}$$

$$t_R = \frac{2.7 \text{ V}}{-13.5 \text{ V}}(100 \times 10^3)(0.1 \times 10^{-6}) = 0.002 \text{ s}$$

$$t_F = \frac{2.7 \text{ V}}{+13.5 \text{ V}}(100 \times 10^3)(0.1 \times 10^{-6}) = 0.002 \text{ s}$$

$$T = 0.002 \text{ s} + 0.002 \text{ s} = 0.004 \text{ s}$$

$$f_{out} = \frac{1}{0.004 \text{ s}} = 250 \text{ Hz}$$

The amplitude of the triangle-wave output will be $\pm V_T$ or V_H. The amplitude of the square-wave output will be $\pm V_{sat}$.

4-4 SINE-WAVE OSCILLATOR

A sine-wave oscillator can generate a single sine wave using a frequency-selective network similar to a narrow bandpass filter. One of the oldest types of sine-wave generators is the Wien bridge oscillator. An application of a Wien bridge oscillator using an op amp is shown in Figure 4-8.

Feedback is applied to both inputs of the op amp. The frequency-selective network consisting of R_1, C_1 and R_2, C_2 provides positive feedback to the noninverting input. Negative feedback is provided to the inverting input via R_3, R_4, and R_5. The positive feedback must be greater than the negative feedback in order to sustain oscillations. Potentiometer R_4 is used

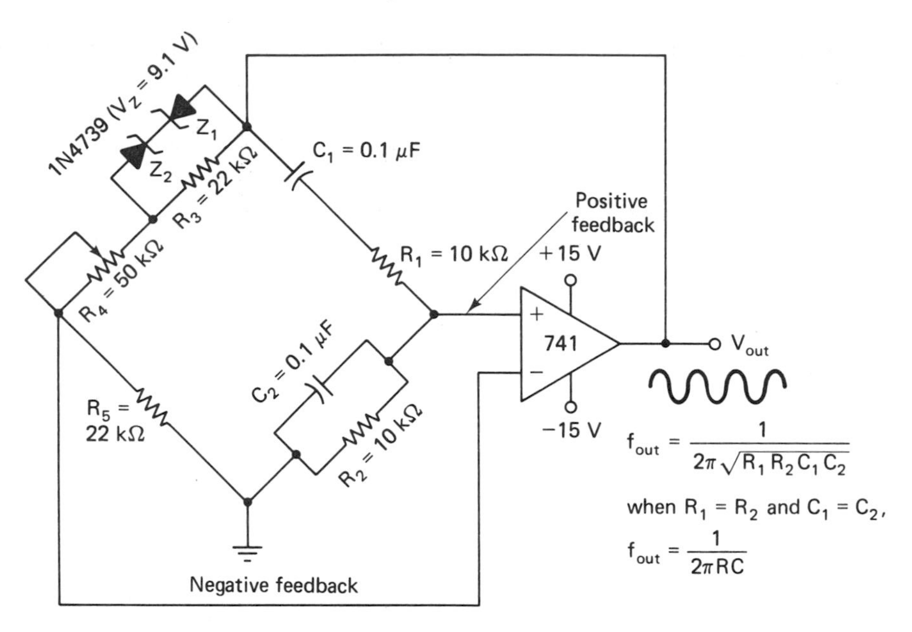

FIGURE 4-8 Wien-bridge oscillator.

to accomplish this by reducing the negative feedback. In effect, it is set to start the circuit oscillating. The frequency-selective network controls the amount of positive feedback, depending upon the frequency. After R_4 is adjusted to start oscillations, the ratio of reactances and resistances determines the proper positive feedback at the noninverting input. If the frequency begins to decrease, the reactance of C_1 becomes greater and the positive feedback decreases. Likewise, if frequency begins to increase, the reactance of C_2 decreases and more positive feedback is shunted to ground. Therefore, the oscillator is forced to operate at the resonant frequency by this network.

Positive feedback causes the output voltage to increase until the op amp locks into saturation. To prevent saturation and have a useful circuit, two zener diodes face to face (or back to back, as it matters very little) are connected across R_3. When the output voltage rises above the zener voltage point, one or the other zener diode conducts, depending on the polarity of the output. The conducting zener diode shunts R_3, causing the resistance of the negative feedback circuit to decrease. More negative feedback is applied to the op amp and the output voltage is controlled at a certain level.

The output frequency can be determined by the formula

$$f_{out} = \frac{1}{2\pi\sqrt{R_1 R_2 C_1 C_2}}$$

or, if $R_1 = R_2$ and $C_1 = C_2$, then

$$f_{out} = \frac{1}{2\pi R_1 C_1}$$

The f_{out} for Figure 4-8 with the components shown is about 160 Hz.

Another type of sine-wave oscillator using two op amps can be produced as shown in Figure 4-9. This circuit consists of a bandpass filter and comparator. One way of generating sine waves is to filter a square wave, which results in the fundamental sine wave at the output. The comparator is fed with a sine wave from the bandpass filter to obtain a square-wave output. The square wave is fed back to the input of the bandpass filter to cause oscillation.

The f_{out} is determined by R_1, R_3, R_4, C_1, and C_2 by the expression

$$f_{out} = \frac{1}{2\pi\sqrt{R_p R_4 C_1 C_2}}$$

where

$$R_p = \frac{R_1 R_3}{R_1 + R_3}$$

Resistor R_2 can be considered part of R_1, but is negligible, because of its extremely small value. It is used only to keep from shorting the feedback to ground.

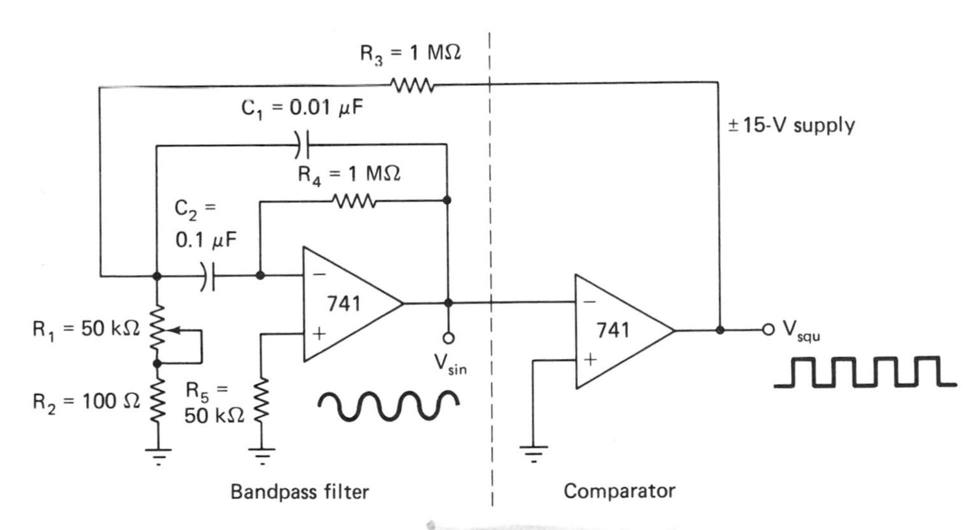

FIGURE 4-9 Two op amp sine wave oscillator.

Since R_1 is variable, the frequency of the oscillator can be changed. The frequency range of the oscillator with the component values shown is about 7 to 1.6 kHz. Other ranges within the audio band can be obtained easily by changing C_1 and C_2.

Two outputs are provided by this circuit, a sine wave (V_{sin}) from the bandpass filter and a square wave (V_{squ}) from the comparator.

4-5 QUADRATURE OSCILLATOR

Sometimes in electronic systems it is necessary to have two sine waves, 90° out of phase, termed quadrature. Figure 4-10 shows a quadrature oscillator, where the outputs are labeled sine and cosine. Basically, the circuit consists of two integrators with positive feedback. The sine output comes from Op-1 and the cosine output comes from Op-2. Since the phase shift of an integrator is 90°, the cosine output is 90° out of phase with the sine output.

Resistor R_1 is usually slightly less in value than R_3 to ensure that the circuit oscillates. If R_1 is too small in value, the outputs will be clipped and resemble square waves. A potentiometer may be used to adjust for minimum distortion of the output voltages. The output voltages may also reach op-amp saturation. If this situation prevails, two zener diodes may be connected face to face across C_3 to limit the output.

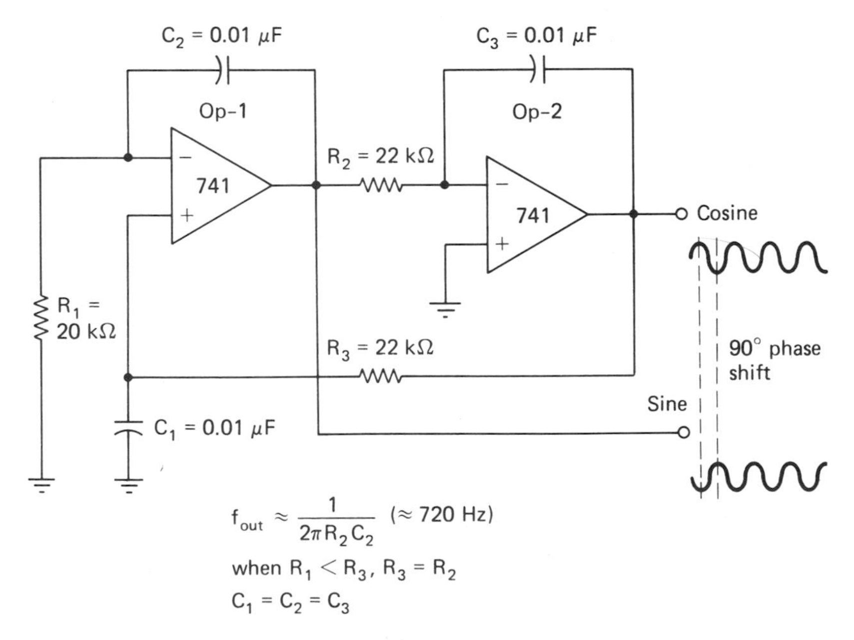

FIGURE 4-10 Basic quadrature oscillator.

If $R_2 = R_3$ with $R_1 < R_3$ and $C_1 = C_2 = C_3$, then f_{out} can be easily found by the expression

$$f_{out} = \frac{1}{2\pi R_2 C_2}$$

4-6 FUNCTION GENERATOR

As mentioned before, a function generator is any signal generator that has two or more different waveform outputs. Various basic circuits can be combined to produce a function generator, as shown in Figure 4-11. A basic

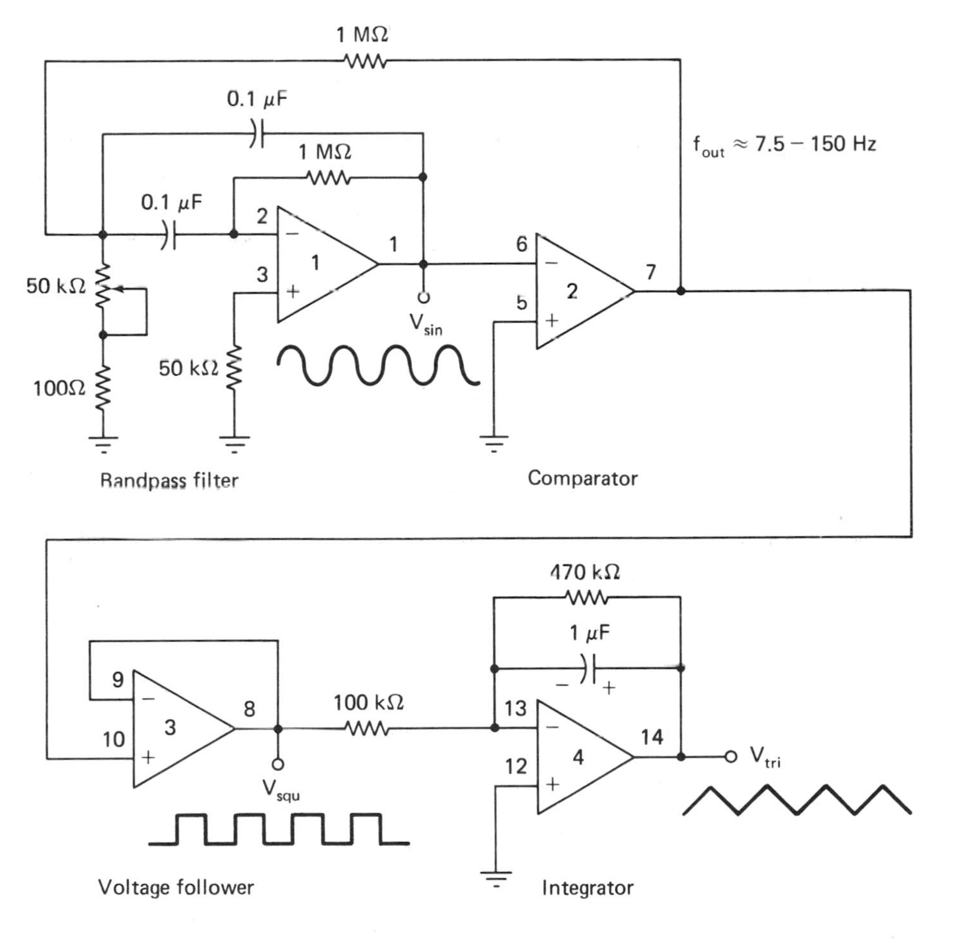

IC = 324 (quad op amps)
pin 4 = +V, pin 11 = −V, ±15-V supply

FIGURE 4-11 Basic low frequency tri-function generator.

sine/square-wave generator (Figure 4-9) is used to establish f_{out} and the sine-wave and square-wave outputs. The output from the comparator is fed to a voltage follower (Figure 2-12a) to prevent loading down the basic oscillator. This also helps to prevent any change in the oscillator's frequency due to loading. The output of the follower is then fed to an integrator (Figure 3-3) to produce the triangle-wave output. Since three outputs are available, the circuit could be called a trifunction generator.

This circuit can be built from a single 14-pin DIP 324 (quad op amp) IC. The pin identification is shown to facilitate construction. A ±15-V supply is used. The amplitudes of the output voltages are about as follows:

$$\text{square wave} = 26 \text{ V p-p}$$

$$\text{sine wave} = 16 \text{ V p-p}$$

$$\text{triangle wave} = 0.3\text{–}6 \text{ V p-p (depending on } f_{out})$$

By adjusting the 50-kΩ potentiometer, f_{out} can vary from about 7.5 to 150 Hz. Decreasing the value of the two capacitors in the bandpass amplifier will increase f_{out}. It is a good practice to keep their values equal, and it also simplifies calculating f_{out} from the formula given with this circuit in Section 4-4.

SUMMARY POINTS

1. Oscillators convert DC voltage to AC voltage or other time-varying DC voltages.
2. Four basic waveforms are the square wave, triangle wave, sawtooth wave, and sine wave.
3. Positive feedback is required for oscillation.
4. The output frequency for op-amp oscillators depends on *RC* time constants.
5. A square-wave generator belongs to the multivibrator family.
6. An integrator is the basic component used in a ramp-voltage generator.
7. A sawtooth-wave generator uses an electronic switch to discharge the capacitor.
8. The switching of some op-amp oscillators is established by ± threshold voltages.

9. The ratio of the resistors (R_F/R_{in}) used with positive feedback in an op-amp circuit establishes the ± threshold voltages.
10. A function generator provides two or more different output waveforms.
11. Bandpass filters are used to select the fundamental sine wave from a square wave in some types of op-amp sine-wave generators.
12. A quadrature oscillator has two outputs 90° out of phase.

TERMINOLOGY EXERCISE

Write a brief definition for each of the following terms:

1. Square-wave voltage
2. Sawtooth-wave voltage
3. Triangle-wave voltage
4. Sine-wave voltage
5. Regenerative feedback
6. Quadrature oscillator
7. Threshold voltages
8. Astable multivibrator
9. Frequency
10. Function generator

PROBLEMS AND EXERCISES

1. Referring to Figure 4-1, find the f_{out} when $R_1 = 47\ \text{k}\Omega$, $R_2 = 100\ \text{k}\Omega$, $R_3 = 86\ \text{k}\Omega$, and $C = 0.002\ \mu\text{F}$.
2. If the saturation voltages in problem 1 are ±10.8 V, what is $+V_T$ and $-V_T$?
3. What is the f_{out} of Figure 4-4 when $V_{ref} = 0.5$ V, $R_3 = 47\ \text{k}\Omega$, $C_f = 0.01\ \mu\text{F}$, and $V_p = 3.5$ V?

4. What is the approximate peak voltage amplitude of the output waveform in problem 3?
5. What is $+V_T$ and $-V_T$ of Figure 4-6 when $R_{in} = 22$ kΩ and $R_F =$ 180 kΩ?
6. What is the f_{out} and voltage amplitude of the triangle-wave output of Figure 4-7 when $R_2 = 22$ kΩ and $R_3 = 150$ kΩ? ($\pm V_{sat} = \pm 13.5$ V.)
7. What is the f_{out} and voltage amplitude for the square-wave output of problem 6?
8. Find the f_{out} for the Wien bridge oscillator of Figure 4-8 when $R_1 = R_2 = 100$ kΩ and $C_1 = C_2 = 0.01$ μF.
9. What is the approximate f_{out} of Figure 4-9 when $R_1 = 100$ kΩ, $R_2 = 100$ Ω, $R_3 = 2.2$ MΩ, $R_4 = 1$ MΩ, and $C_1 = C_2 = 0.02$ μF?
10. Find the f_{out} of Figure 4-10 when $R_2 = R_3 = 47$ kΩ and $C_1 = C_2 = C_3 = 0.1$ μF.

SELF-CHECKING QUIZ

Match each oscillator (generator) in column A with its proper schematic diagram in Figure 4-12.

Column A

1. Square-wave generator - C
2. Triangle-wave generator — A
3. Sawtooth-wave generator — E
4. Wien bridge oscillator — B
5. Quadrature oscillator — D

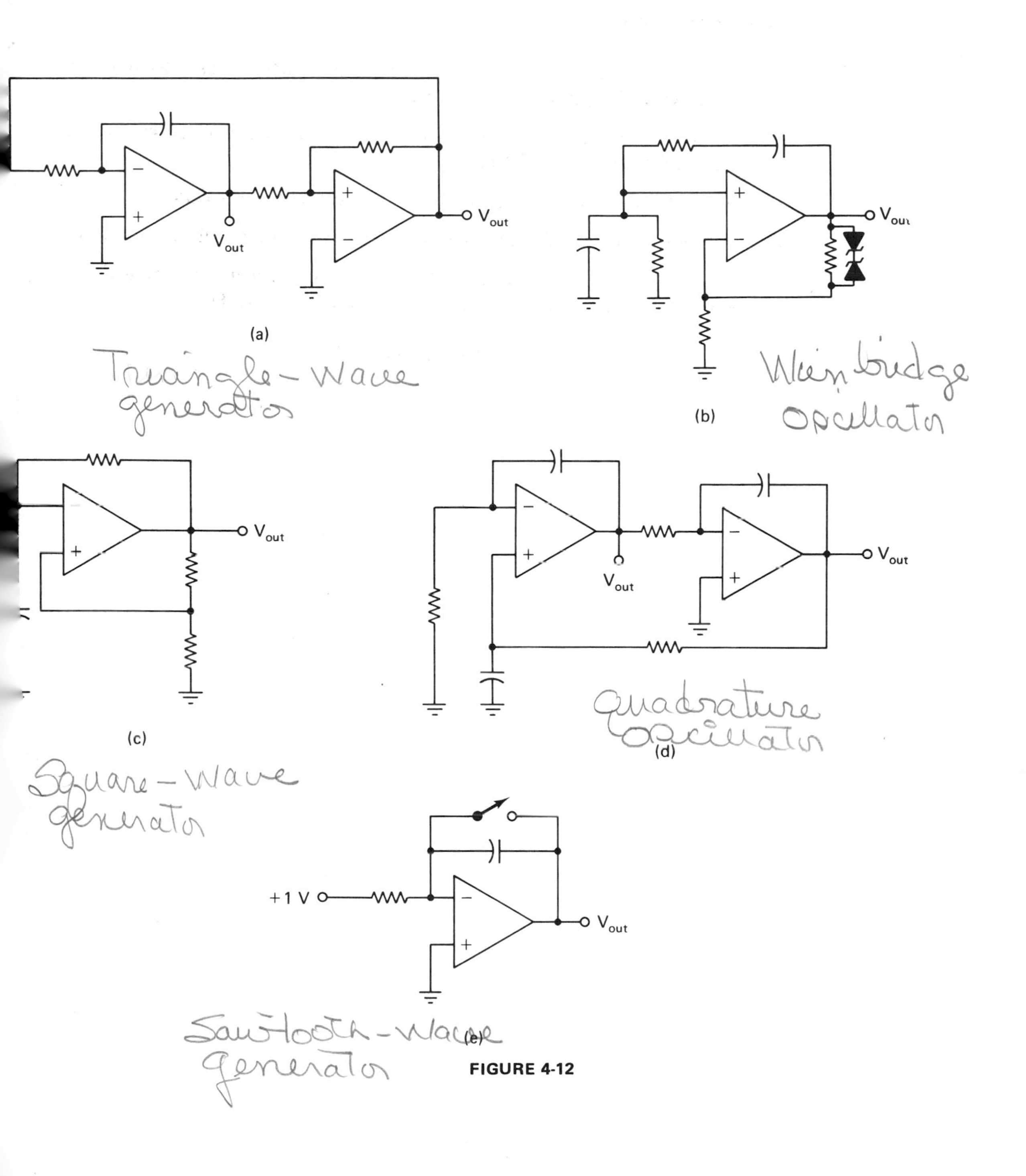

(a)
V_{out}
V_{out}
Triangle-wave generator
(b)
V_{out}
Wein bridge oscillator
(c)
V_{out}
Square-wave generator
(d)
V_{out}
V_{out}
Quadrature oscillator
(e)
+1 V
V_{out}
Sawtooth-wave generator

FIGURE 4-12

True-or-False Questions

6. The output of an op-amp square-wave generator goes positive and negative with reference to ground.
7. If the rise time (t_R) of a sawtooth-wave generator output is 0.02 ms, its frequency is 50 kHz.
8. A quadrature oscillator produces four output waveforms.
9. A Wein bridge oscillator uses both positive and negative feedback.
10. If the rise time (t_R) of the output voltage from a symmetrical triangle wave generator is 0.025 s, its frequency is 40 Hz.

(Answers at back of book)

Chapter 5

Op-Amp Applications to Audio Circuits

Modern semiconductor technology, especially integrated circuits, have made it possible for all of the marvelous consumer products (miniature portable radios, portable color TV sets, cassette and eight-track cartridge tape recorders, etc.) available today. All of these products use audio circuits. The IC op amp has many advantages when employed in audio circuits: small size, lower power consumption (a must for portable devices), minimum associated components, and reliable performance at low cost. Not only is it easier for the engineer to design complex audio systems, but the home experimenter and hobbyist can easily design and construct various audio circuits using IC op amps.

Audio circuits are a specialized area which places unique requirements upon op-amp parameters. The op amp must be able to process complex AC signals in the frequency range 20 to 20 kHz whose amplitudes vary from a few hundred microvolts to several volts. These waveforms are characterized by steep complex wavefronts of a transient nature separated by incalculable periods of absolute silence. The op amp must process these complex AC signals with a minimum of distortion of any kind, either harmonic, amplitude, or phase, and it has to be done as noiselessly as possible. Figure 5-1 illustrates a typical audio waveform. Referencing this figure to musical and voice reproduction, low sustaining frequencies, such as those associated with the string bass, have larger periods of time; medium sustaining frequencies, asso-

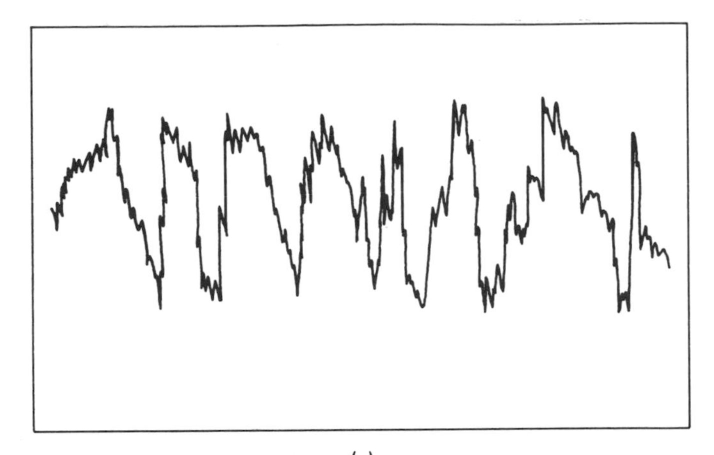

(a)

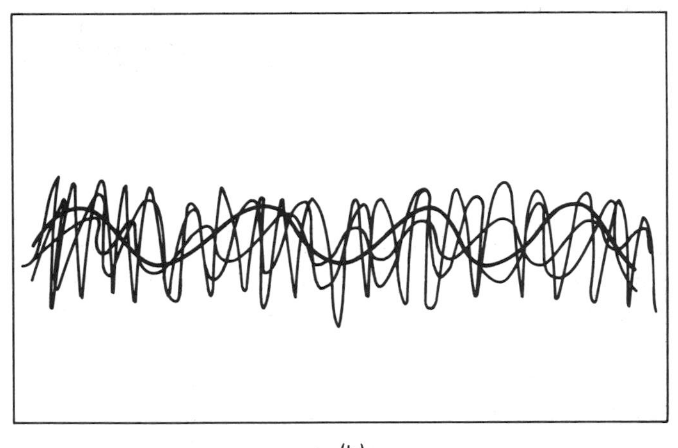

(b)

FIGURE 5-1 Oscilloscope display of typical audio signal: (a) horizontal scan set at about 100 hertz; (b) expanded view.

ciated with voice or instruments (trumpet, saxophone, piano, guitar, etc.), have medium periods of time; and high sustaining frequencies, associated with the higher range of the instruments already listed as well as flute, piccolo, and violins, have narrow periods of time. Percussion instruments, particularly drums and cymbals, have sharp rise times and relatively fast decays. The combinations of these waveforms are, of course, endless, depending on the type of material being produced and the mode of performance.

A word about noise. Noise is particularly objectionable to audio circuits because its effects can be heard in the speaker. As mentioned in Chapter 1, there are two categories of noise, external and internal. External noise originates from operating electrical equipment such as, motors, switches, and lights. This type of noise can be minimized in audio circuits by using shielded cables, shielding various parts of a circuit, proper component layout, and noise decoupling techniques. Internal noise results from offset voltages and currents, charge carriers crossing junctions, and thermal noise in passive resistive elements. Proper biasing and input offset compensation can reduce offset voltage and current noise. Thermal noise can be kept to a minimum by using low-resistance-value resistors, especially in that part of an op-amp circuit that determines gain.

Noise can range from 0.01 Hz to megahertz, and determining its approximate frequency can sometimes be useful in reducing or eliminating it. Low-frequency noise, such as 60 or 120 Hz, comes from AC lines or power supplies, and simple low-pass filters can often reduce these effects. Medium- and high-frequency noise can also be reduced by appropriate filtering.

A capacitor of a few picofarads connected across the feedback resistor in an op-amp amplifier can reduce the gain of high-frequency noise.

General-purpose op amps are suitable for limited audio work in noncritical applications. However, solid-state manufacturers have developed special op amps to meet the stringent requirements of audio work. These op amps will have one or more of the following features that are ideal for audio use.

1. High slew rate.
2. High gain–bandwidth product.
3. High input resistance or impedance.
4. High-voltage and/or power operation.
5. Low-distortion operation.
6. Very low input (voltage and/or current) noise.
7. Very low input current.
8. Easily used with single power-supply source.

This chapter will acquaint you with basic audio circuits using op amps and provide some practical audio circuits, which you may want to construct.

5-1 THE AUDIO VOLTAGE AMPLIFIER

5-1-1 The Inverting Amplifier

An inverting amplifier as applied to audio use is shown in Figure 5-2. It is the same basic circuit used for DC amplifiers, except with capacitor C_i in

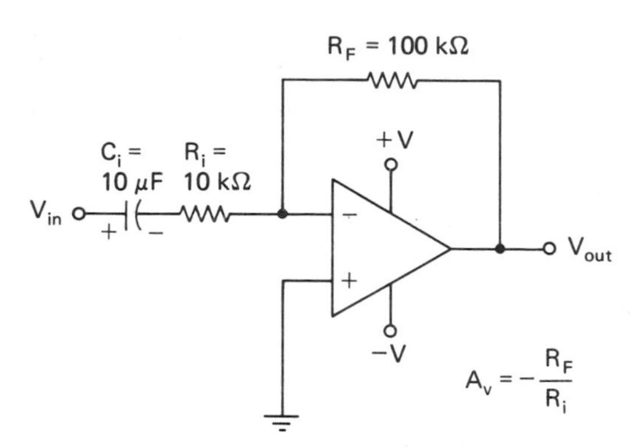

FIGURE 5-2 Inverting AC amplifier.

series with the inverting input. This capacitor serves two important functions. First, it blocks any DC from a previous stage or device that could be amplified, causing the output to go to some unwanted DC level other than zero. This may cause amplifier saturation and distortion, when the audio signal is applied to the input. Second, the capacitor helps to block any low-frequency noise from getting to the input of the amplifier. The low-frequency cutoff is determined by C_i and R_i with the formula

$$f_c = \frac{1}{2\pi R_i C_i}$$

The stage gain is found the same way as with the inverting amplifier in Chapter 2. $A_v = -R_{F/R_i}$ and the input resistance (R_{in}) is the same as R_i.

Single power-supply inverting amplifier. In some instances it may be desirable to operate an op-amp inverting amplifier from a single power supply, as shown in Figure 5-3. The output should be operated at one-half the voltage supply in the quiescent state (V_q) in order to obtain maximum undistorted output. This is accomplished by making biasing resistors R_3 and R_4 equal resistances, which might range from 10 to 100 kΩ. Capacitor C_3 helps to filter out (or decouple) power-supply noise from getting into the noninverting input. Capacitor C_2 will most likely be needed to block the DC output ($\frac{1}{2}$ + V) from the following stages. All capacitors are electrolytic and polarity should be observed when connecting them into the circuit. Values for C_2 and R_L depend upon the input impedance of the following stage.

Although a positive voltage supply is used with the circuit in Figure 5-3, the circuit could operate as well from a negative voltage supply. Resistor R_3 and the positive supply terminal of the op amp would have to be grounded while the bottom of R_4 and the negative supply terminal of the op amp would be connected to the negative voltage supply ($-V$). Also, remember that all capacitors must be reversed to operate with the proper voltage polarity.

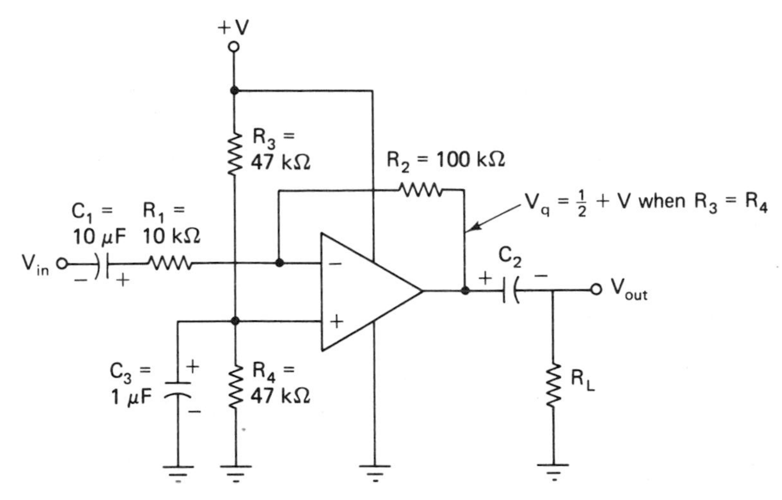

FIGURE 5-3 Inverting amplifier using single power supply.

Some special audio op amps can be used with single or dual power supplies without the need for special biasing resistors.

5-1-2 The Noninverting Amplifier

Since the input impedance of the inverting amplifier equals R_i, matching a high-impedance source (such as a microphone) may present a formidable problem. The noninverting AC amplifier shown in Figure 5-4 is often used in audio circuits to overcome this problem.

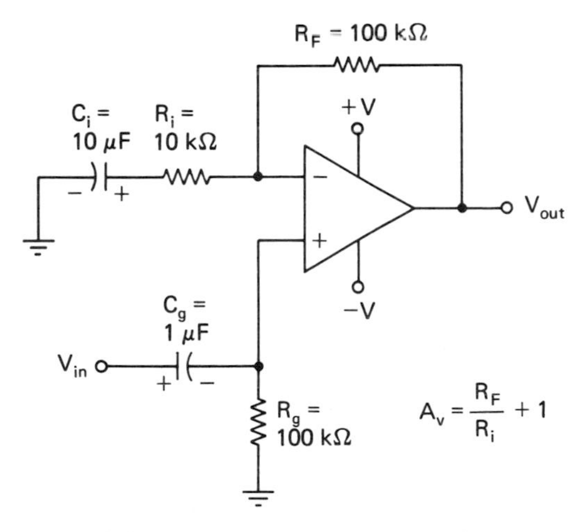

FIGURE 5-4 Noninverting AC amplifier.

Components C_i, R_i, and R_F serve the same function as with the inverting amplifier in Figure 5-2. The noninverting input offers an exceptionally high input impedance and can be matched to the source impedance more readily with the use of C_g and R_g (Fig. 5-4). The rolloff frequency is found the same for C_g and R_g as it is for C_i and R_i. The rolloff frequency at the noninverting input may be up to 10 times lower than the rolloff frequency at the inverting input.

The gain of the AC noninverting amplifier is the same as its DC counterpart. Input impedance is approximately equal to R_g. The rolloff frequencies are found in the same way as with the inverting amplifier.

Single power-supply noninverting amplifier. The noninverting amplifier may also be used with a single power supply, as shown in Figure 5-5. All

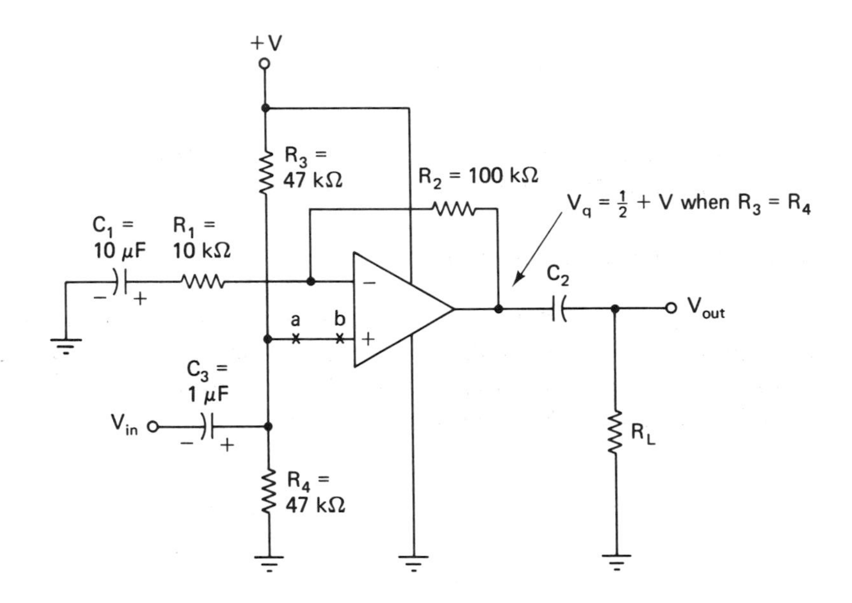

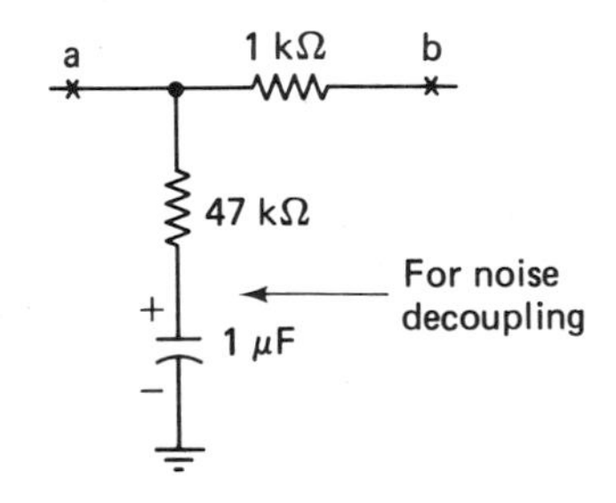

FIGURE 5-5 Noninverting amplifier using single power supply.

components function the same way as with the inverting amplifier of Figure 5-3. However, a noise decoupling capacitor cannot be connected directly to the noninverting input, since this would bypass the signal as well as the noise to ground. An additional circuit for noise decoupling can be inserted at points *a* and *b*, if needed. The 47-kΩ resistor tends to isolate R_3 and R_4 from V_{in}, while the power-supply noise is bypassed to ground by the 1-μF capacitor without affecting the input signal.

Negative power-supply operation is also possible with this circuit by grounding R_3 and the positive supply terminal of the op amp while connecting the bottom of R_4 and the negative supply terminal of the op amp to the negative voltage supply.

5-1-3 Simple Audio-Voltage-Amplifier Applications

Four examples of audio voltage amplifiers are given in the following figures. These are simple, noncritical circuits that perform rather satisfactorily. Each circuit has a voltage gain of 100 (40 dB gain) and any op amp can be used. However, the higher the quality of the op amp, the better the performance.

A high-impedance input microphone preamplifier is shown in Figure 5-6. The rolloff frequency is about 1.5 Hz. By making R_F adjustable, the gain of the circuit can be varied. Resistor R_g might also be made variable, to match the impedance of the microphone. The microphone should have an impedance larger than 600 Ω.

A low-impedance-input differential microphone preamplifier is shown in Figure 5-7. Unbalanced microphone lines are susceptible to common-mode noise signals coupled into the cable. The differential input arrangement of

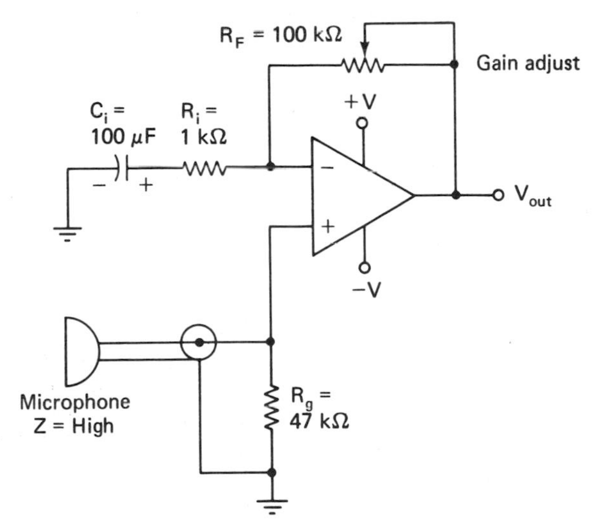

FIGURE 5-6 High-impedance input microphone preamplifier.

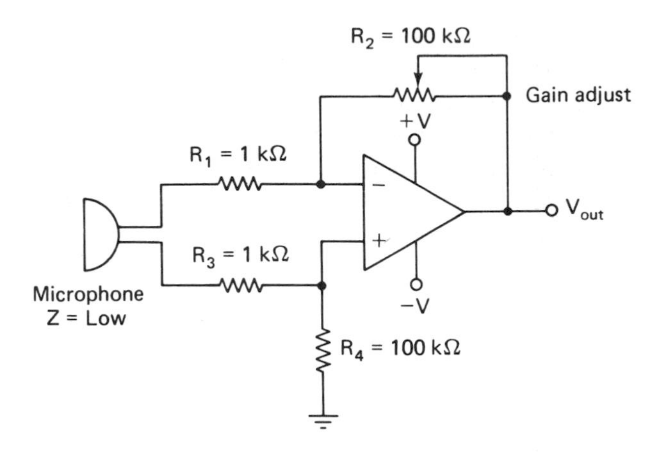

FIGURE 5-7 Low-impedance input, differential microphone preamplifier.

this circuit helps to minimize such induced noise. Matched 1.0% resistors also improve the degree of common-mode rejection. Performance of the circuit is best when used with microphones with impedances of less than 600 Ω (the lower the better).

A single standard op amp can very effectively drive a set of headphones, as shown in Figure 5-8. The headphones should not be lower than 150 Ω. Other headphone sets can be placed in parallel as long as the total load impedance is above 150 Ω. A low-impedance headphone set or speaker can be used if it is coupled by an impedance matching transformer, as shown by the dashed lines. Resistor R_3 serves as a volume control. Components C_3 and R_4 help to bypass high-frequency noise and improve the sound quality at the output.

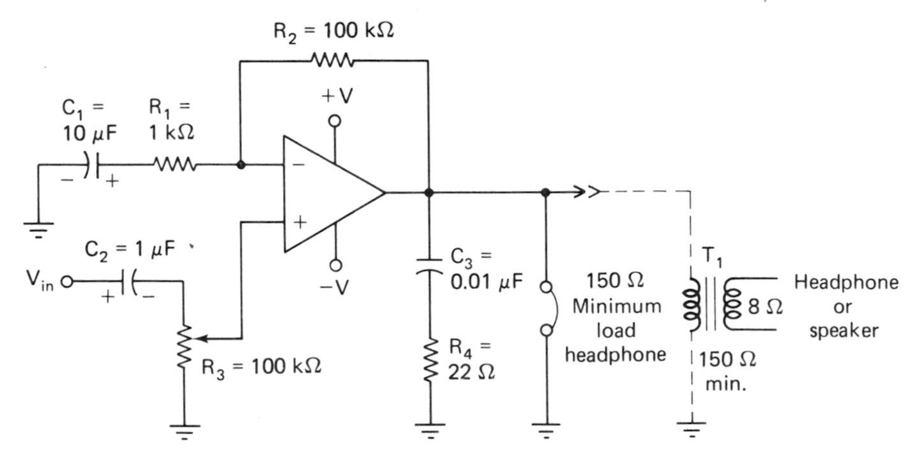

FIGURE 5-8 Basic headphone driver amplifier.

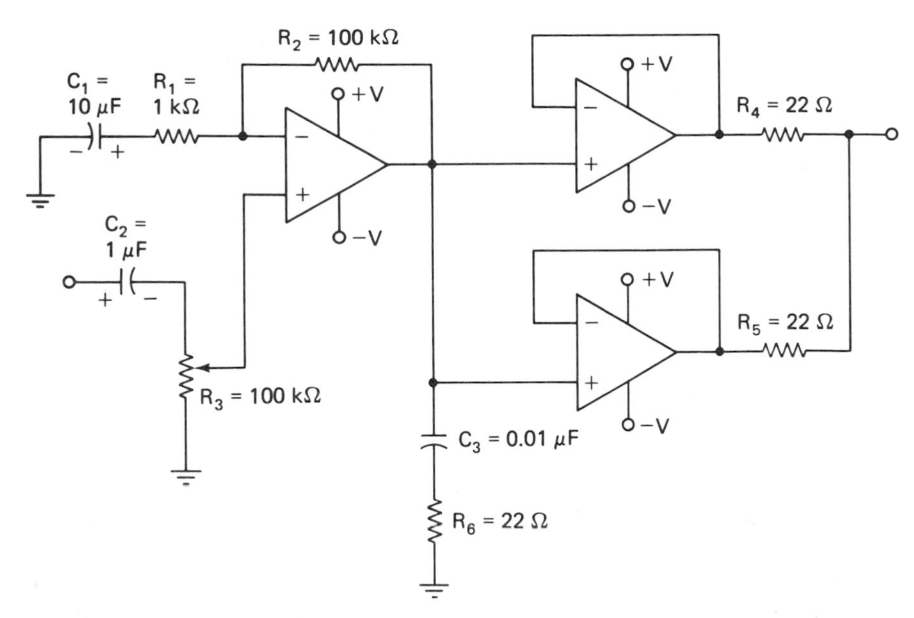

FIGURE 5-9 Amplifier with followers to increase output load current requirements.

Special power op amps with up to a few watts of output can be used to drive headphones and speakers for certain applications. If, however, you need only about twice as much current to drive a load efficiently, you could use the circuit shown in Figure 5-9. The voltage amplifier is connected to two voltage followers in parallel, which in effect doubles the current available to the load. With the dual and quad op-amp ICs now commonly available, it may be advantageous to use one of these packages in a circuit, which would not be considered using single devices.

5-2 EQUALIZATION PREAMPLIFIERS

5-2-1 RIAA Equalization Preamplifier

Producing a phonograph record, commonly called "cutting a record," is a highly complex and technically exacting procedure which is beyond the scope of this book. Basically, however, the grooves of a typical stereo record are cut by a chisel-shaped cutting stylus which vibrates mechanically from side to side. This "lateral cut," as it is termed, is in accordance with the audio signal placed on the cutting-stylus mechanism. This lateral cut is known as groove modulation. The amplitude of the audio signal translates to groove modulation, while the frequency of the audio signal determines the rate of change of the groove modulation. Normally, the groove modulation is set to cut one groove at 1 kHz without affecting adjacent grooves.

Lower frequencies produced by musical instruments have larger amplitudes which will drive the cutting stylus beyond its fixed limit into adjacent grooves. Higher frequencies from the instruments have smaller amplitudes, which will not drive the stylus sufficiently and result in a poor signal-to-noise ratio when the record is played back. Therefore, an electronic procedure known as equalization attenuates the amplitude of the lower frequencies and amplifies the amplitude of the higher frequencies during the recording process.

When a record is played back, the preamplifier of the audio system must reverse the equalization process, as illustrated by the Record Industry Association of America (RIAA) curve shown in Figure 5-10. The turnover frequencies indicate the frequency ranges that need playback equalization. For instance, the preamplifier must have a higher gain for the frequencies from 50 to 500 Hz, since these were attenuated during the recording process, whereas the preamplifier must have less gain for the frequencies from 2 to 20 kHz, since these were amplified in the original recording.

Some phonograph cartridges (playback stylus and device) such as ceramic and crystal, produce from 100 MV to 2 V of output and do not require a preamplifier. These outputs are generally fed to a passive tone network and then directly to a power amplifier, as found in most lo-fi or mid-fi systems. On the other hand, magnetic cartridges produce from 3 to 10 MV output and require a preamplifier with gain. These cartridges are used with hi-fi systems. A preamplifier should have a gain of at least 100, to amplify a

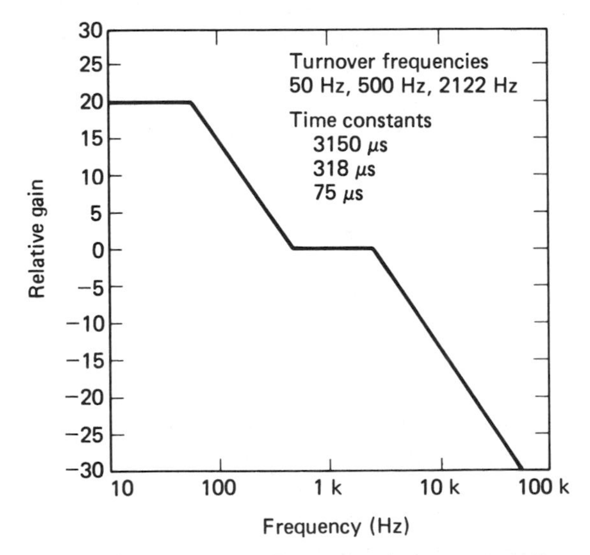

FIGURE 5-10 Standard RIAA equalization curve (Permission to reprint granted by Signetics Corporation, a subsidiary of U.S. Philips Corp., 811 E. Arques Avenue, Sunnyvale, CA 94086.)

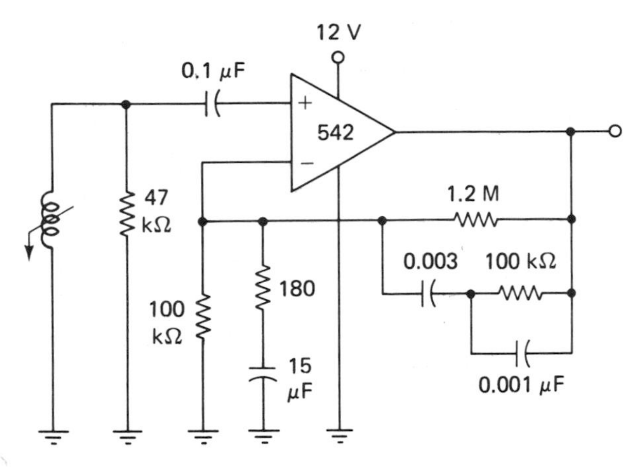

FIGURE 5-11 Typical RIAA preamplifier. (Permission to reprint granted by Signetics Corporation, a subsidiary of U.S. Philips Corp., 811 E. Arques Avenue, Sunnyvale, CA 94086.)

signal of, say, 5 MV from a magnetic cartridge to drive the other circuits in a system.

A typical RIAA preamplifier is shown in Figure 5-11. The 47-kΩ resistor sets the preamplifier input resistance to match the internal resistance of the magnetic cartridge. The 100-kΩ resistor from the inverting input to ground and the 1.2-MΩ feedback resistor set the DC bias. The 180-Ω resistor and the 100-kΩ feedback resistor establish the reference gain for frequencies 500 to 2 kHz (in this case, about 560). At frequencies below 500 Hz, the X_c of the 0.003-μF capacitor is large, causing the feedback impedance to be larger, which results in higher gain for these frequencies. At frequencies above 2 kHz, the X_c of the 0.003-μF capacitor and the 0.001-μF capacitor is low, causing the feedback impedance to be smaller, and the gain falls off for these frequencies.

5-2-2 NAB Equalization Preamplifier

Tape recorder playback preamplifiers require a different type of equalization since the recording process has its own inherent problems with relation to frequencies. The tape head (recording or playback) is an inductive device whose impedance varies directly with frequency (X_L becomes larger when frequency increases). Therefore, the higher the frequency, the larger the impedance and a larger amplitude is produced. Information recorded on the magnetic tape then has low amplitudes for low frequencies and high amplitude for high frequencies. However, because of the amount of magnetic material on the tape, tape magnetic saturation, the speed of the tape, and the

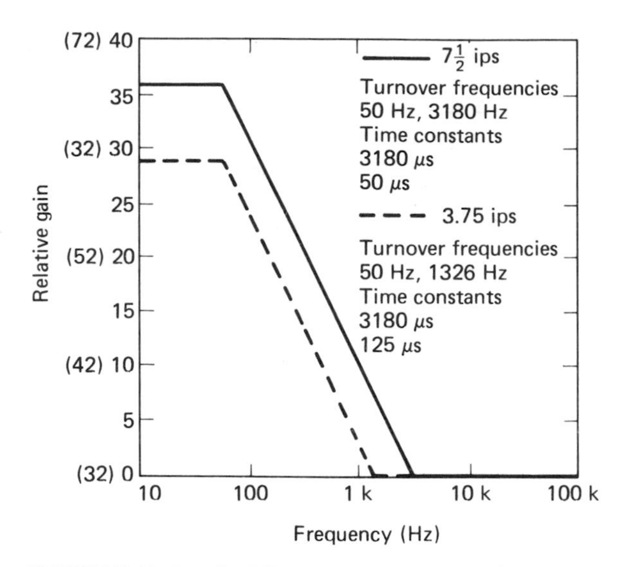

FIGURE 5-12 Standard NAB equalization curve. (Permission to reprint granted by Signetics Corporation, a subsidiary of U.S. Philips Corp., 811 E. Arques Avenue, Sunnyvale, CA 94086.)

width of the tape-head gap, the amplitude of the audio signal can fall off anywhere from 2 to 20 kHz. Nevertheless, a standard tape playback equalization curve does exist, given by the National Association of Broadcasters (NAB), and is shown in Figure 5-12.

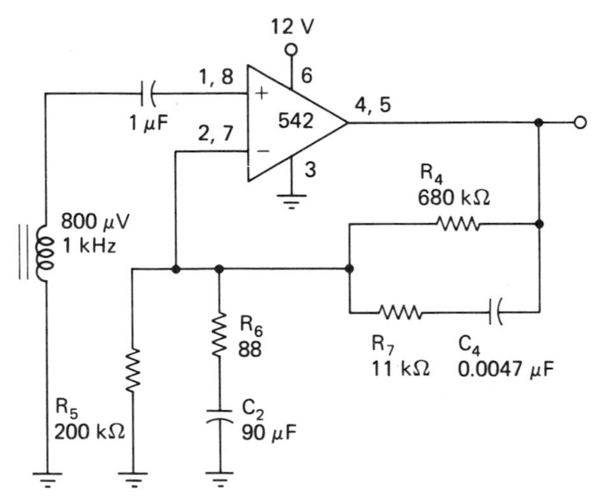

All resistor values are in ohms.

FIGURE 5-13 NAB response amplifier. (Permission to reprint granted by Signetics Corporation, a subsidiary of U.S. Philips Corp., 811 E. Arques Avenue, Sunnyvale, CA 94086.)

The playback preamplifier must be able to amplify low frequencies and attenuate high frequencies. Tape speed also affects the gain, as indicated by the two standard speeds of $7\frac{1}{2}$ ips and $3\frac{3}{4}$ ips.

A typical NAB tape playback preamplifier is shown in Figure 5-13. Similar to the RIAA preamplifier, resistors R_4 and R_5 set the DC bias. The reference gain of the circuit is set by R_6 and R_7. High-frequency attenuation is determined by X_{C4} and R_7, while low-frequency rolloff is determined by X_{C2} and R_6. The tape head (indicated by the iron-core inductor symbol) is fed through a 1-μF capacitor to the noninverting input.

5-3 ACTIVE-TONE-CONTROL CIRCUITS

Many precautions are taken to record music exactly as it is produced, and equalization preamplifiers are designed to reproduce the material exactly (or "flat") as the original. Then why would the user of audio equipment want to alter the frequency response of the material being played? There are several reasons. The output of the amplifier can be affected by speaker response, room acoustics, and other factors, but probably most important is the listener's personal taste. Some persons prefer "bassy" music while others prefer it "trebley." A basic discussion on passive bass and treble tone controls will aid you in understanding active tone controls.

A typical passive bass tone control is shown in Figure 5-14. The term "bass" refers to low frequencies or low audible sounds; therefore, this circuit controls low-frequency amplification or attenuation. Passive tone controls require "audio taper" (logarithmic) potentiometers, since our hearing ability is also logarithmic. When the wiper is set at the halfway point of rotation, the total resistive element is split into two portions, with 90% above the wiper and 10% below the wiper. Basically, when the wiper is placed toward R_1 (bass boost), capacitor C_1 is shorted and there is more resistance from

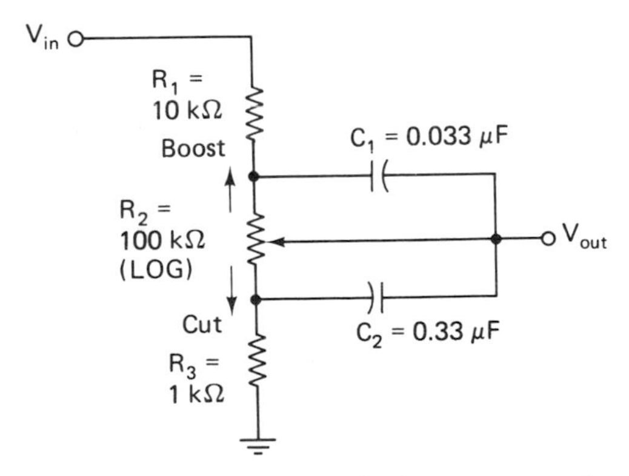

FIGURE 5-14 Typical passive bass tone control.

V_{out} to ground, which produces a larger amplitude. When the wiper is moved toward R_3 (bass cut), there is less resistance between V_{out} and ground and the lower frequencies are allowed to pass through C_1 and C_2 to R_3. Therefore, the amplitude is less.

A typical passive treble tone control is shown in Figure 5-15. The term "treble" refers to high frequencies or high audible sounds; therefore, this circuit controls high-frequency amplification or attenuation. Essentially, the components of the bass circuit have been rearranged for the treble circuit except that the values of the capacitors are changed, since this circuit deals with high frequency. When the wiper of R_2 is placed toward C_1 (boost), there is a larger impedance to ground and more V_{out} is produced. When the wiper is moved toward C_2 (cut), there is less resistance and the signal is shunted to ground.

A passive bass and treble tone control can be combined into a single circuit, as shown in Figure 5-16. Resistor R_4 helps to isolate the two controls and minimizes interaction.

Passive tone controls consume power in the resistors and capacitors used and are referred to as insertion losses. Additional amplification is required to build up the overall signal amplitude to meet the audio system requirements.

The use of an op amp with tone control circuits keeps the output at the overall level of the input or may even result in some gain. This type of circuit is referred to as an active-tone-control circuit, as shown in Figure 5-17.

The addition of a midrange control, which acts to boost or cut the midrange frequencies in a manner similar to the bass and treble controls, offers greater flexibility in tone control. A three-band active tone control is shown in Figure 5-18. The circuit is for the left channel of a stereo amplifier. The

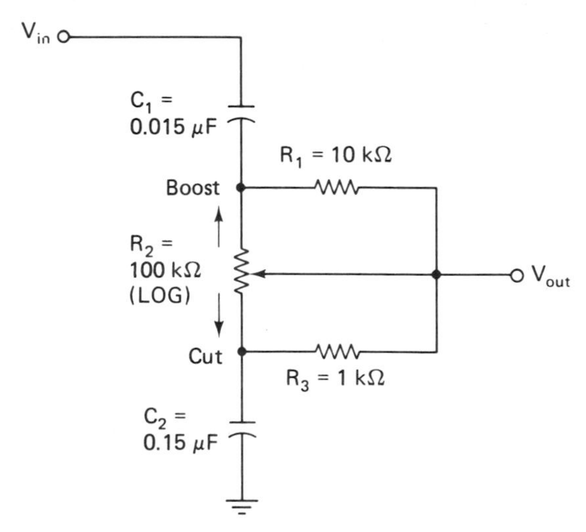

FIGURE 5-15 Typical passive treble tone control.

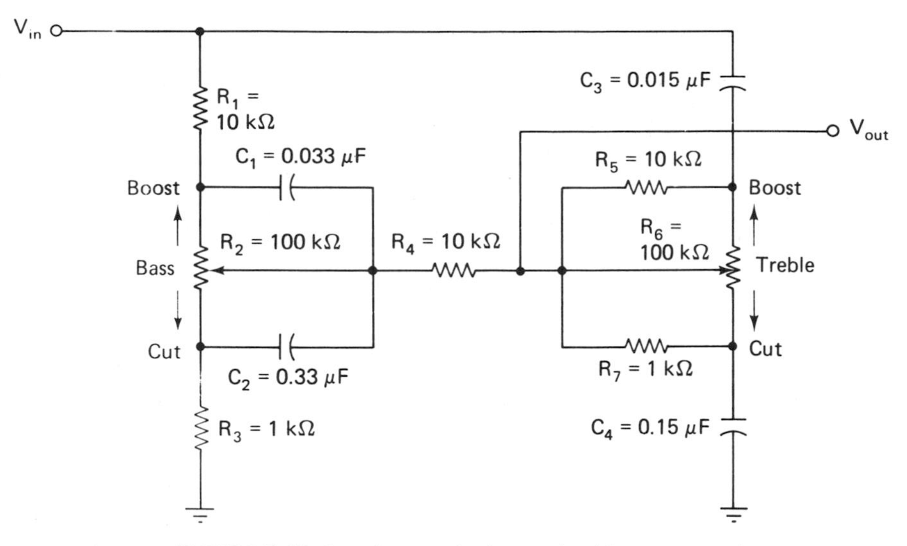

FIGURE 5-16 Complete passive bass and treble tone control.

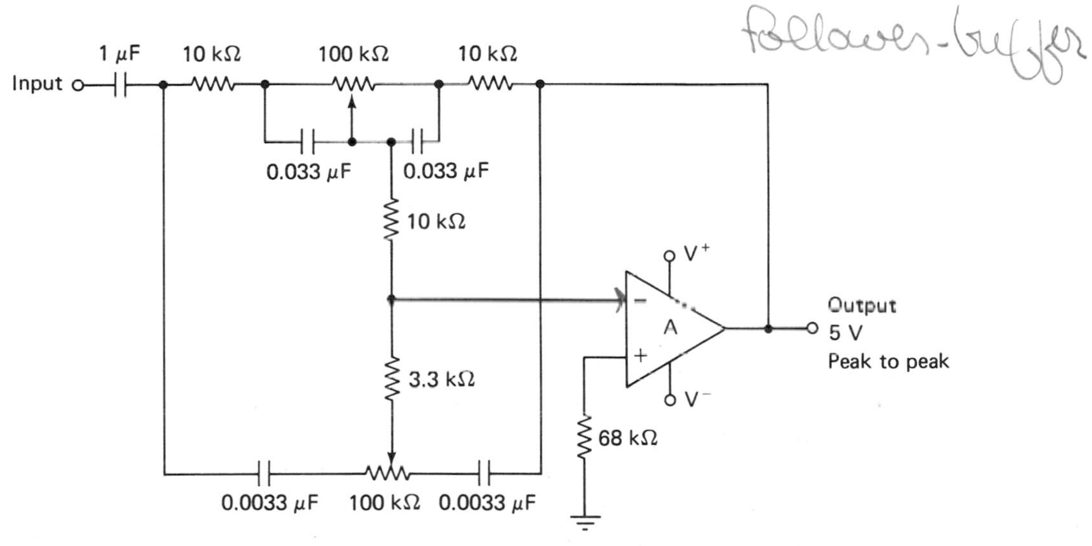

All resistor values are in ohms.

Notes:
1. Amplifier A may be NE531 or 301. Frequency compensation, as for unity-gain noninverting amplifiers, must be used.
2. Turnover frequency — 1 kHz.
3. Bass boost +20 dB at 20 Hz, bass cut −20 dB at 20 Hz, treble boost +19 dB at 20 kHz, treble cut −19 dB at 20 kHz.

FIGURE 5-17 Tone control circuit for operational amplifiers. (Permission to reprint granted by Signetics Corporation, a subsidiary of U.S. Philips Corp., 811 E. Arques Avenue, Sunnyvale, CA 94086.)

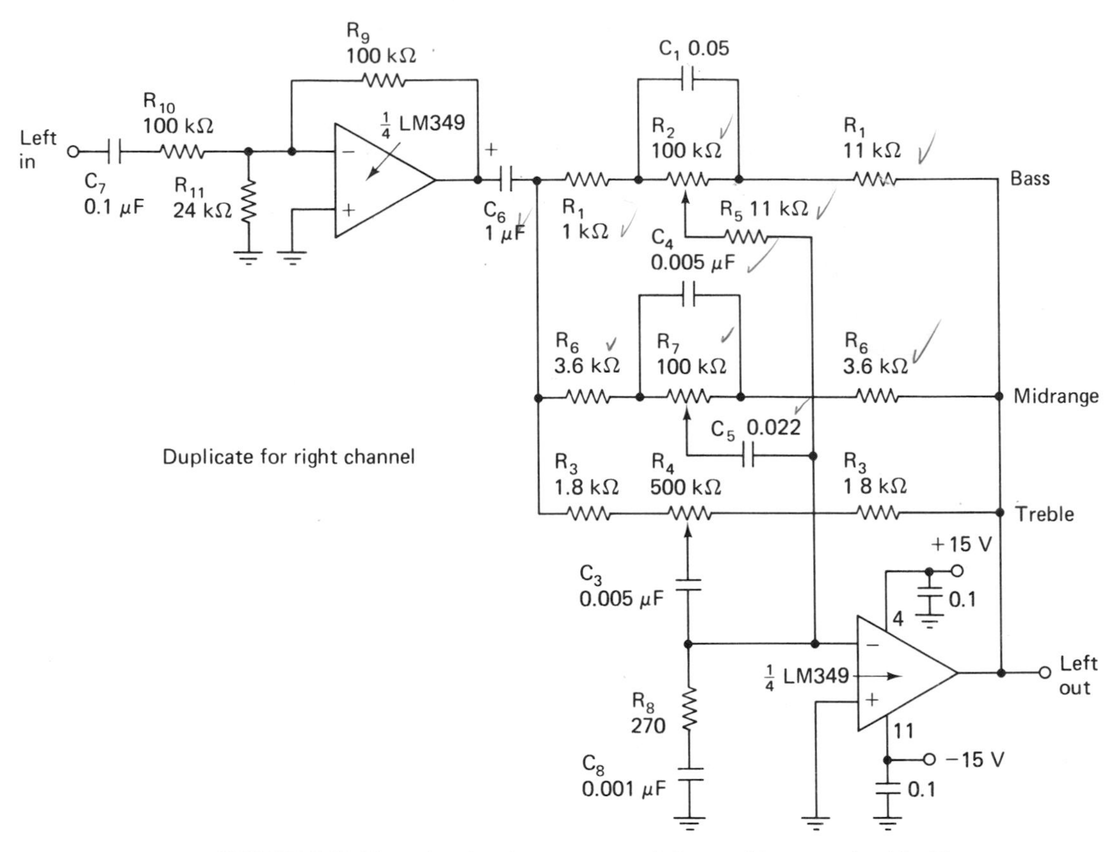

FIGURE 5-18 Three band active tone control—bass, midrange and treble. (Courtesy National Semiconductor Corp.)

input op amp serves as a follower-buffer. If the bass control is a low-pass filter and the treble control is a high-pass filter, the midrange control is a combination of both, or a band-pass filter.

It may be desirable to add more tone-control sections; however, three paralleled sections appear to be the realistic limit as to what can be expected from a single op amp. More tone-control circuits can be used in an audio system, and this is discussed in Section 5-5.

5-4 AUDIO MIXERS

Audio mixers are basically the same as summing amplifiers, discussed in Section 2-5. With a basic audio mixer as shown in Figure 5-19a, each input resistor is made variable. This allows variable gain for each input, in a similar manner as a scaling adder. However, the input resistance to the op amp is

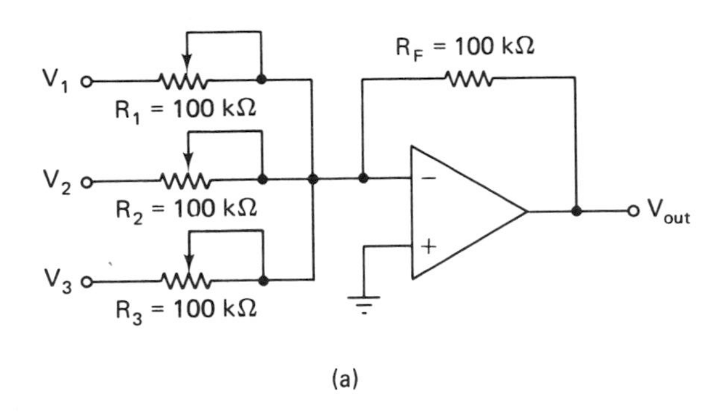

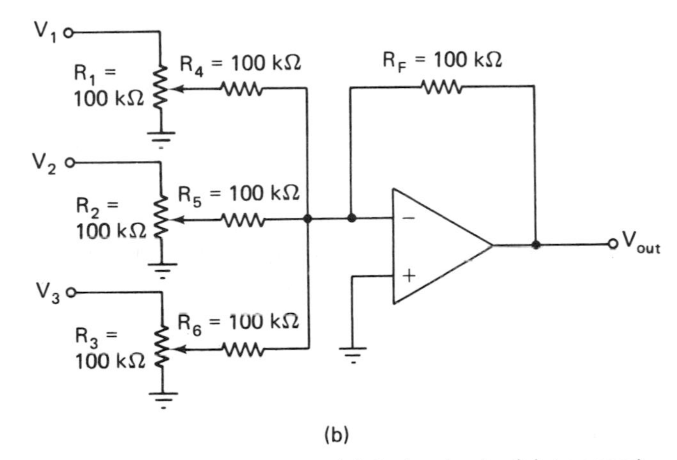

FIGURE 5-19 Audio mixers: (a) basic circuit; (b) improved circuit.

constantly changing, which may prove detrimental to circuit efficiency. An improved audio mixer is shown in Figure 5-19b. In this circuit the potentiometers are used as independent input volume controls. Here the gain for each input is constant and the potentiometers adjust the voltage to each input.

5-5 MISCELLANEOUS AUDIO CIRCUITS

5-5-1 Scratch Filter

A scratch filter is a low-pass filter used to roll off excess high-frequency noise appearing as hiss, ticks, and pops from worn records. The circuit shown in Figure 5-20 has a corner frequency of 10 kHz with a slope of −12 dB octave.

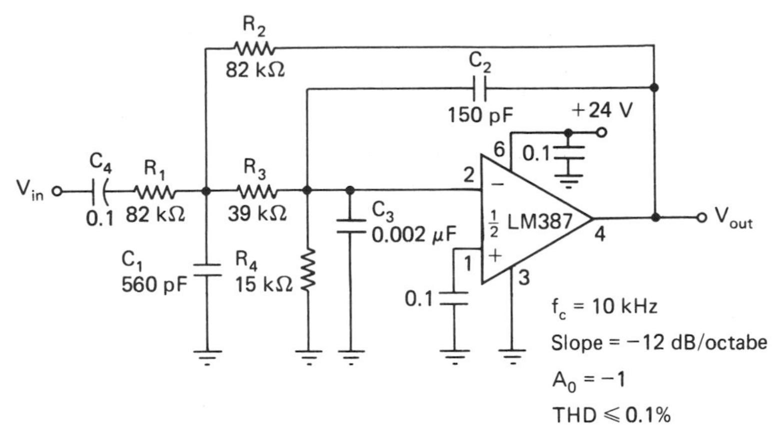

FIGURE 5-20 Scratch filter. (Courtesy National Semiconductor Corp.)

5-5-2 Rumble Filter

A rumble filter is a high-pass filter used to roll off low-frequency noise associated with worn turntable and tape transport mechanisms. Figure 5-21 shows such a circuit with a corner frequency of 50 Hz with a slope of −12 dB/octave.

5-5-3 Speech Filter

In some audio applications only voice is required, where the frequency range is from 300 to 3 kHz. Frequencies above and below this are not needed and are attenuated so that the voice information is as distortionless as possible. The speech filter shown in Figure 5-22 consists of a high-pass filter in

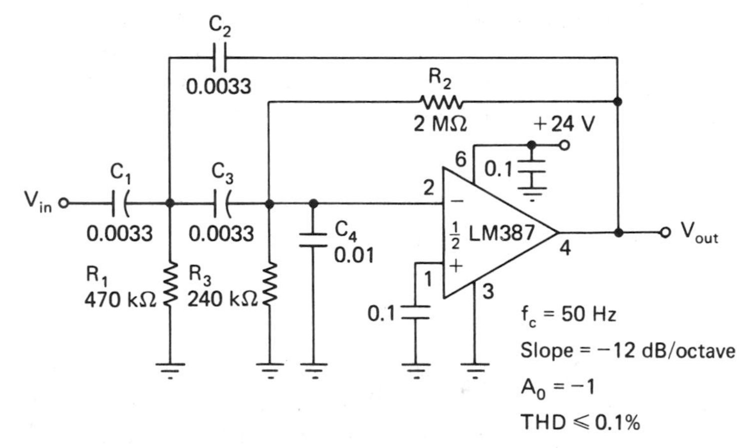

FIGURE 5-21 Rumble filter. (Courtesy National Semiconductor Corp.)

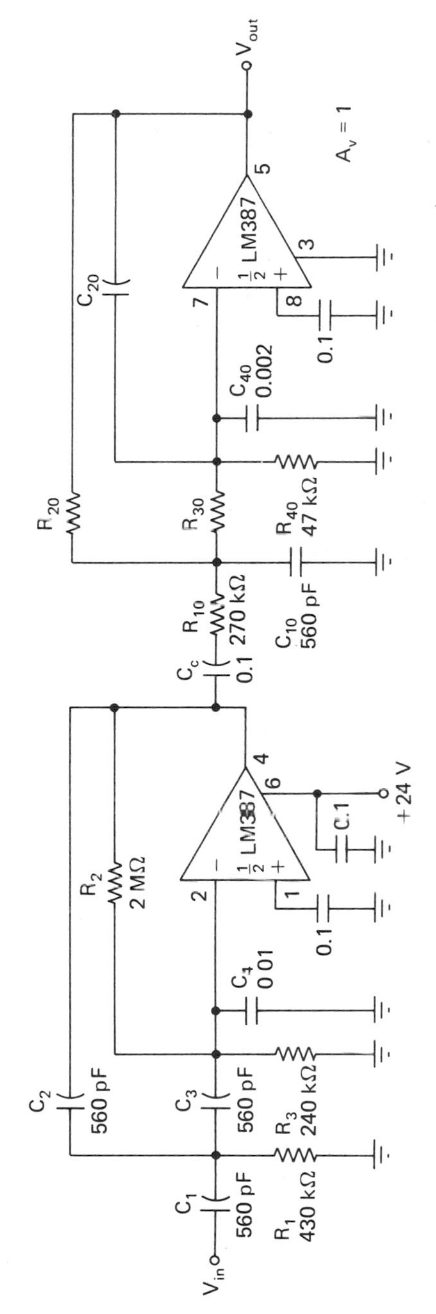

FIGURE 5-22 Speech filter, 300Hz–3KHz bandpass. (Courtesy National Semiconductor Corp.)

cascade with a low-pass filter to produce the necessary bandpass filter. The corner frequencies are 300 and 3 kHz with a rolloff of -40 dB/decade.

5-5-4 Octave Equalizer

Musically speaking, an octave is a note or group of notes that can be twice the frequency or half the frequency of a particular reference note or group of reference notes. Optimum results can be realized in an audio system if the frequency response of each octave can be controlled separately throughout the entire audio-frequency range.

The basic octave equalizer as shown in Figure 5-23 is a simple tone-control circuit. The values of C_1 and C_2 determine the frequency range or octave, which is affected by the circuit. It is designed to compensate for any unwanted amplitude-frequency or phase-frequency characteristics of an audio system.

The basic octave equalizer can be duplicated to produce a 10-section octave equalizer, as shown in Figure 5-24. By using quad op-amp ICs, the entire circuit consists of only three IC packages. The input buffer amplifier provides a low source impedance to drive the equalizer while presenting a high input impedance for the preamplifier. Resistor R_8 is used to stabilize the circuit while retaining its fast slew rate of 2 V/μs. A unity-gain output summing amplifier is used to add each equalized octave of frequencies together again. Resistor R_{20} is scaled such that its in-phase signal is actually subtracted from the inverted (out-of-phase) signals, coming from each equalizer to maintain an overall gain of 1. Capacitor C_3 minimizes the possibility of large DC offset voltages from appearing at the output. Capacitor C_4 in each equalizer section provides more stability. Also, an input and output capacitor may be needed if DC voltages are present.

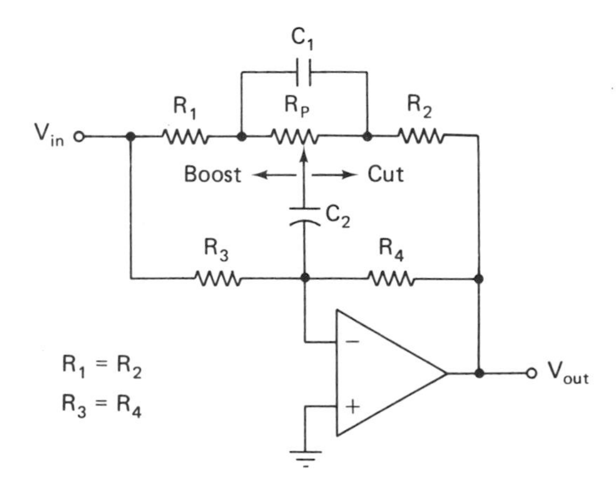

FIGURE 5-23 Typical octave equalizer section.

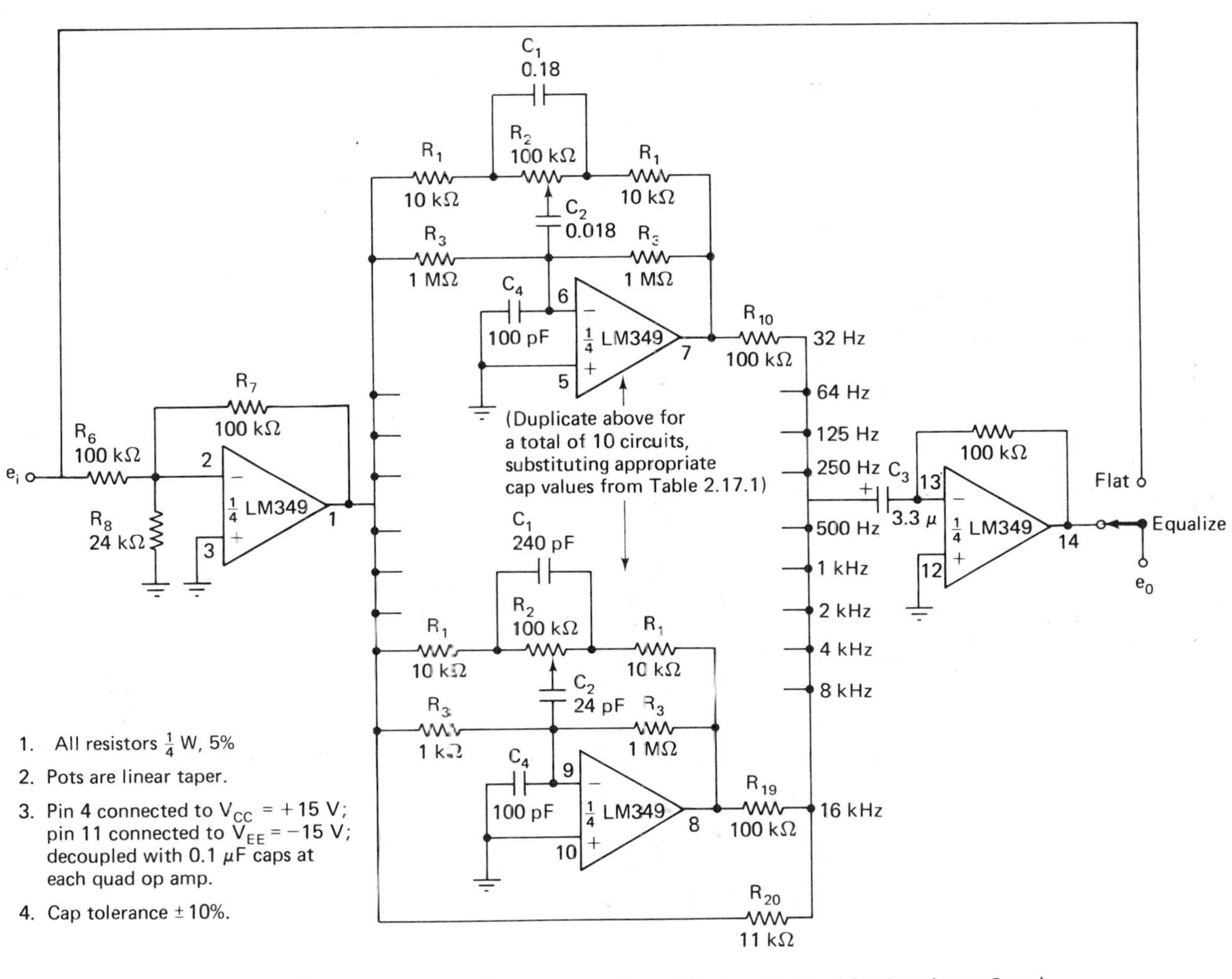

FIGURE 5-24 Complete 10-section octave equalizer. (Courtesy National Semiconductor Corp.)

TABLE 5-1 Values of C_1 and C_2 for appropriate octave. (Courtesy National Semiconductors.)

f_0	C_1	C_2
32 Hz	0.18 μF	0.018 μF
64 Hz	0.1 μF	0.01 μF
125 Hz	0.047 μF	0.0047 μF
250 Hz	0.022 μF	0.0022 μF
500 Hz	0.012 μF	0.0012 μF
1 kHz	0.0056 μF	560 pF
2 kHz	0.0027 μF	270 pF
4 kHz	0.0015 μF	150 pF
8 kHz	680 pF	68 pF
16 kHz	240 pF	24 pF

Table 5-1 lists the values of C_1 and C_2 for each given octave within the audio range.

5-5-5 Active Crossover Network

To achieve maximum speaker efficiency, only high frequencies should be applied to the smaller tweeter speaker, while low frequencies are applied to the larger woofer speaker. Such a circuit to accomplish this is known as

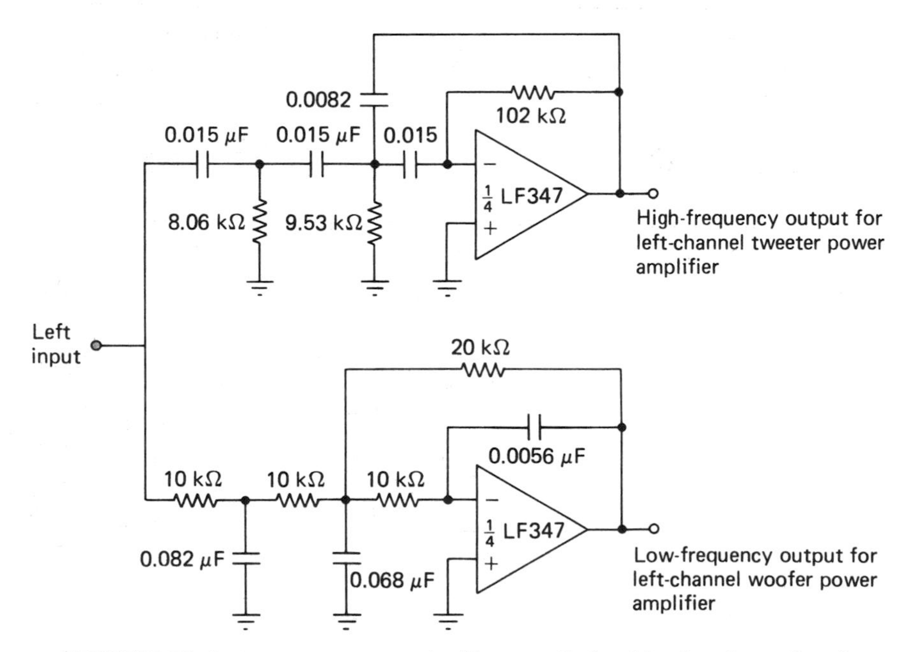

FIGURE 5-25 Active crossover network. (Courtesy National Semiconductor Corp.)

a crossover network. An op-amp crossover network is shown in Figure 5-25. The circuit basically uses a high-pass filter for driving the tweeter power amplifier and a low-pass filter to drive the woofer power amplifier. This circuit uses a quad op-amp IC and shows the left channel of a stereo system, leaving two other op amps within the package for the right channel. Crossover networks using op-amp ICs are considerably less expensive and much less bulky than older types of circuits that use heavy inductors and large capacitors.

5-5-6 Two-Channel Panning Circuit

The circuit shown in Figure 5-26 has the ability of taking a single input signal and moving it to either output via the potentiometer. It can be set to divide the input signal evenly or by any amount between the two output channels. Because of this panoramic control, the circuit is called a "panning" or "pan-pot" circuit. Panning is how an audio engineer can manage to pick up a musical instrument and float the sound over to the other side of the stage and back again. The output of this circuit is required to have unity gain at each extreme of potentiometer travel; that is, one output will have all of the input signal while the other output will be zero. When the potentiometer is centered, each output will be −3 dB down from the input signal.

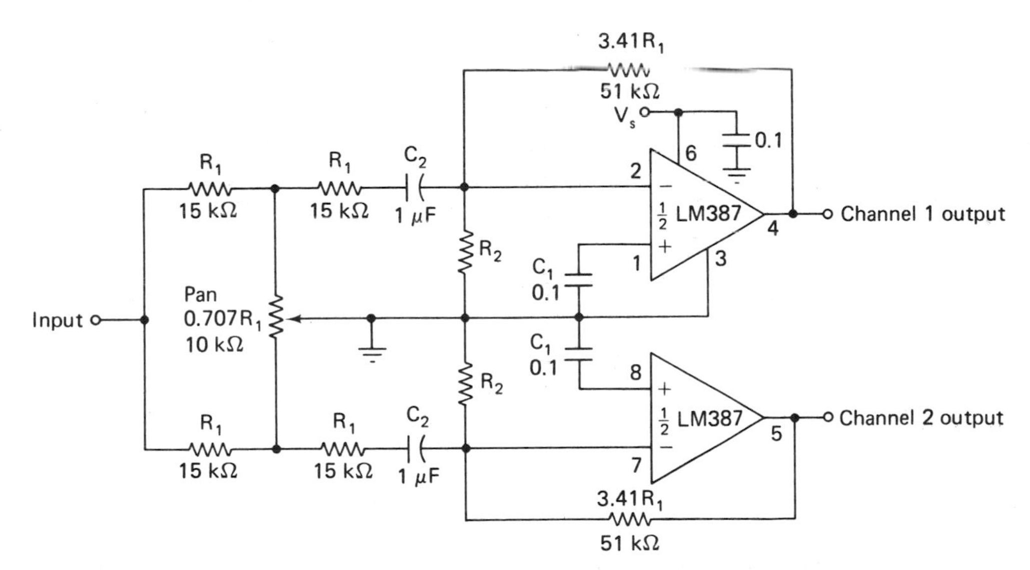

FIGURE 5-26 Two channel panning circuit. (Courtesy National Semiconductor Corp.)

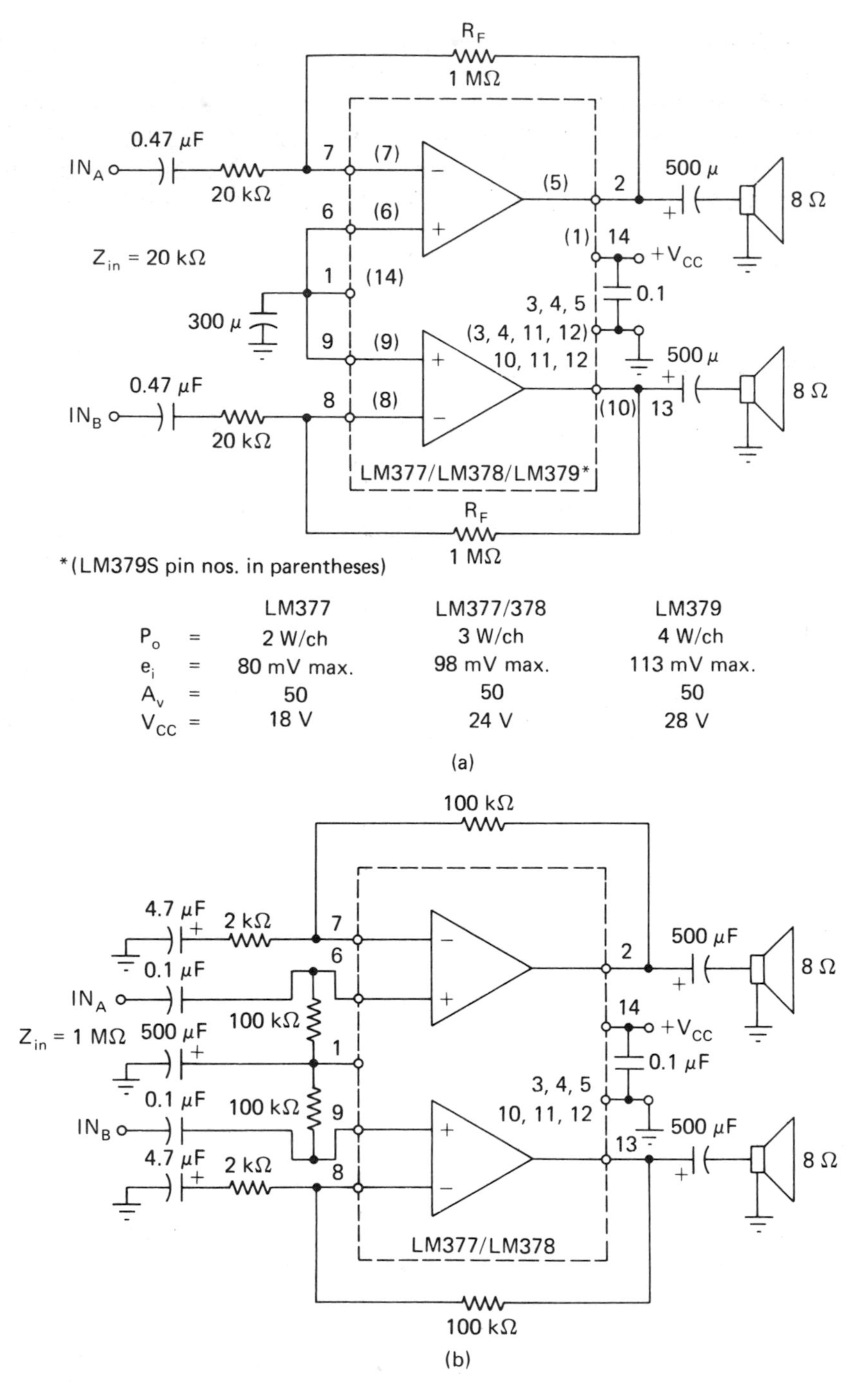

		LM377	LM377/378	LM379
P_o	=	2 W/ch	3 W/ch	4 W/ch
e_i	=	80 mV max.	98 mV max.	113 mV max.
A_v	=	50	50	50
V_{CC}	=	18 V	24 V	28 V

FIGURE 5-27 Stereo amplifiers: (a) inverting amplifier; (b) non-inverting amplifier. (Courtesy National Semiconductor Corp.)

5-6 SIMPLE MEDIUM-POWER AMPLIFIERS

Audio-power-amplifier ICs are used for low- and medium-power applications, such as mono, stereo, or multichannel audio output for phono, tape, or radio. They utilize a power-supply range of 6 to 50 V.

These audio-power-amplifier ICs do not differ significantly in circuit design from traditional op amps. Major design differences appear in the class AB high current output stages and special IC layout techniques to guarantee thermal stability throughout the chip. These IC packages resemble the standard 14-pin DIP, except that several pins on each side are replaced with a fin, which is used for connecting the IC to heat sinks. Normally, the circuits use a single power supply, but dual power supplies may be used without difficulty or degradation in output performance.

Unfortunately, circuits using these ICs are susceptible to picking up noise and self-oscillation at high frequencies. Component layout is extremely critical and extra circuitry may be needed to filter out noise and/or prevent oscillation.

Two circuit configurations using audio-power-amplifier ICs are shown in Figure 5-27. The circuits are used for stereo output. The inverting amplifier requires the least amount of components, but must be driven by a relatively low-impedance circuit. Also given is the power output for each channel when using the rated supply voltage. The input signal voltage e_i should not be exceeded for each circuit to prevent output distortion. For applications requiring a high input impedance, the noninverting amplifier would normally be used, although it has a higher component count.

A typical mono amplifier capable of delivering 5 W of output power is shown in Figure 5-28. For applications where output ripple and high-

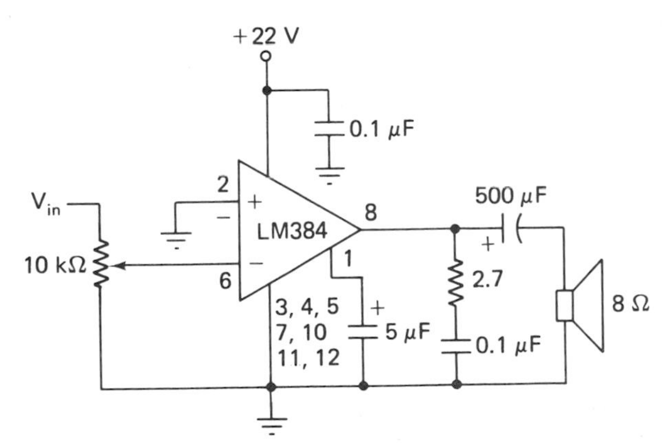

FIGURE 5-28 Typical 5W amplifier. (Courtesy National Semiconductor Corp.)

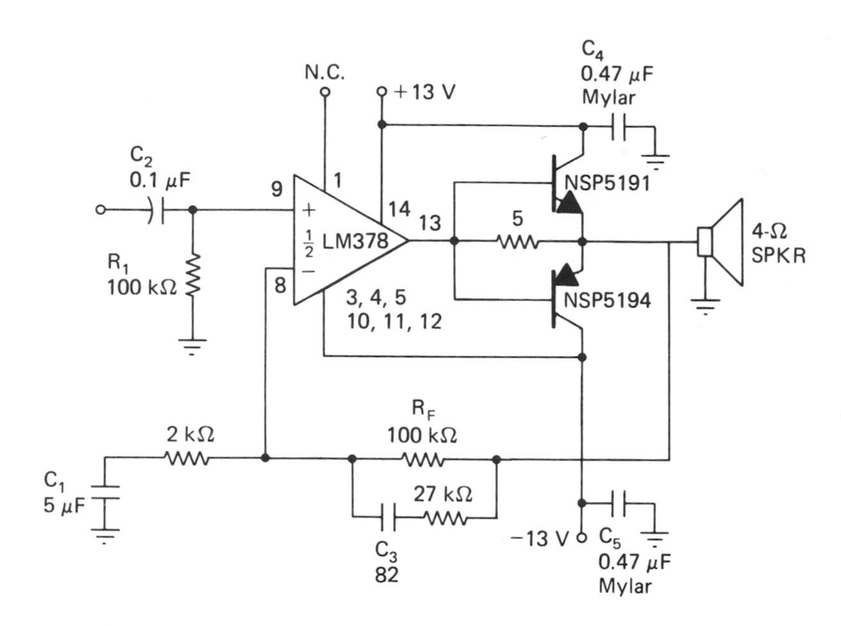

FIGURE 5-29 12-watt low-distortion power amplifier. (Courtesy National Semiconductor Corp.)

frequency oscillations are not a problem, the 2.7-Ω resistor and all capacitors, except the 500-μF output capacitor, may be eliminated. For maximum efficiency the circuit must be driven by a low-impedance source.

When output power requirements exceed the limits of available audio-power-amplifier ICs, the output may be boosted using two external power transistors, as shown in Figure 5-29. This simple circuit uses a complementary emitter-follower stage with a feedback circuit (components R_F, C_3, and the 27-kΩ resistor). Capacitors C_4 and C_5 are used for power-supply decoupling. At signal levels below 20 mW, the IC supplies the speaker directly through the 5-Ω resistor. Above this level, the booster transistors are biased "on" by the same current through the 5-Ω resistor.

Figure 5-30 shows an improved 35-W boosted power amplifier using a specially designed power driver IC. This circuit utilizes current and power limiting, which minimizes distortion and protects the output transistors from possible destruction. Resistors R_1 and R_6 are the actual current-limiting devices, while R_2–R_3 and R_4–R_5 form reference voltages about the transistors, which keep them properly biased under high load conditions. The 1-kΩ potentiometer is used to adjust the bias on the transistors so that 0 V appears between the transistors (at the load point R_L) during the quiescent state.

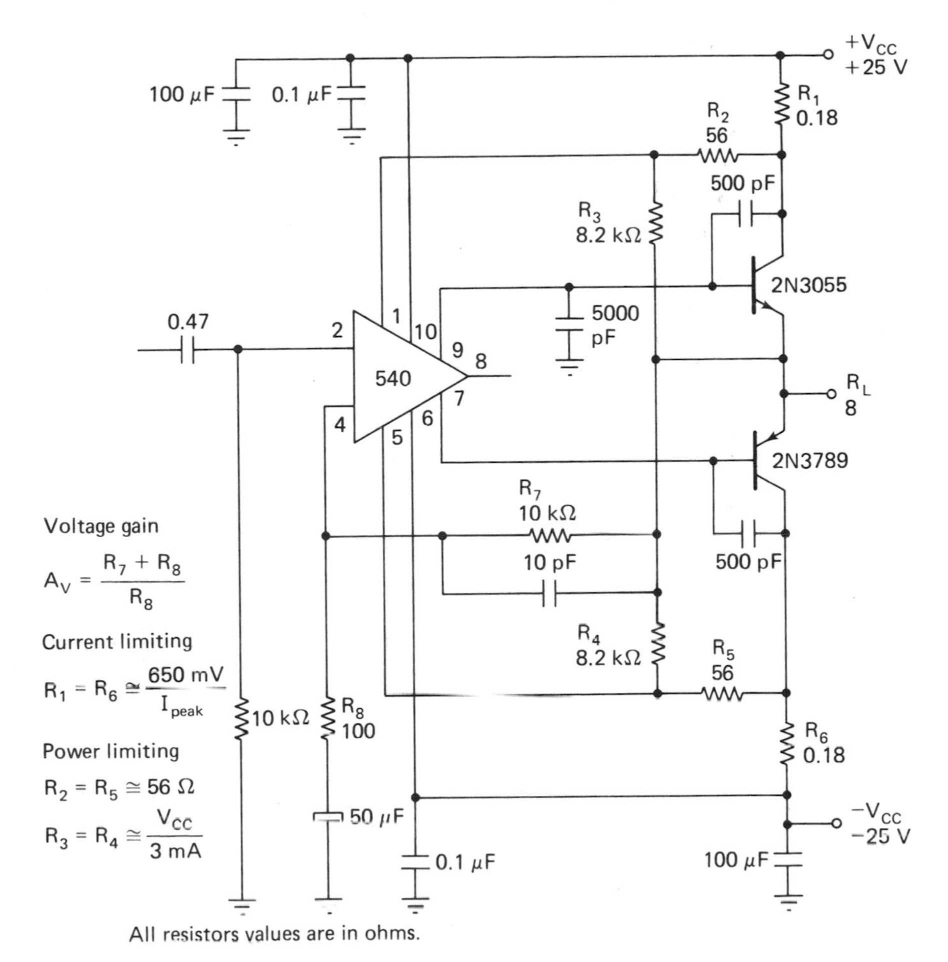

FIGURE 5-30 35-watt power amplifier. (Permission to reprint granted by Signetics Corporation, a subsidiary of U.S. Philips Corp., 811 E. Arques Avenue, Sunnyvale, CA 94086.)

SUMMARY POINTS

1. Inverting audio op amps are usually used with low-impedance sources.
2. Noninverting audio op amps are used with high-impedance sources.
3. Often, op-amp audio applications use a single power supply by biasing the noninverting input at $\frac{1}{2} V_{cc}$ (+ or −).

4. Special audio op amps are designed to operate from a single or a dual power supply.
5. RIAA equalization preamplifiers are designed to amplify frequencies below 500 Hz and to attenuate frequencies above 2 kHz.
6. NAB equalization preamplifiers are designed to have a rather linear attenuation for frequencies above 50 Hz.
7. Active-tone-control circuits maintain the input signal amplitude and may even have some gain.
8. Audio mixers are basically summing amplifiers.
9. A scratch filter is a low-pass filter.
10. A rumble filter is a high-pass filter.
11. A speech filter is a bandpass filter of 300 to 3 kHz.
12. Octave equalizers are basically tone controls that provide greater flexibility over the entire audio range than simple bass and treble controls.
13. A crossover network directs low frequencies to the larger "woofer" speaker and directs the higher frequencies to the smaller "tweeter" speaker.
14. A panning circuit enables the operator to move a single audio source completely or proportionately from one audio channel to another.
15. Specially designed audio op amps are capable of delivering up to several watts of power directly to a speaker.
16. When more audio power is required, an audio op amp may drive booster output transistors.

TERMINOLOGY EXERCISE

Write a brief definition for each of the following terms:

1. Audio frequency range
2. External noise
3. Internal noise
4. Single power-supply operation

5. RIAA equalization
6. RIAA equalization amplifier
7. NAB equalization
8. NAB equalization amplifier
9. Passive tone controls
10. Active tone controls
11. Audio mixers
12. Scratch filter
13. Rumble filter
14. Speech filter
15. Octave equalizer
16. Crossover networks
17. Panning circuit
18. Power amplifier

PROBLEMS AND EXERCISES

1. Referring to Figure 5-2, what is the low-frequency cutoff when $C_i = 4.7\ \mu F$ and $R_i = 2.2\ k\Omega$?
2. Draw an inverting amplifier and a noninverting amplifier using a single power supply.
3. Draw a noninverting preamplifier with a gain of 250, followed by an inverting amplifier with a gain of 50, which is driving an 8-Ω speaker.
4. Explain the function of the components that make up the RIAA preamplifier shown in Figure 5-11.
5. Explain the function of the components that comprise the NAB preamplifier shown in Figure 5-13.
6. Draw an active-tone-control (bass and treble) circuit.
7. List and explain three types of filters used in audio applications.

8. Explain the operation of an octave equalizer. How does it perform compared to normal tone-control circuits?

9. Explain the difference between an active crossover network and a panning circuit.

10. Draw a simple boosted power output amplifier and define the function of each component used. (*Hint:* See Figures 5-29 and 5-30.)

SELF-CHECKING QUIZ

Referring to Figure 5-31, identify the op-amp component or circuit nomenclature for questions 1 through 6.

1. Audio mixer ____
2. Output driver ____
3. Bass-tone-control section ____
4. Volume control ____
5. Preamplifier ____
6. Treble-tone-control section ____

Multiple-Choice Questions

7. A circuit capable of selecting sound from one of two speakers or both is known as:

   a. A crossover network
   b. A panning circuit
   c. A tone control
   d. An octave equalizer

8. The best circuit used to eliminate 60-Hz hum is:

   a. A scratch filter
   b. A speech filter
   c. A rumble filter
   d. An octave equalizer

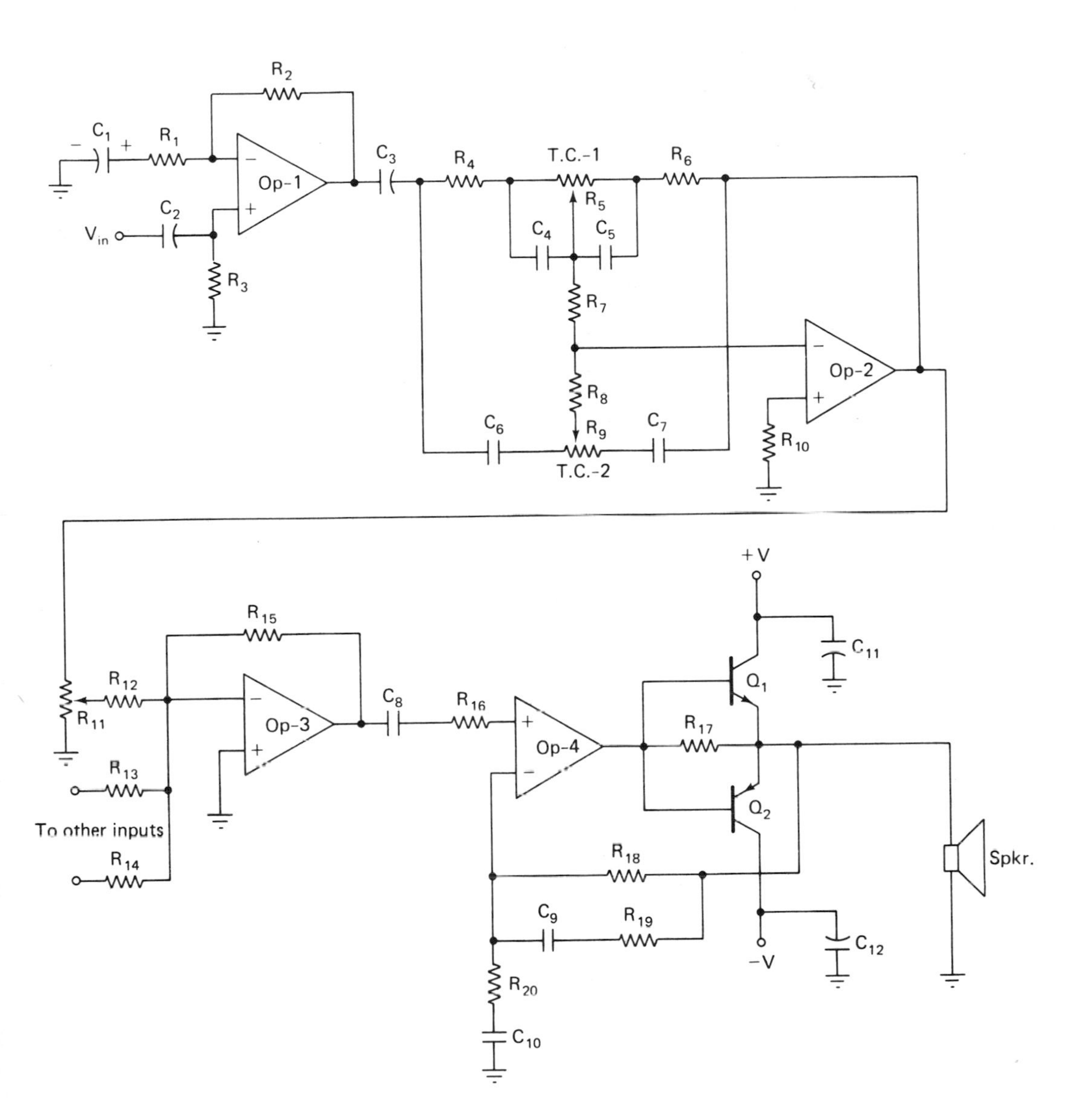

R2
C1
R1
Op-1
C2
Vin
R3
C3
R4
T.C.-1
R6
R5
C4
C5
R7
Op-2
R8
C6
R9
C7
R10
T.C.-2
+V
R15
C11
Q1
R12
C8
R16
R11
Op-3
Op-4
R17
R13
Q2
To other inputs
R18
Spkr.
R14
C9
R19
C12
-V
R20
C10

FIGURE 5-31

9. An audio system requires a preamplifier with a gain of 200 to match a 50-kΩ impedance microphone. The best amplifier configuration to use is:

   a. Noninverting

   b. Inverting

   c. Single follower

   d. None of the above

10. A 10-section octave equalizer can adjust the frequency response of:

    a. Low frequencies

    b. High frequencies

    c. Middle frequencies

    d. All of the above

(Answers at back of book)

Chapter 6

Op-Amp Protection, Stability, and Testing

It is inevitable with any device that breakdown will occur sooner or later. Reliable manufacturers or persons constructing their own circuits will desire dependable trouble-free operation for their efforts. Certain precautions can be taken during the design and construction stages of op-amp circuits that will ensure proper operation even before the circuits are powered up. This chapter will present practical applications for overvoltage protection, circuit stability, and testing of op amps.

6-1 INPUT PROTECTION

As with any solid-state device, exceeding the manufacturer's voltage rating will probably destroy the op amp or alter its characteristics. Failure at the input can result from exceeding the differential input rating or the common-mode rating. The differential input transistors of an op amp form an equivalent circuit that resembles two zener diodes back to back. If a voltage across the input exceeds the zener breakdown point, sufficient reverse current could flow to damage the input transistors (usually a short). Current-limiting resistors and diodes may be used at the inputs of op amps to provide input protection, as shown in Figure 6-1.

Resistors R_1 and R_2 should be equal and may range up to 10 kΩ with-

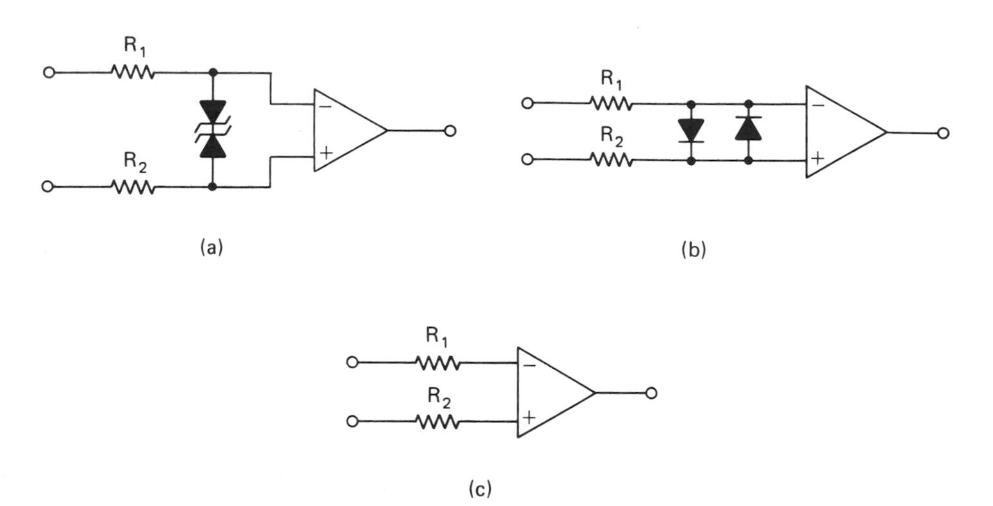

FIGURE 6-1 Input overvoltage/overcurrent protection methods: (a) zener diodes; (b) regular diodes; (c) internally protected.

out any effect on circuit performance. The internal emitter-to-base zener breakdown point of the transistors within the op amp is about 7 V. Zener diodes with a lower breakdown voltage can be used to protect the op-amp input (Figure 6-1a). Very often, because it is less expensive, regular diodes are used in a reverse-connected parallel arrangement (Figure 6-1b). Where some op amps have internal input protection, only the resistors may need to be added (Figure 6-1c). In a practical working circuit, the input ends of the resistors may be considered the actual inputs of the op amp.

6-2 OUTPUT PROTECTION AND LATCH-UP

Most op amps currently manufactured have output short-circuit protection incorporated within the IC. You may encounter older types of op amps that require output short-circuit protection as shown in Figure 6-2. The output

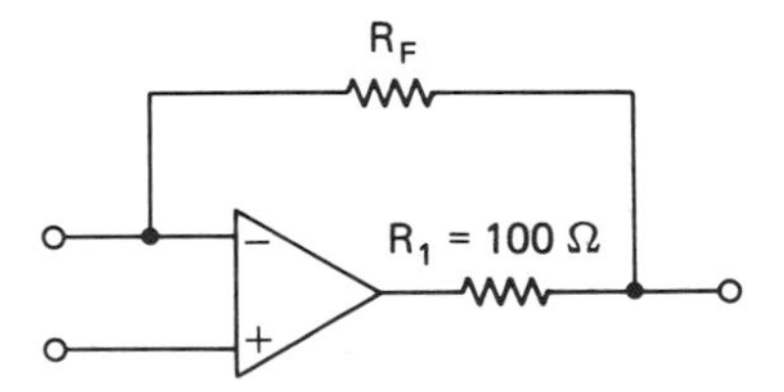

FIGURE 6-2 Output short circuit protection.

current is limited during a short circuit by a low-value resistor in series with the output. During normal operation the drop in voltage at the output is insignificant and performance is otherwise not impaired. Moreover, the circuit stability is improved, especially with capacitive loads.

A problem sometimes occurring in voltage-follower stages using op amps is called latch-up. If the input signal peak-to-peak voltage swing is greater than the input bias levels, the op amp can saturate. In the saturated condition, the op amp no longer has negative feedback, but goes into positive feedback. Positive feedback, of course, keeps the op amp in saturation, and the output will be latched up to a high voltage level (near $+V$ or $-V$ supply voltages).

The use of a high-value feedback resistor can reduce or eliminate the possibility of latch-up but may impair the input characteristics of an op amp. One commonly used method of preventing latch-up is shown in Figure 6-3a. A diode is placed from the output to one of the frequency-compensation terminals. This prevents the output voltage from rising above the potential on the terminal due to diode clamping action.

Since latch-up usually occurs from extreme limits of input voltage swing, a circuit that clamps the input voltage to a specific level will also prevent latch-up, as shown in Figure 6-3b. The input signal will not be able to swing greater than supply voltages $+V$ or $-V$.

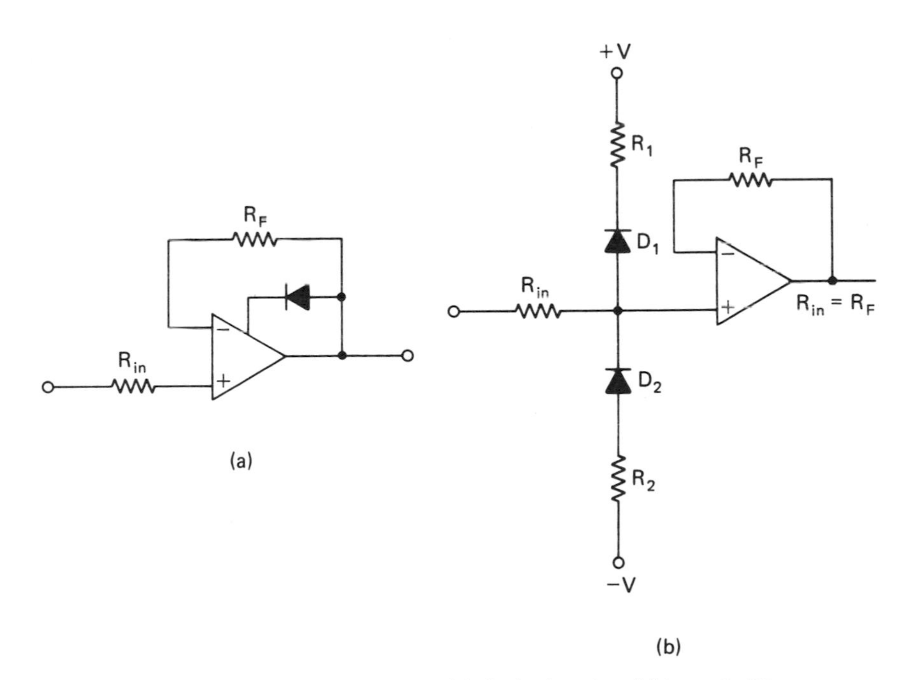

FIGURE 6-3 Preventing latch-up: (a) diode clamping; (b) input limiting.

6-3 POWER-SUPPLY PROTECTION

Op-amp ICs are constructed in such a manner that they must always be operated with the correct power supply polarity, and the maximum power-supply voltage rating must not be exceeded. If the power-supply voltage ever becomes reversed, even for a moment, destructive currents will flow and the IC will be useless. A single amplifier can be reverse-polarity-protected by placing a diode in series with the $-V$ power supply, as shown in Figure 6-4a. Reverse-polarity protection for a group of amplifiers may be provided by connecting a pair of power diodes in reverse across the power supply, as shown in Figure 6-4b. The diodes should be capable of handling more current than the fuse or short-circuit current limit of the power supplies. If a polarity reversal occurs, D_1 and D_2 will clamp the power supplies to the limit or draw enough current to blow the fuse, thereby protecting the amplifiers.

General-type IC op amps usually have a maximum power-supply operating voltage of ±18 V (36 V total) given by the manufacturer. If this limit is exceeded, even momentarily, the IC could be destroyed. Two power-supply overvoltage protection circuits are shown in Figure 6-5. A single zener diode with a V_Z of 36 V can be placed across the power-supply terminals. If this is not available, two 18-V zener diodes in series can be used.

Normally, the actual operating voltages will be lower than the rated maximum supply voltages and the zener diodes will not interfere with circuit operation. Just remember that the V_Z of the zener diodes for any power-supply overvoltage protection must be equal to or a little less than the rated maximum operating voltage of the op amp.

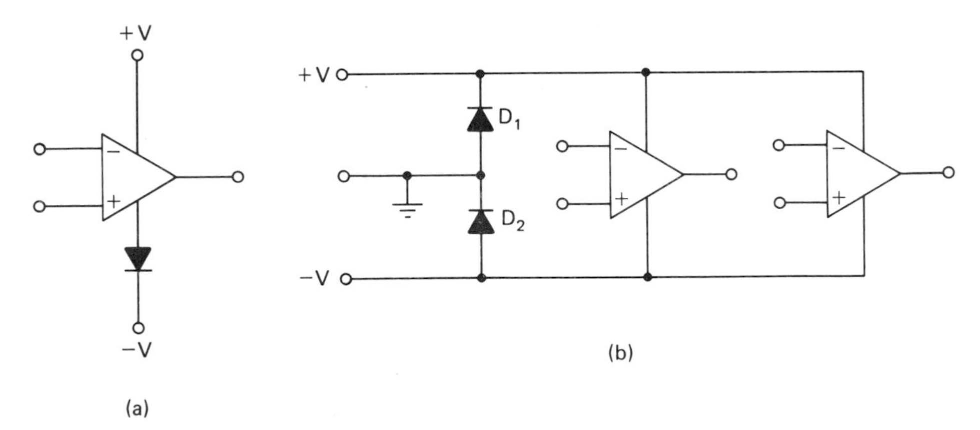

FIGURE 6-4 Power supply reverse polarity protection: (a) single amplifier; (b) group of amplifiers.

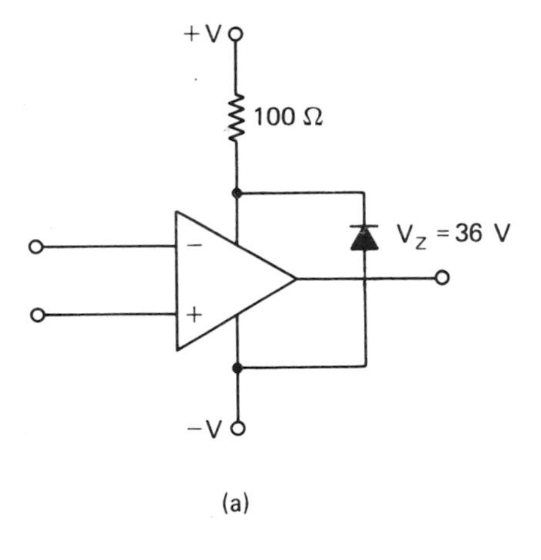

(a)

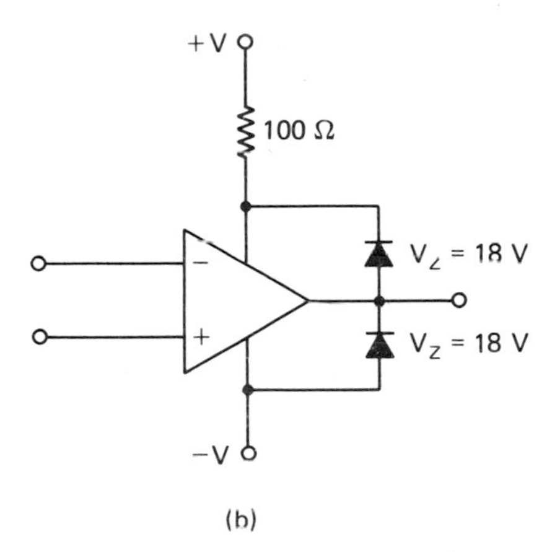

(b)

FIGURE 6-5 Power supply overvoltage protection methods: (a) using a single zener diode; (b) using two zener diodes.

6-4 BASIC CIRCUIT STABILITY APPLICATIONS

Stabilizing a feedback amplifier means to keep it from oscillating, maintain the constant gain for which it was designed, and reduce noise to a minimum. Proper circuit layout enhances the stability of a circuit. Component lead length should be kept to a minimum and, ideally, circuits be connected as directly as possible with short conductor lengths. Ground paths should have low resistance and low inductance.

It is important to keep the power-supply voltages constant for good stability. Power-supply decoupling as shown in Figure 6-6 can help accomplish this. The capacitors bypass power-supply variations to ground. These capac-

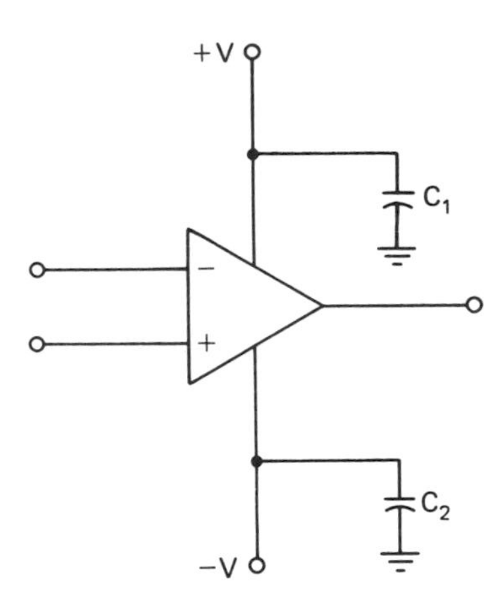

FIGURE 6-6 Power supply decoupling.

itors are usually 0.1-μF disk ceramic or 1.0-μF tantalum types. It is good practice to bypass each printed circuit board or at least every five op-amp circuits.

Stray input capacitance due to amplifier input capacitance and wiring capacitance can affect the stability of a circuit. Signal phase shifts can occur, which may even cause oscillation. A small feedback capacitor in the order of 3 to 10 pF placed in parallel with the feedback resistor can reduce or eliminate the problems caused by stray input capacitance. Examine Figure 6-7a.

Stray output capacitance can also cause stability problems. By adding a small output resistor in series with the output, the capacitance is isolated from the amplifier, as shown in Figure 6-7b.

High-frequency gain (and noise) is reduced by the feedback capacitor C_F. The reactance of C_F should be $\frac{1}{10}$ (or less) that of R_F at the unity-gain frequency.

6-5 TESTING THE OP AMP

There are numerous tests that can be performed on op amps to determine their operating abilities. These exacting tests are comprised of bias current, offset voltage, offset current, slew-rate limiting, transient response, frequency response, voltage gain, CMRR, and other more specialized testing. These tests are normally performed by the manufacturer. But what about testing op amps from the functional standpoint? Will it work or not work in a circuit? Unlike other solid-state devices, ICs, particularly op amps, are too complex to perform simple ohmmeter tests for determining their "go–no go" ability.

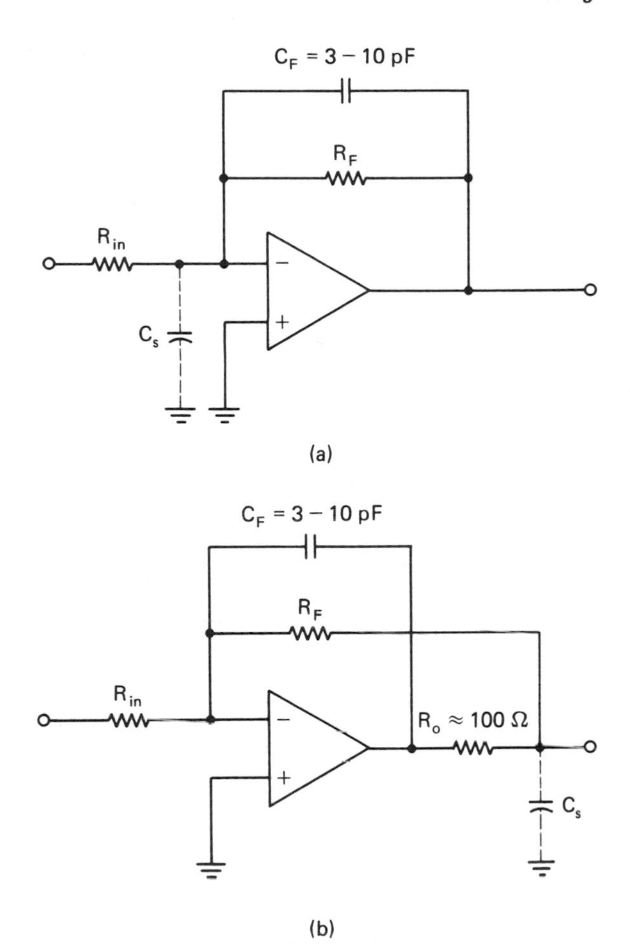

FIGURE 6-7 Stabilizing circuits with stray capacitance: (a) input capacitance; (b) stray output capacitance.

This section gives four simple op-amp test circuits that determine if the output of the op amp can swing positive and negative and indicate some relative gain. Battery power supplies can be used for these circuits or you may want to construct an AC line-operated power supply as presented in Section 1-4.

These circuits can be constructed on a perforated board and are easily stored in a small space. Figure 6-8 illustrates a general layout for the op-amp testers. Connecting terminals are required for attaching the power supplies and having the flexibility of wiring op amps with various lead identification. Some of the terminal-lead methods that you might use are:

mini-pin plug and jack

mini-banana plug and jack

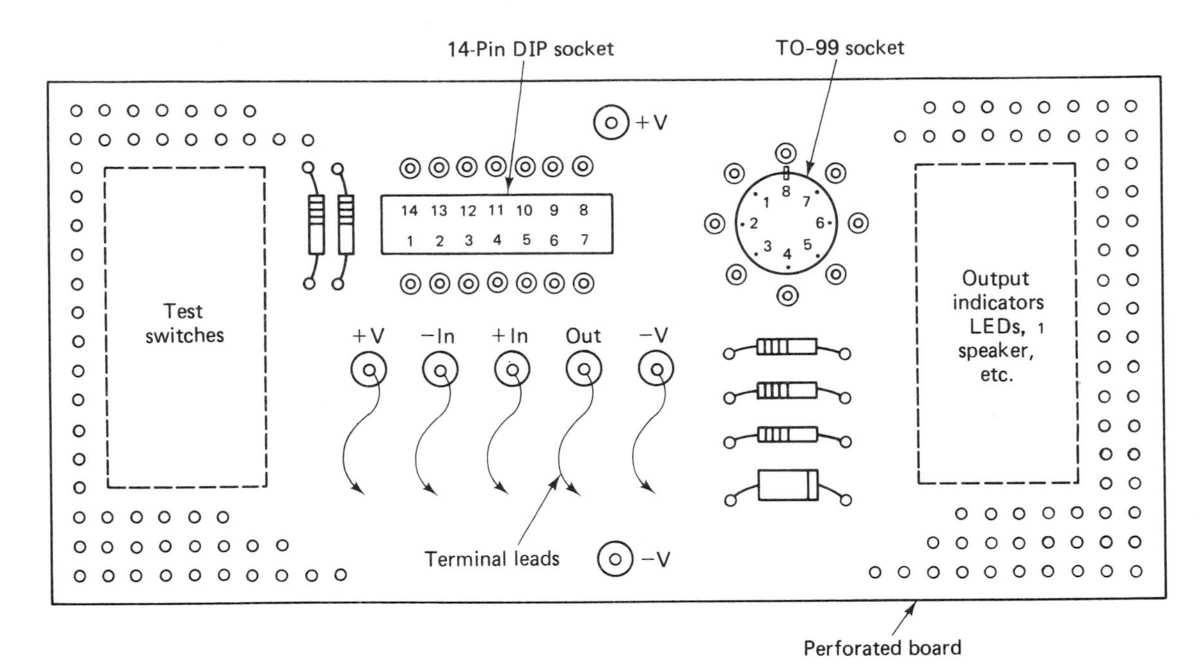

FIGURE 6-8 General layout for op-amp testers.

spring connector and tinned wires

mini-alligator clip lead

Two IC test sockets can be used: 14-pin DIP and 8-pin TO-99. All components are mounted on the board. Five terminals, labeled +V, −IN, +IN, OUT, and −V, have wires soldered to them which are used to wire the appropriate terminals of the IC socket.

Special care should be used when installing and removing the ICs from the socket. All power must be removed from the circuit. Make sure that the IC is properly aligned to the socket terminals. With the DIP socket, insert the pins on one side of the IC first and then with gentle pressure insert the other side. With the TO-99 socket, insert pin 1 first and then with a pointed object, such as a pencil or screwdriver, align each pin with a socket hole as you gently press down on the IC.

You may want to buy an IC puller or make one of your own from a pair of long tweezers by bending the tips $\frac{1}{8}$ in. inward 90 degrees, facing each other.

A very small screwdriver can be used to pry an IC out of a socket. Just remember to raise the IC evenly so that any pins are not overly bent as the IC comes free of the socket.

The circuit shown in Figure 6-9 uses LEDs (light-emitting diodes) to indicate the output swing. When the input polarity switch is in the positive posi-

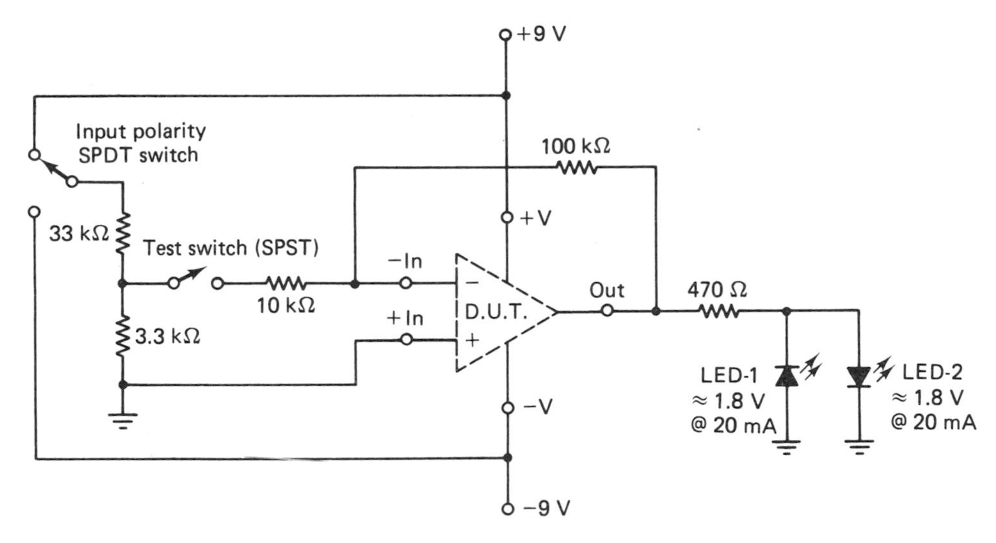

FIGURE 6-9 LED indicator GO-NO GO op-amp checker.

tion and the test switch is closed, the output of the "device under test" (D.U.T.) will go negative and LED-1 will light. Conversely, when the input polarity switch is set to negative and the test switch is closed, the output will swing positive, causing LED-2 to light.

If one or both of the LEDs fail to light, the op amp has an internal problem, which may be an open circuit. If the LEDs are on simultaneously anytime during the test, there could possibly be a short circuit within the op amp.

An automatic LED go–no go op-amp checker is shown in Figure 6-10. This circuit is a square-wave generator such as that discussed in Section 4-1. It proves the same test as the circuit in Figure 6-9. The output of the D.U.T. swings negative and positive, causing LED-1 and LED-2 to light, respectively.

By changing the values of the capacitor and resistor in the inverting input circuit of the square-wave generator you can produce an audible op-amp checker, as shown in Figure 6-11. The 1-kΩ load and AC voltmeter can test the op-amp output by giving a relative output reading. The output should be swinging nearly $+V$ and $-V$; therefore, the output reading should be 50% or greater of the $\pm V$ supply. When SI is closed, there should be a 1-kHz audible sound in the loudspeaker. The meter reading should decrease drastically due to the speaker loading down the output. No sound, of course, would indicate a defective op amp.

A basic op-amp dc gain tester is shown in Figure 6-12. This tester uses a ±15-V supply but may be increased to ±18 V if it is decided to use batteries. The input resistors and diodes drop the voltage down for the input to the op amp. The diodes help to regulate the voltage at the input. Their forward voltage drop is about 0.7 V. The 470-Ω and 100-Ω resistors form a divider

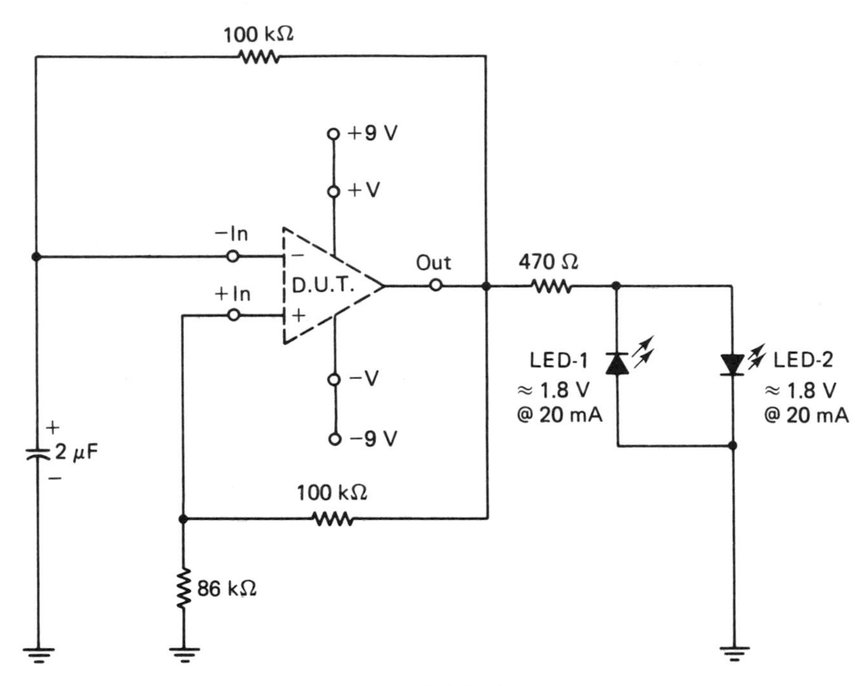

FIGURE 6-10 Automatic LED GO-NO GO op-amp checker.

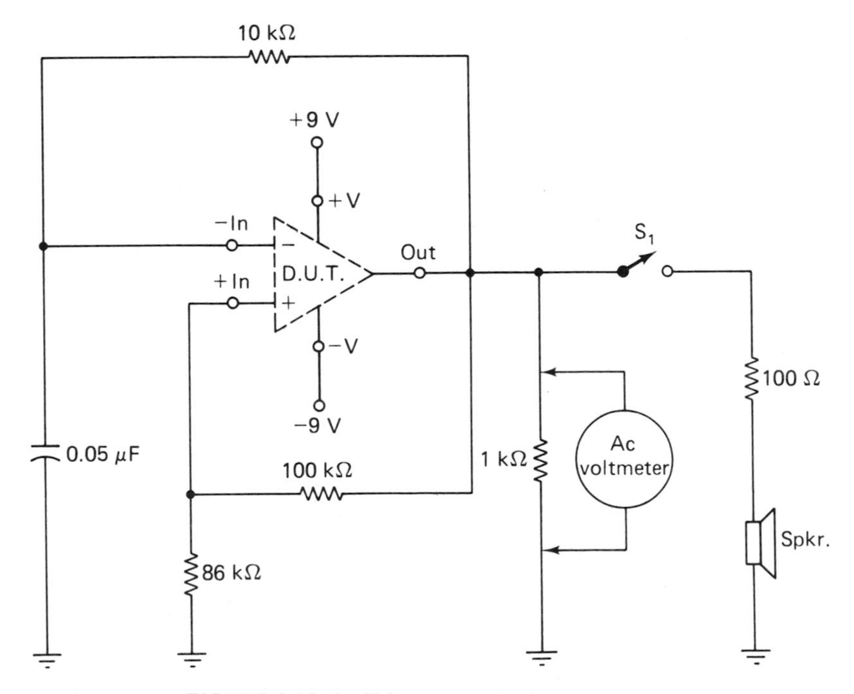

FIGURE 6-11 Audible op-amp checker.

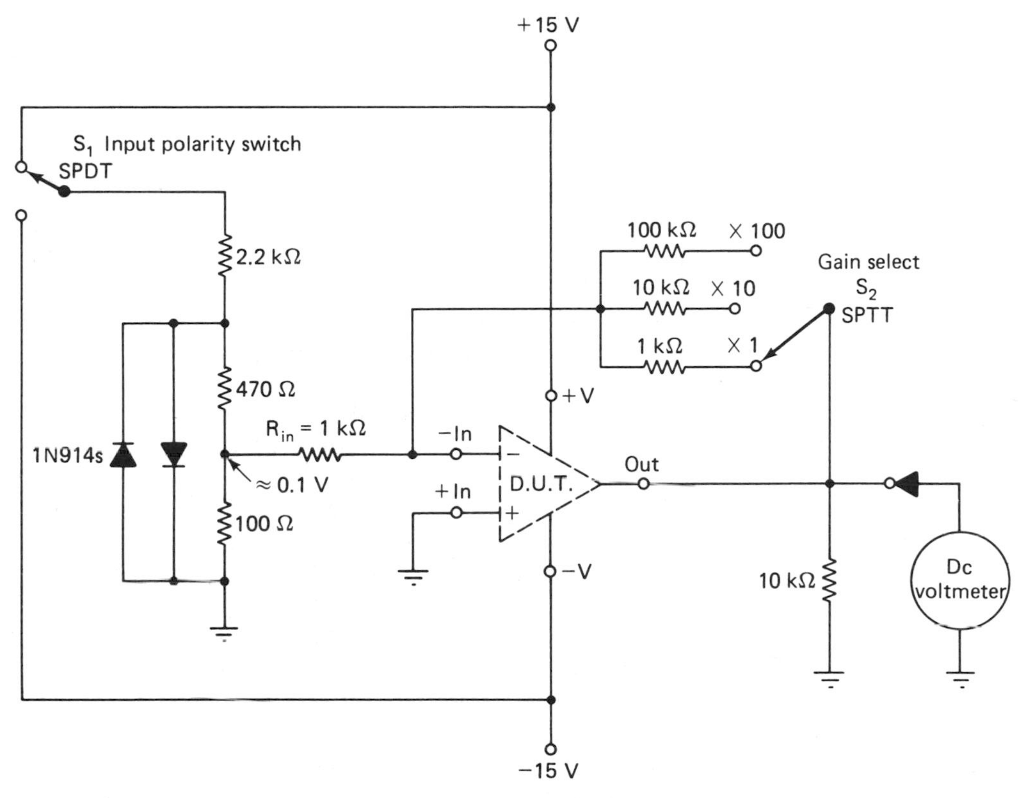

FIGURE 6-12 Basic op-amp DC gain tester.

which provides approximately 0.1 V at the inverting input. The gain select switch, S_2, controls the gain of the circuit for ×1, ×10, and ×100. The output voltages should be 0.1, 1.0, and 10 V, respectively for each setting of the gain select switch. When the input polarity switch, S_1, is in the positive position, the output voltages will be negative, and vice versa, since the circuit is a basic inverting amplifier. Any op amp that cannot produce the required gain for this tester should be discarded.

6-6 IN-CIRCUIT TESTING WITH A VOLTMETER

The voltmeter is basically used to test and measure the DC voltages in an electric circuit. As with any electronic servicing, it is always good practice to check power-supply voltages for proper values. With op amps, this voltage should be checked right at the supply-voltage pins of the IC. Improper readings may indicate that the IC is bad; however, the power supply could have a problem or another defective circuit could be upsetting the power to all circuits. Do not overlook faulty overvoltage protection or decoupling circuits, which may affect the power-supply voltages. Leaky diodes and capacitors can lower the supply voltages. Once it is determined that the power-supply voltages are correct, other dc voltage measurements can be made to ascertain where a problem is occurring.

Component aging can cause DC voltages to change and upset circuit stability. A check should be made of DC balance controls or null adjustments. These controls may be adjusted to restore circuits to their proper working condition, thereby eliminating the need for further troubleshooting.

When changes in DC levels occur, the problem is to tell whether these changes are caused by defects in the op amp or in the external circuit. One method of determining this is by removing the op amp and rechecking the voltage readings. If the voltage readings become normal, the IC (op amp) is probably defective. If the readings are still incorrect, the problem is most likely in the external circuitry. As an example, refer to Figure 6-13. The voltage at the noninverting input should read +4.8 V; however, our meter indicates about +2.0 V. If upon removing the op amp this voltage becomes normal, then the op amp seems to be defective. If the voltage remains at +2.0 V, R_1 has probably increased in value and R_2 is dropping less voltage.

Another troubleshooting technique is to short together the two inputs of the op amp, as shown in Figure 6-14. The differential input voltage becomes zero; therefore, the output voltage should also drop to zero. If this does not occur, the op amp is defective.

There are a few precautions to be observed with this technique. First, make sure that the IC is an op amp. Many other ICs cannot tolerate a short circuit across the inputs. Second, make sure that the pins being shorted are

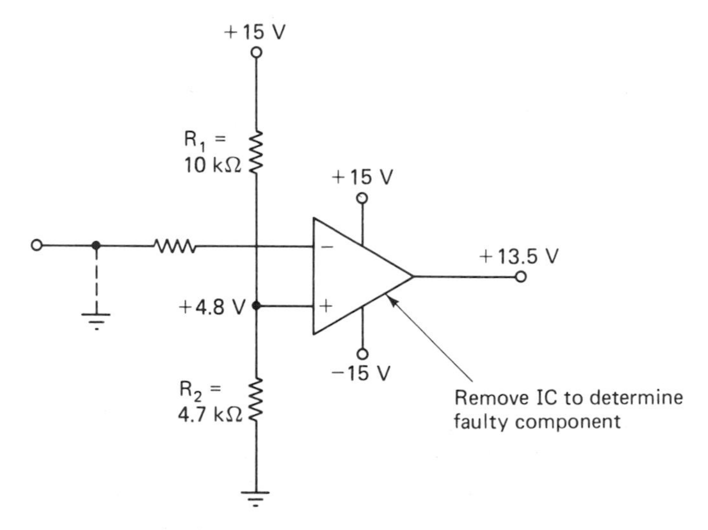

FIGURE 6-13 Testing DC input voltage.

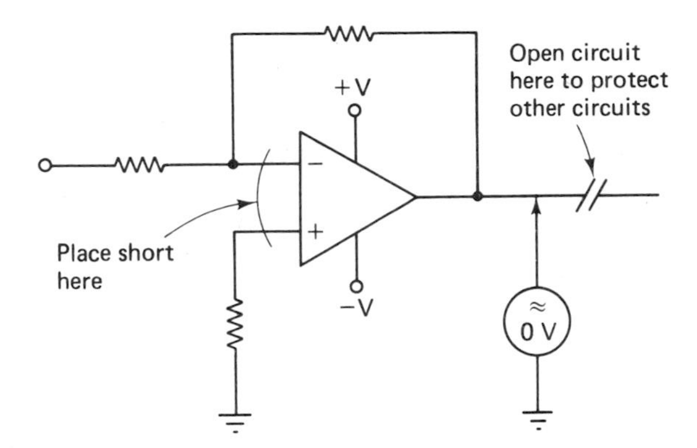

FIGURE 6-14 Testing for 0-volts output when inputs are shorted.

the inputs, because shorting other pins will destroy the IC. Third, do not use this method or at least open the output circuit if there is direct coupled circuitry following the op amp. Fourth, this test should be used only on circuits using dual polarity power supplies.

If the loss of gain is a problem within a system, individual amplifiers may be tested as shown in Figure 6-15. This is a DC gain test and it may be necessary to isolate the amplifier from other circuits by opening the input and output leads. A small 1.5-V battery and 10-kΩ potentiometer can be adjusted to provide the input voltage. Resistors R_{in} and R_F should be measured for accuracy and their values applied to the gain formula for determining the output voltage expected. Also, remember to observe the input/output voltage polarity relationship of an inverting amplifier and noninverting amplifier.

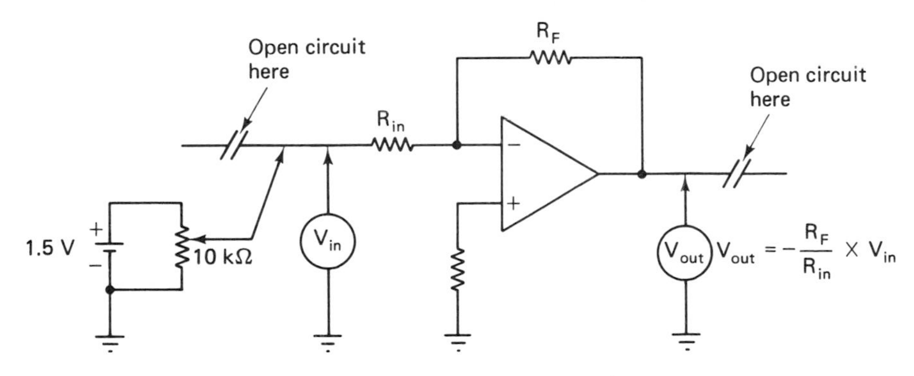

FIGURE 6-15 Testing stage DC gain.

6-7 IN-CIRCUIT TESTING WITH AN OSCILLOSCOPE

The oscilloscope is probably the best test instrument to use on any electronic circuit, since it can measure DC, signal, transient, and noise voltages. Its greatest advantage is in signal-tracing a system, as shown in Figure 6-16. Generally, a signal is placed or "injected" at the input to the system. An oscilloscope is used to check for the signal or any distortion from stage to stage. When a defective stage is located, the troubleshooting is concentrated in finding the faulty component or components. A blocking capacitor in series with the signal generator and the input to the first stage prevents any dc component from the generator getting to the unit under test. This extraneous DC component could upset the voltage measurements and cause distortion.

Testing a single stage with an oscilloscope is done in a similar manner, as shown in Figure 6-17. The signal generator is placed at the input and the oscilloscope detects the signal at the output. The gain of an amplifier can be determined and tested by using peak-to-peak voltage measurements.

6-7-1 Noise Problems

Since op amps are used in critical measuring or detecting circuits, noise presents a particularly bothersome condition. The oscilloscope is invaluable in locating noise in op-amp circuits as well as any electronic circuit.

Hum and ripple is low-frequency noise that usually comes from the power-supply voltages feeding an amplifier. One method of localizing a hum or ripple problem is to short the input pins of an op amp as was done in Fig-

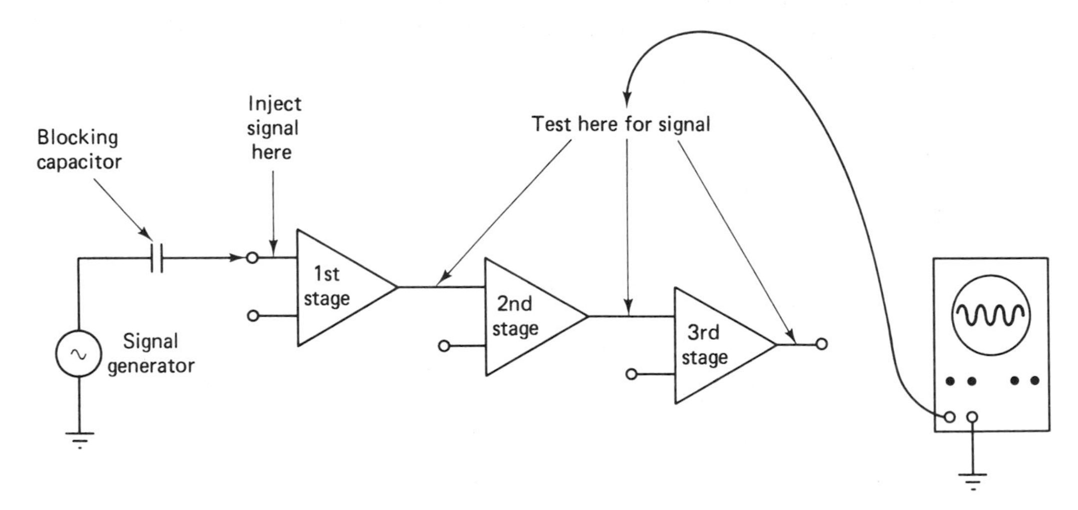

FIGURE 6-16 Signal tracing with an oscilloscope.

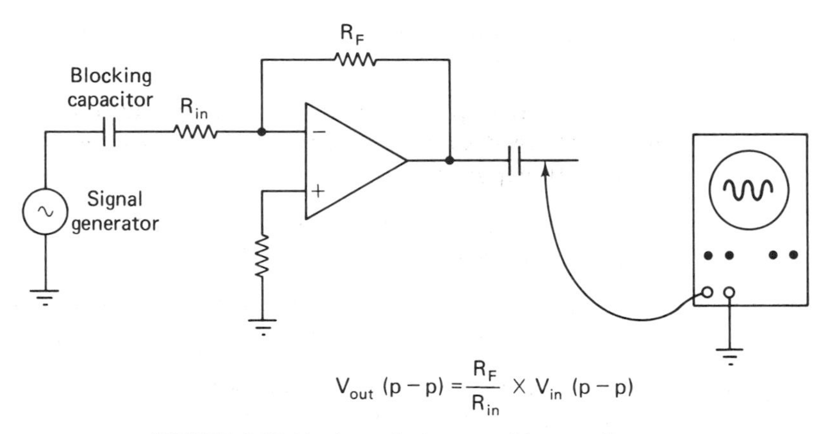

FIGURE 6-17 Testing a single stage with an oscilloscope.

ure 6-14. The oscilloscope monitors the output. If the hum or ripple disappears, it is probably being picked up by the leads or at the terminals of the op amp. Look for cold-solder joints on lead shielding, loose shields, and loose ground terminals.

If the hum or ripple does not disappear with the op-amp input terminals shorted, the problem is probably in the power supply. Check the power-supply terminals of the op amp for an excessive amount of ripple. Remember that the amplifier may have considerable gain, and the ripple monitored at the output of the amplifier may be much greater than at the power-supply terminals.

Higher-frequency noise can be generated by aging components, such as leaky diodes and capacitors and resistors that have increased in value. These components are usually in the protection and stabilizing circuits associated with op amps. Leakage in these components is extremely difficult or impossible to detect with in-circuit testing. Replacement with a component known to be good and then observing the oscilloscope for noise seems to be the best possible remedy.

SUMMARY POINTS

1. Op amps may need input protection in the form of limiting resistors, zener diodes, and regular diodes.
2. Output protection and preventing latch-up also utilizes resistors and diodes.

3. Op amps should have power-supply protection from overvoltage and polarity reversal.
4. Stabilizing op-amp circuits involves counteracting stray capacitance with small feedback capacitors and decoupling power-supply terminals with capacitors.
5. Functional testing an op amp shows that its output can swing positive and negative and that it is capable of gain.
6. Check dc adjustments and null controls when testing op-amp circuits to see if the circuit can be restored to normal operation.
7. Check for leaky diodes and capacitors and resistors that are out of tolerance when testing op-amp circuits.
8. A circuit may be tested by removing the op amp and making voltage measurements.
9. An op amp may be tested by shorting the inputs and measuring the output for zero volts.
10. An in-circuit gain test can be performed by applying the formula $V_{out} = R_F / R_{in} \times V_{in}$.
11. The voltmeter is used to check DC power-supply voltages and DC voltages at the terminals of op amps.
12. The oscilloscope is used for signal tracing and can detect transient voltages and noise.

TERMINOLOGY EXERCISE

Write a brief definition for each of the following terms:

1. Input protection
2. Output protection
3. Latch-up
4. Power supply protection
5. Stability
6. GO-NO GO test

PROBLEMS AND EXERCISES

1. List and explain three types of op-amp input protection.
2. Explain one way to have output short-circuit protection for an op amp.
3. Explain what is meant by latch-up and what causes it.
4. List and explain two methods of preventing latch-up.
5. List two reasons power-supply protection is needed.
6. What is meant by power-supply decoupling, and why is it needed?
7. Draw and explain a stabilizing circuit to reduce input stray capacitance.
8. Draw and explain a stabilizing circuit to reduce output stray capacitance.
9. Explain three methods of DC voltage testing for an op-amp circuit.
10. List the uses that an oscilloscope can have when testing op-amp circuits.

SELF-CHECKING QUIZ

Match each circuit in Figure 6-18 with its proper description in column A.

Column A

1. Input protection
2. Output short-circuit protection
3. Latch-up protection
4. Reverse-polarity protection
5. Overvoltage protection
6. Power-supply decoupling
7. Stray capacitance stabilization

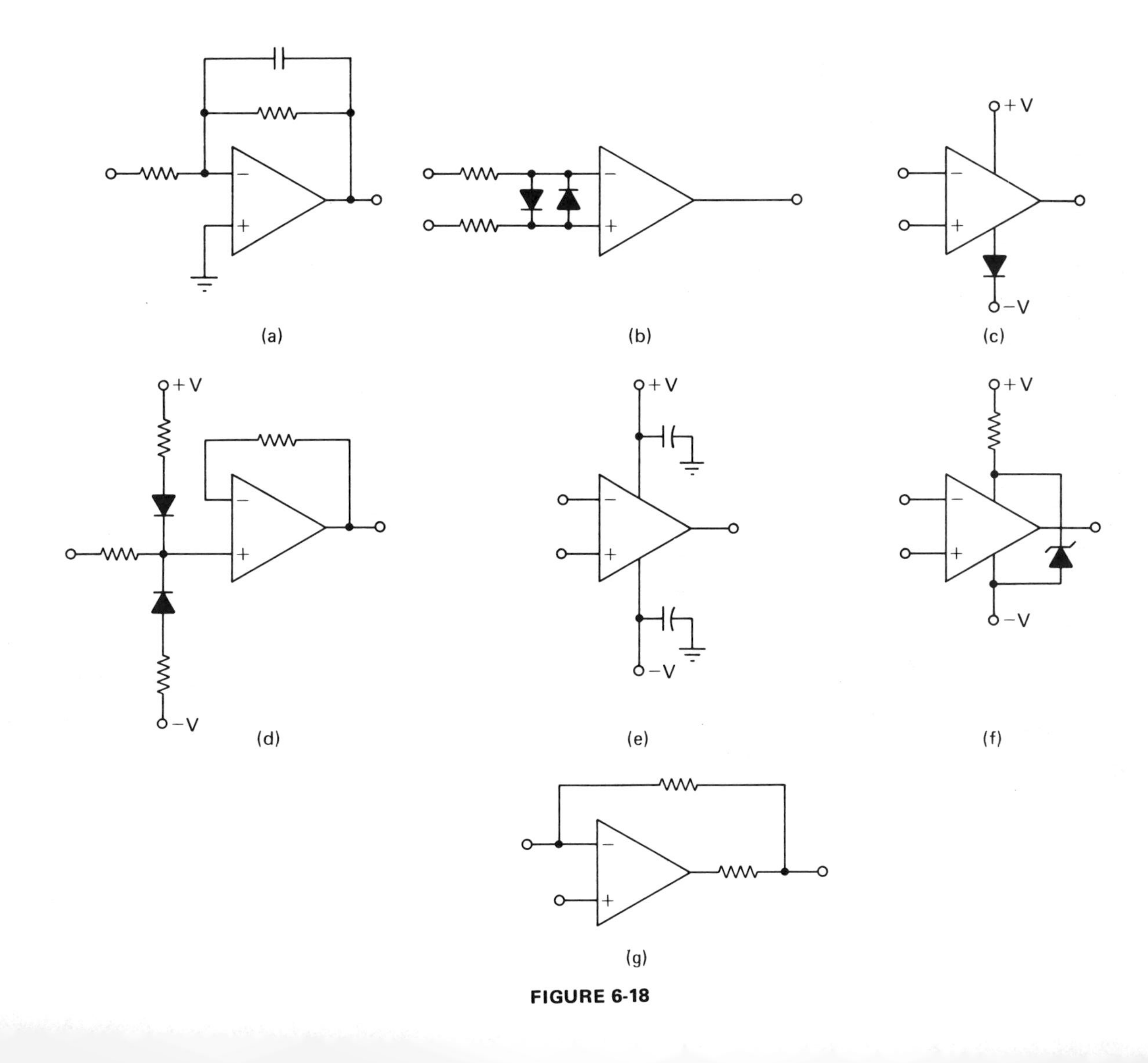

FIGURE 6-18

Multiple-Choice Questions

8. If both inputs of an op amp are at the same voltage potential, the output should be at:

    a. $+V$

    b. $-V$

    c. Depends on A_v factor

    d. Zero volts

9. Low voltage at a power-supply terminal of an op amp could be caused by:

    a. An open decoupling capacitor

    b. A leaky protection diode

    c. An open feedback resistor

    d. A shorted stabilizing feedback capacitor

10. The gain of an op-amp inverting amplifier with an input of 0.25 V and an output of 17.5 V is:

    a. 4.375

    b. 17.75

    c. 17.25

    d. 70

(Answers at back of book)

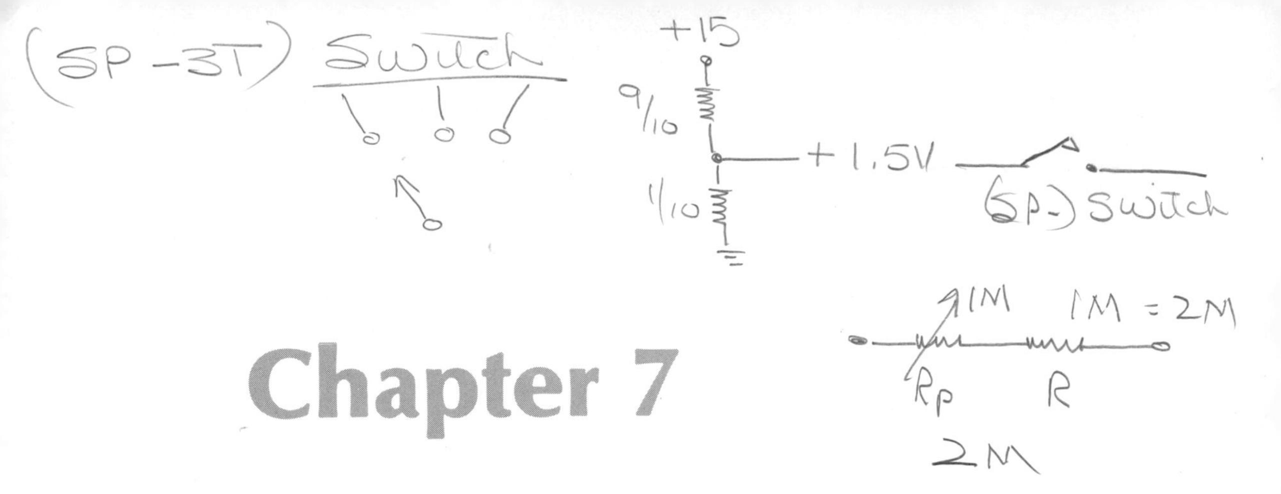

Chapter 7

Experiments to Determine Op-Amp Characteristics and Parameters

This chapter presents some basic experiments to familiarize you with op-amp characteristics and provide working skills in their use. Some of the experiments show how to determine specific op-amp parameters. The experiments are arranged in order from basic operation to more complex procedures. You are not compelled to follow this sequence, but a newcomer to op amps may find it very helpful to do so.

Step-by-step procedures and directions are used to ensure success of the experiments and minimize the chances of damaging the ICs and equipment. There are a few questions at the end of each experiment to aid you in proving that you understand the specific characteristic or parameter.

The popular 741 op-amp IC is used in all the experiments. Be sure you use one of high quality (no surplus rejects) to ensure accurate results in the experiments. No pin identification is shown on the schematic diagrams, but you can refer to Figure 1-14 for proper numbering, depending on the style of package you use. Other types of op amps may be used provided that you have a ''spec'' sheet and pin identification guide for the one you select.

Circuit connections can be temporarily ''tack''-soldered, perhaps on a perforated board, or better yet, some sort of breadboard can be used. Regardless of the method you use, be sure to keep the leads as short as possible to reduce stray pickup and provide good results.

General Procedures and Test Equipment to Use

All power supplies should be turned off or not be connected to the test circuit when you are constructing, changing components, or disassembling the circuit. This prevents any surge currents from destroying the sensitive IC or damaging test equipment. Each circuit should be constructed and then double checked before applying power.

The power supply shown in the schematic diagrams is ±15 V; however, you may elect to use ±12 V or ±9 V with the same success. The power supply used can be similar to those shown in Section 1-4 but should have very good regulation. If you decide to use batteries (±12 V or ±9 V), be sure to bypass them to ground with 0.1-μF capacitors (see Figure 6-6). When checking the power-supply voltages, be sure that they are available at the corresponding pins of the IC.

A DC input voltage (signal) is required in several experiments. This signal can be supplied by 1.5-V cells, such as a fresh flashlight battery of size AA, C, or D.

Specifications and descriptions of the test equipment used in the experiments are as follows:

1. *Meters.* Voltmeters should have 20 kΩ/V or higher for measuring DC and be capable of measuring millivolts and microamperes. Electronic voltmeters, such as vacuum-tube, transistor, or FET, are the best to use and are an absolute must for measuring critical low values of voltage and current. A good-quality digital multimeter would be excellent for these experiments. The meters can be used to measure DC voltage and current as well as RMS values of AC voltage. An oscilloscope can be used to measure peak-to-peak voltages.

2. *Oscilloscope.* A basic oscilloscope can be used for most of the experiments; however, one having a voltage-calibrated vertical axis and a time-calibrated horizontal axis is easier to use and speeds up testing time. A dual-trace oscilloscope is even more advantageous, permitting you to look at input and output signals simultaneously.

3. *Signal generators.* The ac signals required for most of the experiments are sine waves or square waves within the audio-frequency range and can be supplied by an audio generator. An RF generator can be used for frequencies above this range provided that it can produce a signal up to 1-V rms output. These generators should have low output impedance (the lower the better). A function generator is well suited for these experiments.

4. *Components.* Resistors can be $\frac{1}{4}$- or $\frac{1}{2}$-W composition types with 5% or 10% tolerance. Capacitors are nonelectrolytic types, such as ceramic or Mylar, with a minimum 50-V rating. Potentiometers are standard composition type except where specified as wirewound (WW).

Since the same op amp, power-supply voltages, and test equipment is used throughout the experiments, only the components needed for a specific experiment will be given.

Experiment 7-1 Output Polarity

OBJECTIVE: To show how the op-amp output can swing positive and negative with respect to ground and is 180° out of phase with the input.

Required Components:

1 10-kΩ resistor (R_{in})

1 three-position switch (S_1)

Test Procedure

1. With the power supply off, construct the circuit shown in Figure 7-1a.
2. Set the voltmeter to the next highest range above 15 V.
3. Turn the power supply on.
4. Record the V_{out} reading. −1.05 V +1 (Meter may deflect due to offset.)
5. Place S_1 in position A.
6. Record V_{out}, indicating polarity. −12.5V +14.7V
7. Place S_1 in position B.
8. Record V_{out}, indicating polarity. +14.5V −12.5V
9. Turn the power supply off and remove the batteries.
10. Rearrange the circuit to that shown in Figure 7-1b. (Note the meter lead polarity.)
11. Repeat steps 3 through 9.

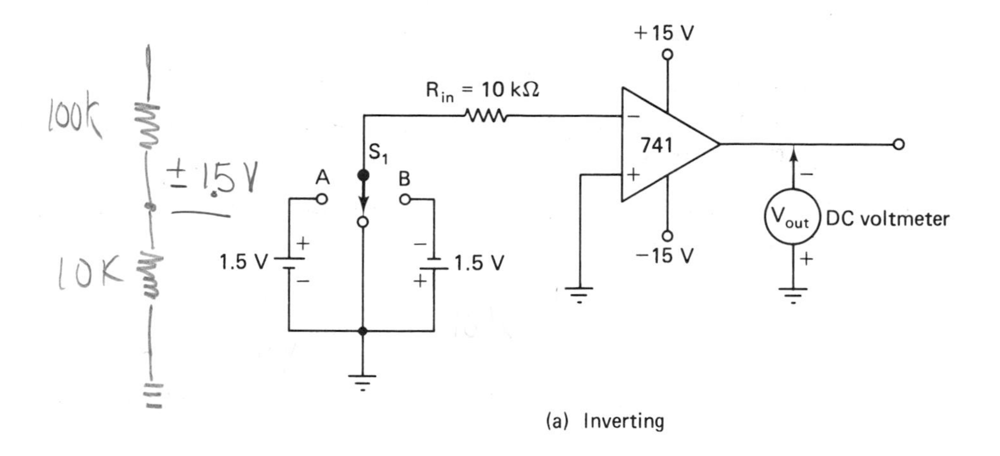

(a) Inverting

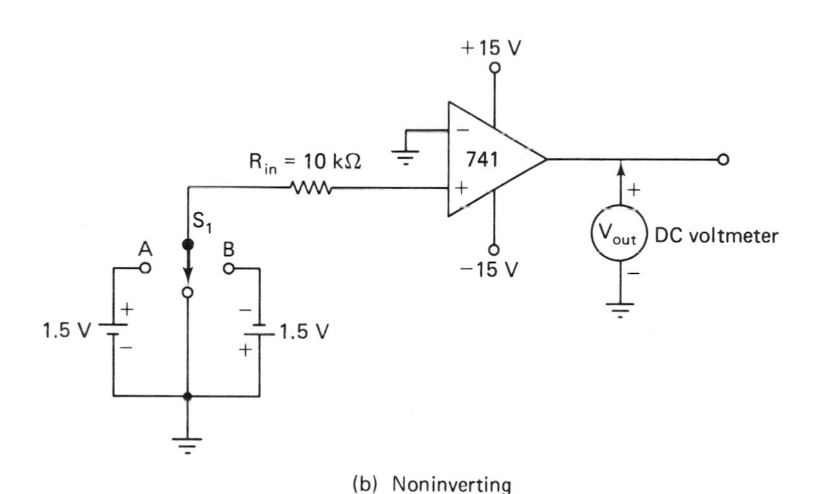

(b) Noninverting

FIGURE 7-1 Output polarity.

QUESTIONS FOR FIGURE 7-1

1. What is the V_{out} of Figure 7-1a when S_1 is in position A? -12.5V
2. What is the V_{out} of Figure 7-1a when S_1 is in position B? +14.5V
3. What polarity will V_{out} be when V_{in} is positive for an inverting amplifier? - V
4. What polarity will V_{out} be when V_{in} is negative for an inverting amplifier? +V

5. What is the V_{out} of Figure 7-1b when S_1 is in position A? +14.7V
6. What is the V_{out} of Figure 7-1b when S_1 is in position B? −12.5V
7. What polarity will V_{out} be when V_{in} is positive for a noninverting amplifier? +V
8. What polarity will V_{out} be when V_{in} is negative for a noninverting amplifier? −V

Experiment 7-2 Closed-Loop DC Voltage Gain (Inverting)

OBJECTIVE: To demonstrate how circuit voltage gain depends upon R_{in} and R_F.

Required Components:

1 10-kΩ wirewound potentiometer (R_p)

1 10-kΩ resistor (R_{in})

1 100-kΩ resistor (R_F)

1 SPDT switch (S_1)

Test Procedure

1. With the power supply off, construct the circuit shown in Figure 7-2.
2. Place S_1 to position A.

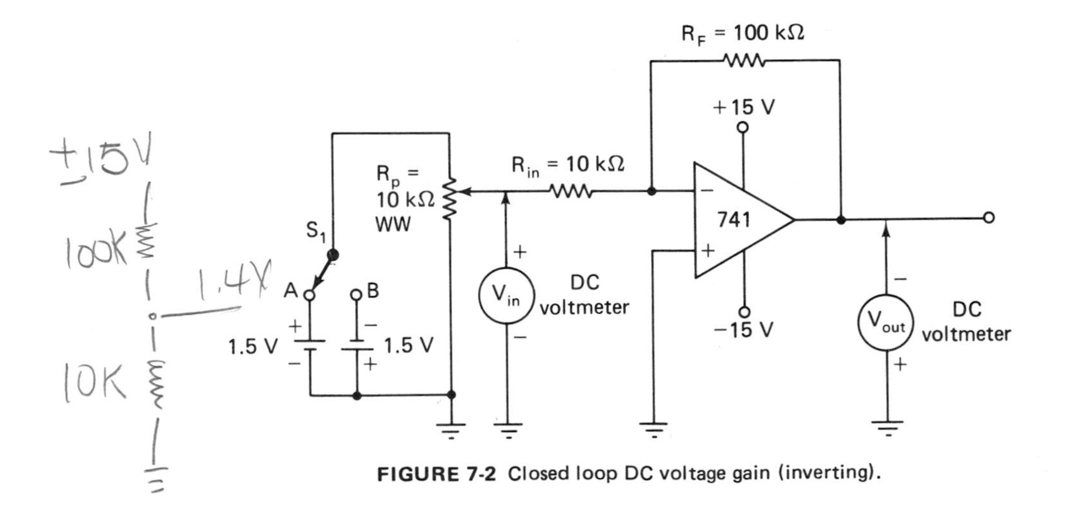

FIGURE 7-2 Closed loop DC voltage gain (inverting).

3. Turn on the power supply.
4. Adjust R_p for a V_{in} of +0.5 V.
5. Record V_{out}, indicating polarity. +5.2V
6. Turn off the power supply.
7. Reverse the meter leads.
8. Place S_1 to position B.
9. Turn on the power supply.
10. Adjust R_p for a V_{in} of −0.8 V.
11. Record V_{out}, indicating polarity. +8V
12. Turn off the power supply and disconnect the batteries used for V_{in}.
13. Calculate the gain of the circuit from the formula

$$A = -\frac{R_F}{R_{in}}$$

100K / 10k = −10V

14. Calculate V_{out} from the formula

$$V_{out} = -\left(\frac{R_F}{R_{in}}\right)(V_{in})$$

100K / 10K · .8V = 8V

15. Calculate the voltage gain from the formula

$$A_v = \frac{V_{out}}{V_{in}}$$

16. Repeat the experiment several times for different gains by changing the value of R_F.

QUESTIONS FOR FIGURE 7-2

1. What is V_{out} for a V_{in} of +0.5 V? (Remember polarity.) −5V
2. What is V_{out} for a V_{in} of −0.8 V? (Remember polarity.) +8V
3. What is the gain of the circuit? 10
4. What is the gain of the circuit if R_F = 120 kΩ? 12
5. What is V_{out} when R_F = 56 kΩ and V_{in} = +0.6 V? −3.36V
6. What is V_{out} when R_F = 220 kΩ and V_{in} = −0.4 V? +8.8V

Experiment 7-3 Closed-Loop DC Voltage Gain (Noninverting)

OBJECTIVE: To show how circuit gain is greater with the noninverting amplifier than with the inverting amplifier.

Required Components:

1 10-kΩ wirewound potentiometer (R_p)

1 10-kΩ resistor (R_{in})

1 100-kΩ resistor (R_F)

1 SPDT switch (S_1)

Test Procedure

1. With the power supply off, construct the circuit shown in Figure 7-3.
2. Place S_1 to position A.
3. Turn on the power supply.
4. Adjust R_p for a V_{in} of +0.5 V.
5. Record V_{out}, indicating polarity. +5.6V
6. Place S_1 to position B.
7. Adjust R_p for a V_{in} of −0.8 V.

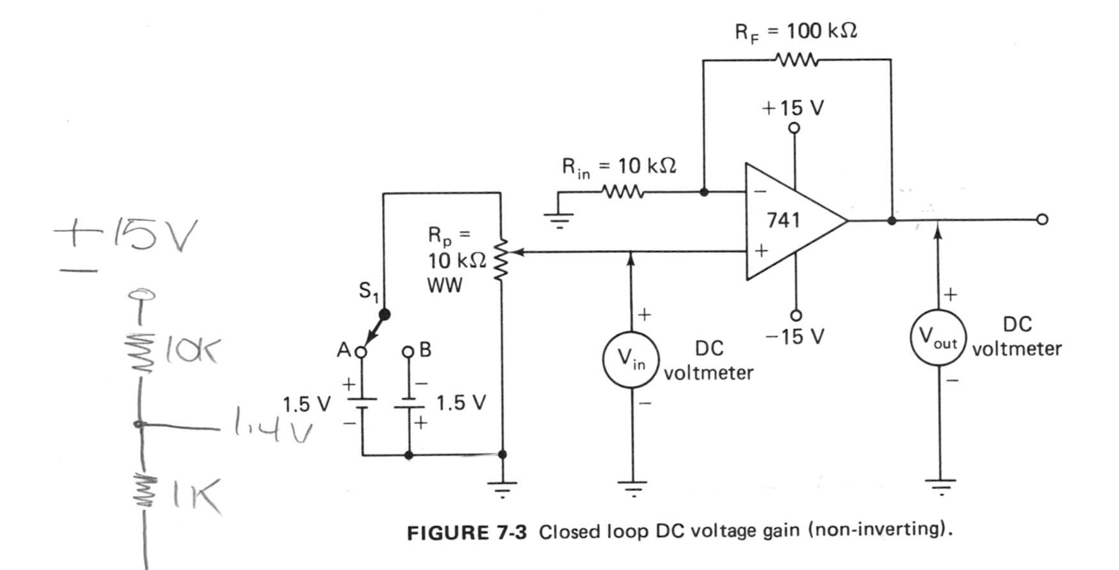

FIGURE 7-3 Closed loop DC voltage gain (non-inverting).

8. Record V_{out}, indicating polarity. -8.9V

9. Turn off the power supply and disconnect the batteries used for V_{in}.

10. Calculate the gain of the circuit from the formula

$$A = \frac{R_F}{R_{in}} + 1 \qquad \frac{100K}{10K} + 1 = 11$$

11. Calculate V_{out} from the formula

$$V_{out} = \left(\frac{R_F}{R_{in}} + 1\right)(V_{in}) \qquad \left(\frac{100K}{10K} + 1\right)(.5) = 5.5$$

12. Calculate the voltage gain from the formula $\left(\frac{100K}{10K} + 1\right)(.8) = 8.8$

$$A_v = \frac{V_{out}}{V_{in}} \qquad \frac{5.5}{.5} = 11 \qquad \frac{8.9}{.8} = 11.1$$

13. Repeat the experiment several times for different gains by changing the value of R_F. Gain of 11 consistent to Vsat where increasing Vin then no effect on Vout

QUESTIONS FOR FIGURE 7-3

$\frac{12.6}{1.14} = 11$

1. What is V_{out} for a V_{in} of +0.5 V? (Remember polarity.) +5.6
2. What is V_{out} for a V_{in} of −0.8 V? (Remember polarity.) -8.9
3. What is the gain of the circuit? 11
4. What is the gain of the circuit if R_F = 120 kΩ? 13
5. What is V_{out} when R_F = 56 kΩ and V_{in} = +0.6 V? + 3.96 (4)
6. What is V_{out} when R_F = 220 kΩ and V_{in} = −0.4 V? 9.2

$\frac{56K}{10K} + 1 = 6.6 \times .6 = 3.96$

$\frac{220}{10} + 1 = 23$

Experiment 7-4 Current Gain

OBJECTIVE: To prove that a large current gain is possible with an op amp.

Required Components:

1 100-kΩ resistor (R_F)

1-2-MΩ potentiometer (WW preferred)

1 SPDT switch (S_1)

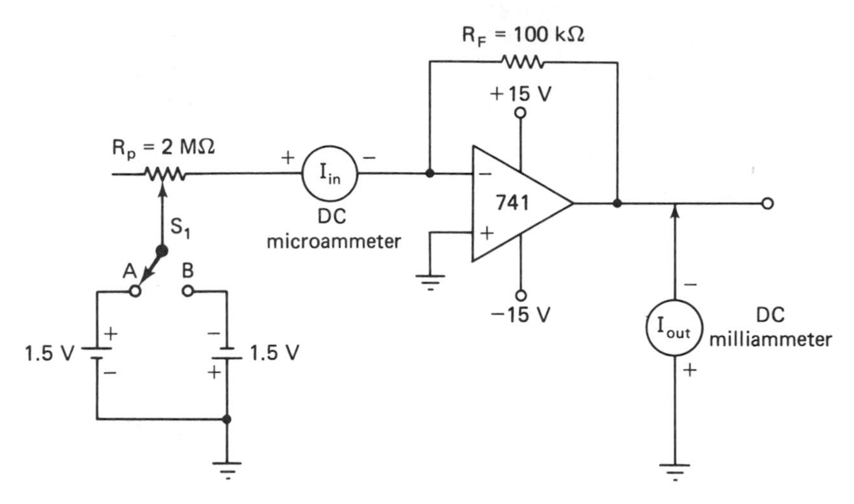

FIGURE 7-4 Current gain.

Test Procedure

1. With the power supply off, construct the circuit shown in Figure 7-4.
2. Place S_1 to position A.
3. Set R_p for maximum resistance.
4. Turn on the power supply.
5. Carefully adjust R_p until $I_{in} = 1\ \mu A$.
6. Record I_{out}; indicate polarity in reference to op-amp output. + 111 µA
7. Turn off the power supply and remove the batteries.
8. Calculate the current gain from the formula

$$A_i = \frac{I_{out}}{I_{in}} \quad \frac{111\ \mu A}{1\ \mu A} = 111$$

9. Place S_1 to position B and reconnect the batteries.
10. Reverse the meter leads.
11. Repeat steps 3 through 8.

QUESTIONS FOR FIGURE 7-4

1. What is I_{out} when $I_{in} = +1\ \mu A$? − 111 µA
2. What is I_{out} when $I_{in} = -1\ \mu A$? + 111 µA

3. What is the current gain of the circuit? ///

4. Is the current gain the same for both the + and - inputs? No

I_{IN} = 1 μA I_{out} = 17.3 mA = Gain 17,300

Experiment 7-5 Maximum DC Output-Voltage Range

OBJECTIVE: To show the output voltage saturation of an op amp.

Required Components:

1 10-kΩ resistor (R_{in})

1 100-kΩ resistor (R_F)

1 10-kΩ wirewound potentiometer (R_p)

1 SPDT switch (S_1)

Test Procedure

1. With the power supply off, construct the circuit shown in Figure 7-5.
2. Place S_1 to position A.
3. Turn on the power supply.
4. Using data log-1, adjust R_p for the $+V_{in}$ values shown and record each corresponding $-V_{out}$ reading.
5. Turn off the power supply.
6. Place S_1 to position B.
7. Reverse the meter leads.
8. Turn on the power supply.
9. Using data log-2, adjust R_p for the $-V_{in}$ values shown and record each corresponding $+V_{out}$ reading.
10. Turn off the power supply and remove the batteries.
11. Observe the data logs.

The point where a further increase in V_{in} causes no change in V_{out} is called saturation. The lowest point at which $+V_{out}$ and $-V_{out}$ saturation occurs is the maximum DC output-voltage range.

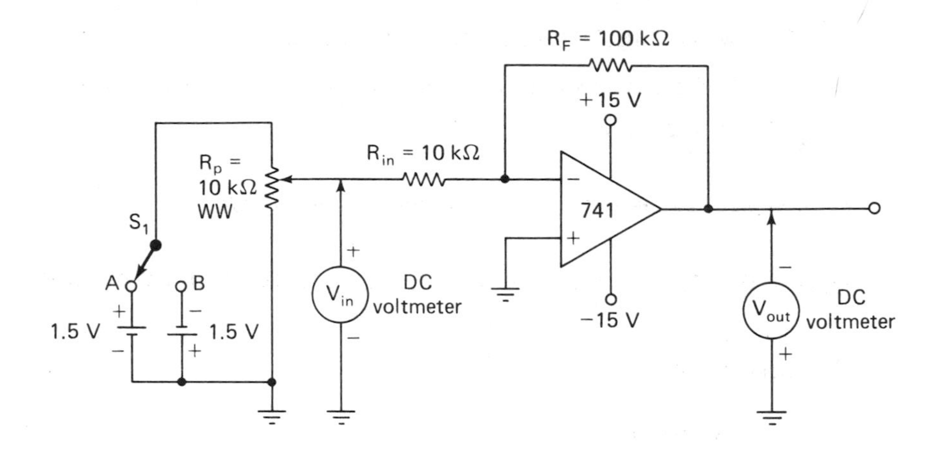

Data log 1

$+V_{in}$	$-V_{out}$
0.2	2.0
0.4	4.1
0.6	6.1
0.8	8.2
1.0	10.1
1.2	12.1
1.4	12.1
1.5	12.1

Data log 2

$-V_{in}$	$+V_{out}$
0.2	2.1
0.4	4.1
0.6	6.2
0.8	8.2
1.0	10.5
1.2	12.3
1.4	13.9
1.5	13.9

FIGURE 7-5 Maximum DC output-voltage range.

QUESTIONS FOR FIGURE 7-5

1. What is the positive saturation voltage ($+V_{sat}$)? 12.1V
2. What is the negative saturation voltage ($-V_{sat}$)? −13.9V
3. What is the maximum DC output-voltage range? 26V

   12.1V − (−13.9V) = 26V

Experiment 7-6 Maximum DC Input-Voltage Range

OBJECTIVE: To demonstrate how input voltage causes output voltage saturation.

Required Components:

1 10-kΩ resistor (R_{in})

1 100-kΩ resistor (R_F)

1 10-kΩ wirewound potentiometer (R_p)

1 SPDT switch (S_1)

Test Procedure

1. With the power supply off, construct the circuit shown in Figure 7-6.
2. Place S_1 to position A.
3. Turn on the power supply.
4. Adjust R_p until V_{out} reaches the value of $-V_{sat}$ from Experiment 7-5. Record the polarity and value of V_{in}. ________
5. Turn off the power supply.
6. Place S_1 to position B.
7. Reverse the meter leads.
8. Turn on the power supply.

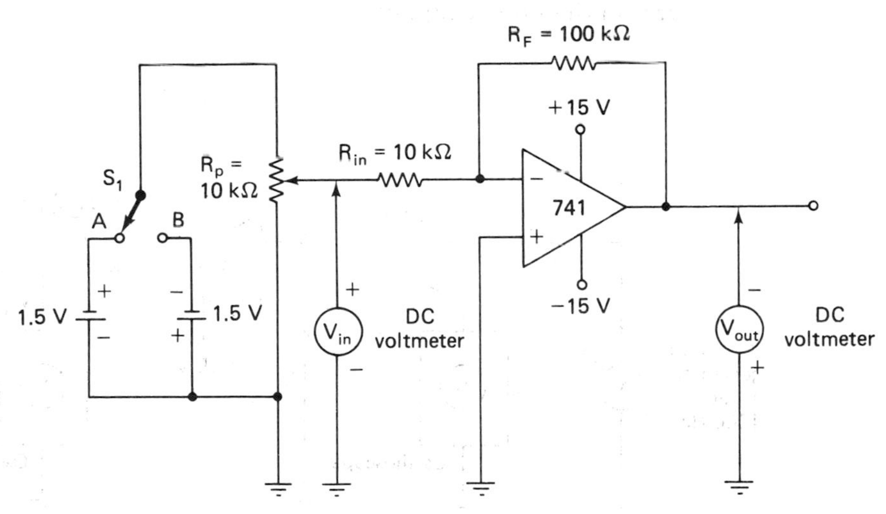

FIGURE 7-6 Maximum DC input-voltage range.

9. Adjust R_p until V_{out} reaches the value of $+V_{sat}$ from Experiment 7-5. Record the polarity and value of V_{in}. ______________

10. Turn off the power supply and remove the batteries.

QUESTIONS FOR FIGURE 7-6

1. What is the minimim V_{in} for $-V_{sat}$? ______________
2. What is the minimum V_{in} for $+V_{sat}$? ______________
3. What is the total (maximum) V_{in} DC range? ______________

Experiment 7-7 Closed-Loop AC Voltage Gain

OBJECTIVE: To show how circuit gain remains constant.

Required Components:

1 10-kΩ resistor (R_{in})

1 100-kΩ resistor (R_F)

Test Procedure

1. With the power supply off, construct the circuit shown in Figure 7-7.
2. Turn on the power supply.

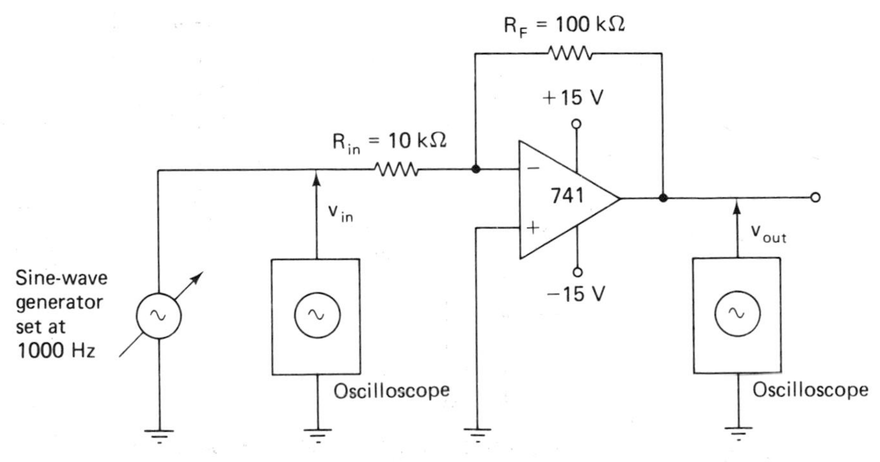

FIGURE 7-7 Closed-loop AC voltage gain.

3. Set the sine-wave generator for 1000 Hz.
4. Adjust the output amplitude of the sine-wave generator for a maximum undistorted sine wave at v_{out}.
5. Record the peak-to-peak amplitude of v_{out}. 20 V
6. Record the peak-to-peak amplitude of v_{in}. 2 V
7. Adjust the output amplitude of the sine-wave generator for one-half of the value in step 5.
8. Record the peak-to-peak amplitude of v_{out}. 10 V
9. Record the peak-to-peak amplitude of v_{in}. 1 V
10. Turn off the power supply.
11. Using the values in steps 5 and 6, calculate the ac voltage gain from the formula

$$\frac{20}{2} = 10 \quad \frac{10}{1} = 10 \quad A_v = \frac{v_{out}}{v_{in}} = 10\,V$$

12. Using the values in steps 8 and 9, calculate the ac voltage gain.

QUESTIONS FOR FIGURE 7-7

1. What is the maximum peak-to-peak voltage of v_{out}? 20 V
2. What is the maximum peak-to-peak voltage of v_{in}? 2 V
3. What is the AC voltage gain of the circuit? 10 V
4. Does the AC voltage gain change when v_{in} is reduced by one-half?
NO decrease prop.

Experiment 7-8 Two-Stage Op-Amp Amplifier

OBJECTIVE: To prove how the gain of one circuit is multiplied by the gain of another circuit and show signal phase relationships.

Required Components:

2 10-kΩ resistors (R_1, R_3)

1 22-kΩ resistor (R_2)

1 100-kΩ resistor (R_4)

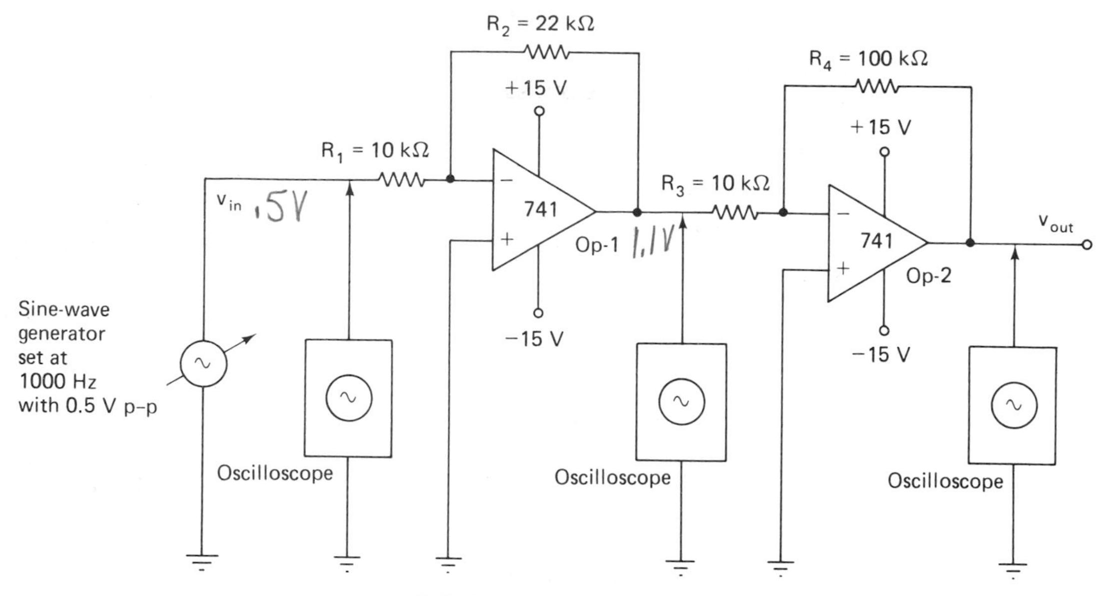

FIGURE 7-8 Two stage op amp amplifier.

Test Procedure

1. With the power supply off, construct the circuit shown in Figure 7-8.
2. Turn on the power supply.
3. Set the sine-wave generator at 1000 Hz with an amplitude of 0.5 V p–p.
4. Record the peak-to-peak voltage at the output of OP-1. 1.1V
5. Record the peak-to-peak voltage at the output of OP-2. 11V
6. Turn off the power supply.

QUESTIONS FOR FIGURE 7-8

1. What is the v_{out} of OP-1? 1.1V
2. What is the A_v of OP-1? 2.2 V
3. What is the v_{out} of OP-2? 11 V
4. What is the A_v of OP-2? 10 V
5. What is the overall voltage gain of the circuit? 22 V

How gain of first circuit is multiplie[d] by gain of Second circuit, OP-2 is 10 X greater then OP-1. (2.2V)(10V) = 22V

Experiment 7-9 Input Impedance

OBJECTIVE: A practical approach to finding op amp input impedance.

Required Components:

1 10-kΩ resistor (R_{in})

1 100-kΩ resistor (R_F)

1 50-kΩ wirewound potentiometer (R_p)

1 SPDT switch (S_1)

Test Procedure

1. With the power supply off, construct the circuit shown in Figure 7-9.
2. Turn on the power supply.
3. Set the sine-wave generator to 1000 Hz.
4. Adjust the generator output amplitude for a maximum undistorted signal at v_{out}.
5. Record the peak-to-peak value of v_{in}. 6V

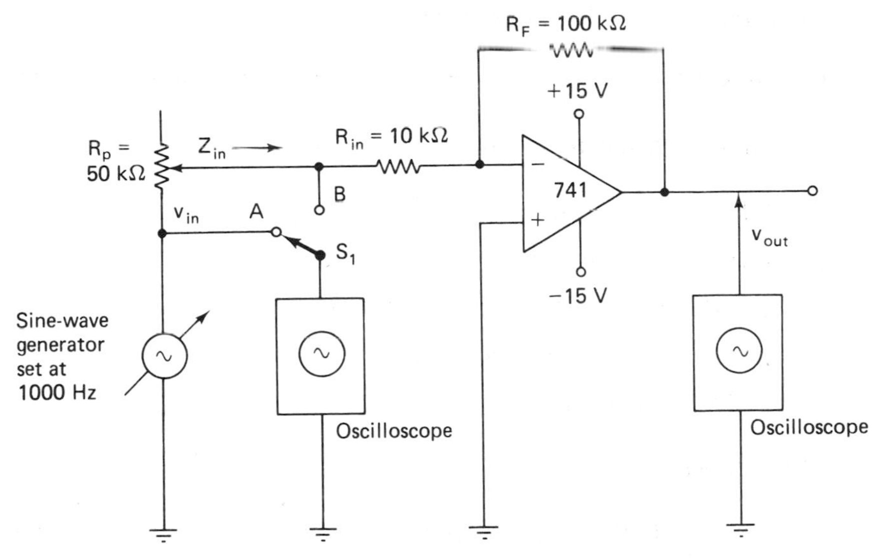

FIGURE 7-9 Input impedance.

6. Place S_1 to position B.
7. Adjust R_p until the peak-to-peak value of v_{in} is one-half the value recorded in step 5. ______
8. Turn off the power supply.
9. Record the ohmic value of R_p from the wiper to the end connected to the generator. ______ This resistance value is equal to the input impedance of the op amp.
10. Repeat steps 2 through 9 for the following frequencies: 100 Hz, 10 kHz, 50 kHz, and 100 kHz.

QUESTIONS FOR FIGURE 7-9

1. What is the value of v_{in} in step 5? ______
2. What is this value divided by 2? ______
3. What is the ohmic reading of R_p from the wiper to the end connected to the generator? ______
4. What is the input impedance of the op amp? ______
5. Does the input impedance change with different input signal frequencies? ______. Why?

Experiment 7-10 Output Impedance

OBJECTIVE: A practical approach to finding op-amp output impedance.

Required Components:

1 10-kΩ resistor (R_{in})

1 100-kΩ resistor (R_F)

1 1-kΩ wirewound potentiometer (R_p)

1 1-μF capacitor (nonpolarized) (C_1)

1 SPST switch (S_1)

Test Procedure

1. With the power supply off, construct the circuit shown in Figure 7-10.

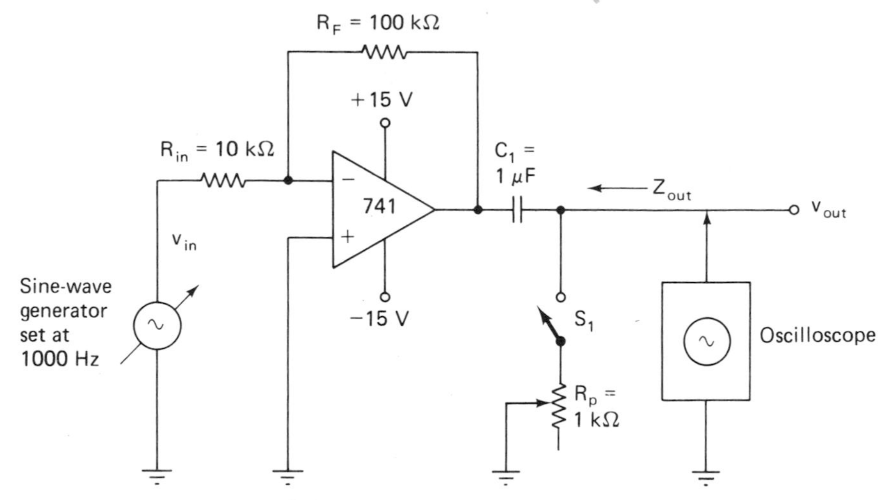

FIGURE 7-10 Output impedance.

2. Set the sine-wave generator at 1000 Hz.
3. Turn on the power supply.
4. Adjust the generator output amplitude for a maximum undistorted signal at v_{out}.
5. Record the peak-to-peak value of v_{out}. ____________
6. Close S_1. Observe that v_{out} is affected.
7. Adjust R_p so that v_{out} is the same as the original reading obtained in step 5.
8. Turn off the power supply.
9. Record the ohmic value of R_p from the end connected to S_1 and the wiper. ____________ This resistance value is equal to the output impedance of the op amp.
10. Repeat steps 2 through 9 for the following frequencies: 100 Hz, 10 kHz, 50 kHz, and 100 kHz.

QUESTIONS FOR FIGURE 7-10

1. What is the value of v_{out} in step 5? ____________
2. What is the ohmic reading of R_p from the end connected to S_1 and the wiper? ____________

3. What is the output impedance of the op amp? ____________

4. Does the output impedance change with different input signal frequencies? ____________ Why?

Experiment 7-11 Input-Offset Current

OBJECTIVE: To show how input-offset current affects output voltage.

Required Components:

1 22-Ω resistor (R_1)

1 100-kΩ resistor (R_F)

1 10-kΩ wirewound potentiometer (R_p)

1 SPDT switch (S_1)

Test Procedure

1. With the power supply off, construct the circuit shown in Figure 7-11.
2. Set the dc voltmeter at output to low range.
3. Turn on the power supply.

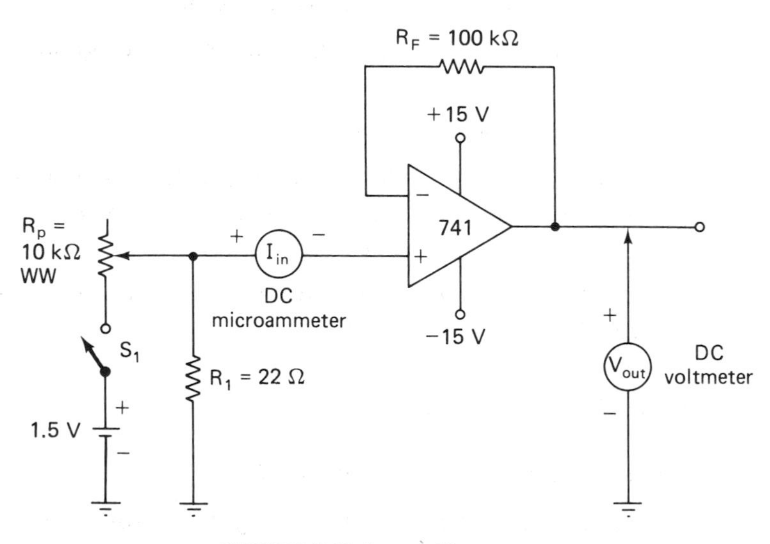

FIGURE 7-11 Input offset current.

4. Record V_{out}. ______________
5. Record I_{in}. ______________
6. Close S_1 and adjust R_p until V_{out} is zero.
7. Record I_{in}. ______________ This value of current is the input offset current. An op amp may be offset-null-adjusted by this method.
8. Turn off the power supply and remove the battery.

QUESTIONS FOR FIGURE 7-11

1. What is V_{out} when S_1 is open? ______________
2. What is the value of I_{in} when S_1 is closed and V_{out} is zero? ________
3. What is the value of input offset current? ______________

Experiment 7-12 Input-Offset Voltage

OBJECTIVE: To show how input-offset voltage affects output voltage.

Required Components:

1 22-Ω resistor (R_1)

1 100-kΩ resistor (R_F)

1 10-kΩ wirewound potentiometer (R_p)

1 SPST switch (S_1)

Test Procedure

1. With the power supply off, construct the circuit shown in Figure 7-12.
2. Set the DC voltmeter to low range.
3. Turn on the power supply.
4. Record V_{out}. ______________
5. Record V_{in}. ______________
6. Close S_1 and adjust R_p until V_{out} is zero.
7. Record V_{in}. ______________ This value of voltage is the input offset voltage.
8. Turn off the power supply and remove the battery.

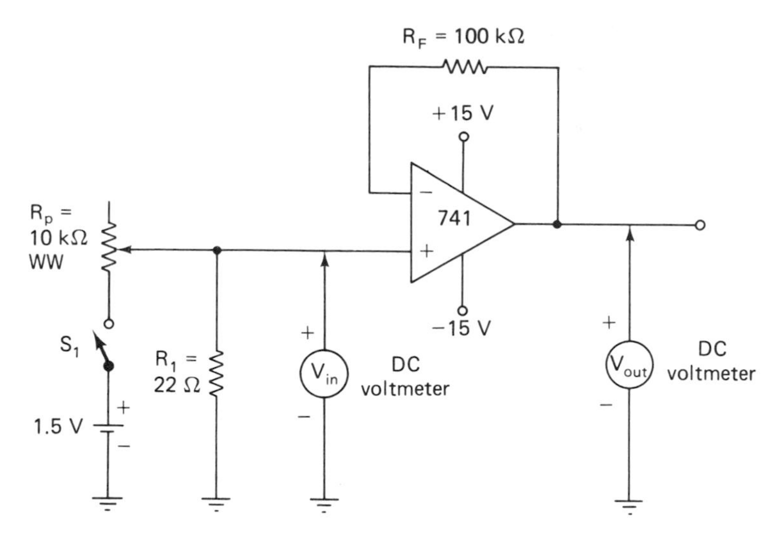

FIGURE 7-12 Input offset voltage.

QUESTIONS FOR FIGURE 7-12

1. What is V_{out} when S_1 is open? ____________
2. What is V_{in} when S_1 is open? ____________
3. What is V_{in} when S_1 is closed and R_p is adjusted for a V_{out} of zero? ____________
4. What is the value of input offset voltage? ____________

Experiment 7-13 Output-Offset Voltage

OBJECTIVE: To prove that output-offset voltage exists in an op amp.

Required Components:

1 10-kΩ resistor (R_{in})

1 100-kΩ resistor (R_F)

Test Procedure

1. With the power supply off, construct the circuit shown in Figure 7-13.
2. Set the DC voltmeter to low range.

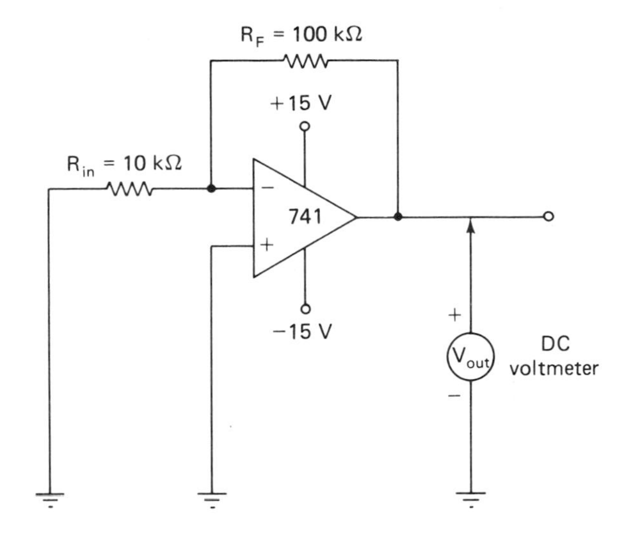

FIGURE 7-13 Output offset voltage.

3. Turn on the power supply.
4. Record V_{out}. ______________ This is the output offset voltage and may be either positive or negative, depending on the type of unbalance inside the op amp.
5. Turn off the power supply.

QUESTIONS FOR FIGURE 7-13

1. What is the value of V_{out}? ______________
2. What is the value of output-offset voltage? ______________

Experiment 7-14 Offset Null Adjustment

OBJECTIVE: To show a method of correcting offset voltage errors.

Required Components:

1 10-kΩ resistor (R_{in})

1 10-kΩ resistor (R_F)

1 10-kΩ wirewound potentiometer (R_n)

1 SPST switch (S_1)

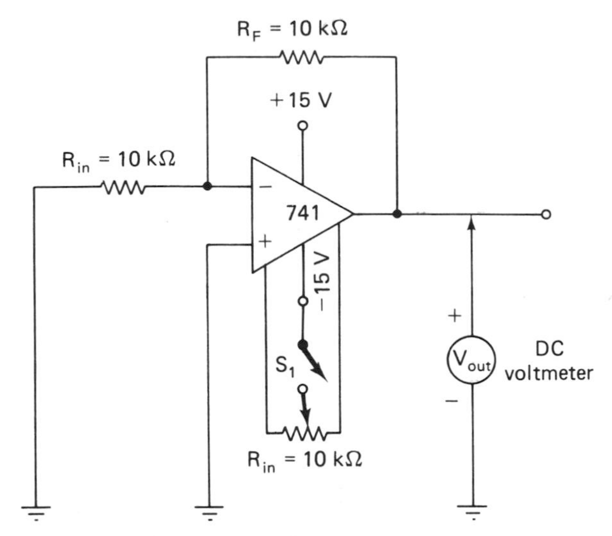

FIGURE 7-14 Offset null adjustment.

Test Procedure

1. With the power supply off, construct the circuit shown in Figure 7-14.
2. Set the DC voltmeter to low range.
3. Turn on the power supply.
4. Record V_{out}. ______________
5. Close S_1 and adjust R_n for minimum V_{out} (hopefully zero). This method of offset null adjustment uses the internal connections of the op amp. Other methods of external offset null adjustment were shown in Experiments 7-11 and 7-12.
6. Turn off the power supply.

QUESTIONS FOR FIGURE 7-14

1. What is V_{out} when S_1 is open? ______________
2. With S_1 closed and R_n properly adjusted, what is V_{out}? __________.

Experiment 7-15 Slew Rate

OBJECTIVE: To demonstrate the limitations of the op amp due to slew rate.

Required Components:

1 10-kΩ resistor (R_{in})

1 100-kΩ resistor (R_F)

Test Procedure

1. With the power supply off, construct the circuit shown in Figure 7-15.
2. Set the square-wave generator at 10 kHz.
3. Adjust the output amplitude of the generator to 1 V p–p.
4. Turn on the power supply.
5. Observe v_{in} and draw an accurate voltage waveform, indicating amplitude and time. ____________
6. Observe v_{out} and draw an accurate voltage waveform, indicating amplitude and time. ____________
7. Calculate the slew rate from the formula SR = $\Delta V/\Delta t$. ____________
8. Repeat steps 2 through 7 for the following frequencies: 100 Hz, 1 kHz, 50 kHz, and 100 kHz.
9. Turn off the power supply.

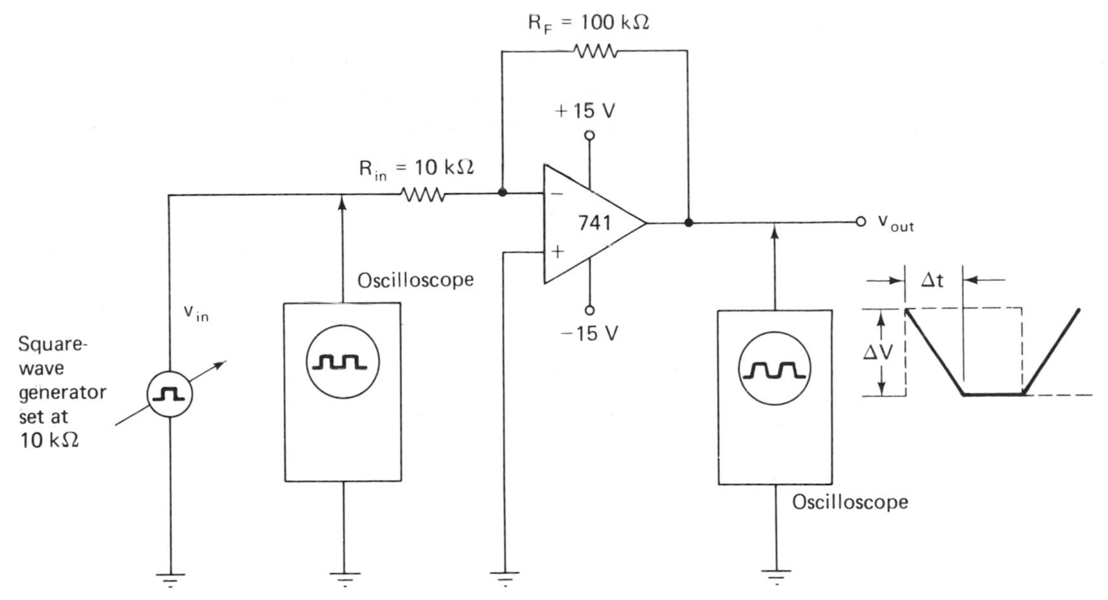

FIGURE 7-15 Slew rate.

QUESTIONS FOR FIGURE 7-15

1. What is the ΔV of the circuit at 10 kHz? ______________
2. What is the Δt of the circuit at 10 kHz? ______________
3. What is the slew rate of this circuit? ______________
4. Does the slew rate increase, decrease, or remain the same when the input frequency decreases? ______________
5. Does the slew rate increase, decrease, or remain the same when the input frequency increases? ______________

Experiment 7-16 Open-Loop Frequency Response

OBJECTIVE: To show the limited frequency response and instability in the open-loop mode.

Required Components:

None

Test Procedure

1. With the power supply off, construct the circuit shown in Figure 7-16. Because the open-loop gain of the op amp is extremely high, the circuit is extremely sensitive and all leads should be kept as short as possible. Measurements will be difficult to perform and care should be taken to ensure accuracy.
2. Turn on the power supply.
3. Adjust the generator for an undistorted signal at the op-amp output. Use the various frequencies given in the data log and record v_{in} and v_{out}.
4. Turn off the power supply.
5. Calculate the voltage gain for each frequency given from the formula

$$A_v = \frac{v_{out}}{v_{in}}$$

Record the answer in the data log.

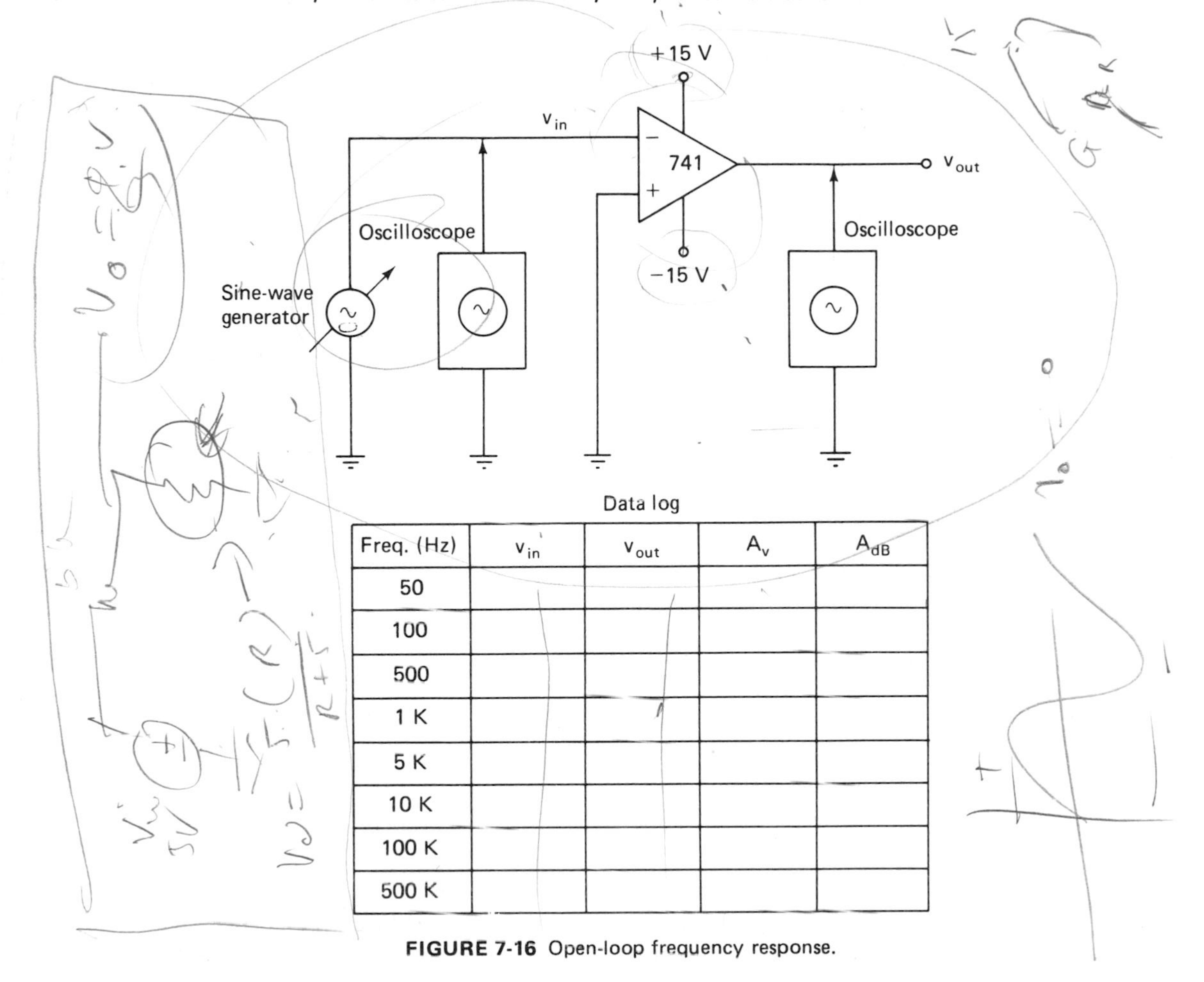

Data log

Freq. (Hz)	V_{in}	V_{out}	A_v	A_{dB}
50				
100				
500				
1 K				
5 K				
10 K				
100 K				
500 K				

FIGURE 7-16 Open-loop frequency response.

6. Calculate the dB gain for each frequency given from the formula

$$A_{dB} = 20 \log \frac{v_{out}}{v_{in}}$$

Record the answers in the data log.

7. Sketch a graph of A_v versus frequency from the results shown in the data log. (Try to use semilog graph paper.)

QUESTIONS FOR FIGURE 7-16

1. What frequency had the highest gain? ____________
2. What frequency had the lowest gain? ____________

3. As frequency increases, what happens to A_v? ______________

4. What is the dB loss from 1 to 10 kHz? ______________

Experiment 7-17 Unity-Gain Frequency

OBJECTIVES: To prove at what frequency v_{in} is equal to v_{out}.

Required Components:

None

Test Procedure

1. With the power supply off, construct the circuit shown in Figure 7-17.
2. Turn on the power supply.
3. Set the generator to 100 Hz and adjust the amplitude of the generator to 100 mV or less, so as to produce an undistorted signal at the output of the op amp. A single generator capable of producing frequencies above 2 MHz should be used, or an audio generator and RF generator will be needed. Since this circuit is in the open-loop mode, the same problems will be encountered as with the previous experiment.
4. While keeping v_{in} constant, run the generator through its frequencies until v_{out} is equal to v_{in}. This will mean monitoring both input and output simultaneously or taking measurements many times.

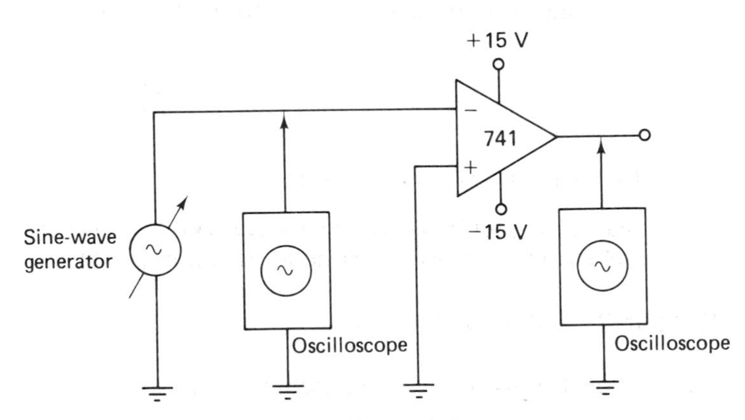

FIGURE 7-17 Unity gain frequency.

5. When v_{out} is equal to v_{in}, record the frequency setting of the generator. ______________ This is the unity-gain frequency of the op amp.
6. Turn off the power supply.

QUESTIONS FOR FIGURE 7-17

1. At what frequency does A_v equal 1 ($v_{in} = v_{out}$)? ______________
2. What is the unity-gain frequency of the op amp? ______________

Experiment 7-18 Closed-Loop Frequency Response

OBJECTIVE: To show how closed-loop mode extends the frequency response and improves the stability of an op amp.

Required Components:

1 10-kΩ resistor (R_{in})

1 100-kΩ resistor (R_F)

1 1-kΩ resistor (R_{in}) to be used for second test run.

Test Procedure

1. With the power supply off, construct the circuit shown in Figure 7-18.
2. Turn on the power supply.
3. Set the generator for an undistorted output signal for each of the frequencies given in the data log and record v_{in} and v_{out}. It will be convenient to keep the amplitude of v_{in} constant, say, 1 V p-p.
4. Turn off the power supply.
5. Calculate the voltage gain for each frequency from the formula

$$A_v = \frac{v_{out}}{v_{in}}$$

Record the answers in the data log.

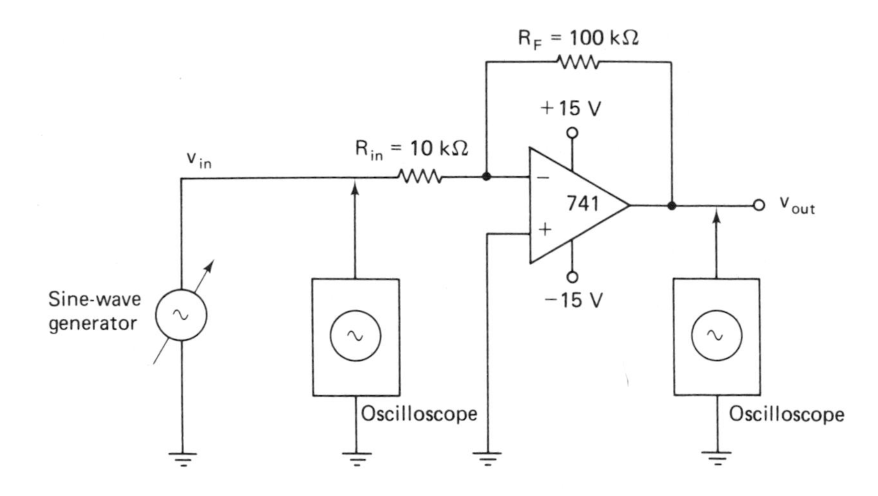

Data log

Freq. (Hz)	V_{in}	V_{out}	A_v	A_{dB}
50				
100				
500				
1 K				
5 K				
10 K				
100 K				
500 K				

FIGURE 7-18 Closed-loop frequency response.

6. Calculate the dB gain for each frequency from the formula

$$A_{dB} = 20 \log \frac{v_{out}}{v_{in}}$$

Record the answers in the data log.

7. Sketch a graph of A_v versus frequency from the results shown in the data log. (Try to use semilog graph paper.)

8. Change R_{in} to 1 kΩ and repeat steps 2 through 7.

QUESTIONS FOR FIGURE 7-18

1. With a circuit gain of 10, at approximately what frequency did the output drop to the half-power point ($0.707V_{out}$)? ______________
2. With a circuit gain of 10, what is the approximate bandwidth of the circuit? ______________
3. With a circuit gain of 100, at approximately what frequency did the output drop to the half-power point ($0.707V_{out}$)? ______________
4. With a circuit gain of 100, what is the approximate bandwidth of the circuit? ______________
5. What is the gain–bandwidth product of the circuit with a gain of 10? ______________
6. What is the gain–bandwidth product of the circuit with a gain of 100? ______________

Experiment 7-19 Common-Mode Rejection

OBJECTIVE: To demonstrate the effectiveness of common-mode rejection.

Required Components:

1 1-kΩ resistor (R_1)

1 1-μF capacitor (nonpolarized) (C_1)

Test Procedure

1. With the power supply off, construct the circuit shown in Figure 7-19.
2. Turn on the power supply.
3. Set the generator for an undistorted output signal of 100 Hz.
4. Record v_{in_1} and v_{out_1} in the data log as a first reading for 100 Hz.
5. Reduce the generator amplitude and record v_{in_2} and v_{out_2} in the data log as a second reading for 100 Hz.
6. Repeat steps 3 through 5 for frequencies 1 kHz, 10 kHz, and 100 kHz, as shown in the data log.
7. Turn off the power supply.

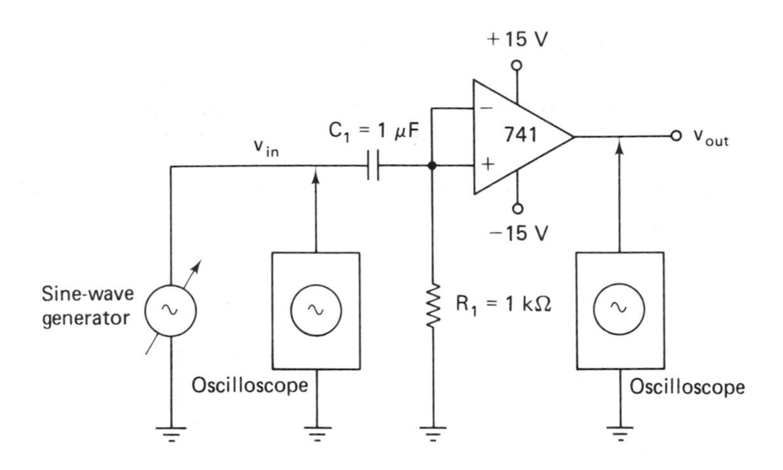

Data log

Freq. (Hz)	v_{in_1}	v_{in_2}	v_{out_1}	v_{out_2}	$\Delta v_{in}/\Delta v_{out}$	dB
100						
1 K						
10 K						
100 K						

FIGURE 7-19 Common-mode rejection.

8. Calculate the small change in v_{in} ($\Delta v_{in} = v_{in_1} - v_{in_2}$) for each frequency given.
9. Calculate the small change in v_{out} ($\Delta v_{out} = v_{out_1} - v_{out_2}$) for each frequency given.
10. Calculate the ratio $\Delta v_{in}/\Delta v_{out}$ for each frequency and record in the data log.
11. Convert this ratio to −dB for each frequency from the formula

$$-\text{dB} = 20 \log \frac{\Delta v_{in}}{\Delta v_{out}}$$

Record the answers in the data log.

QUESTIONS FOR FIGURE 7-19

1. What is the −dB reading for 1 kHz? ____________
2. What is the −dB reading for 10 kHz? ____________

3. What happens to the common-mode rejection ratio when frequency increases? ____________

Experiment 7-20 Power-Supply Rejection Ratio

OBJECTIVE: To show the effects of power-supply instability on the op amp.

Required Components:

1 22-Ω resistor (R_1)

1 100-kΩ resistor (R_F)

1 10-kΩ wirewound potentiometer (R_p)

Test Procedure

1. With the power supply off, construct the circuit shown in Figure 7-20.
2. Set the DC voltmeters to low range.
3. Turn on the power supply.
4. Adjust R_P for a V_{out} of zero.
5. Record the reading of V_{in} as V_{in_1} in the data log.
6. Reduce the $+V$ and $-V$ supply voltages to 12 V.
7. If the output is not zero, readjust R_P for zero output.
8. Record V_{in} as V_{in_2} in the data log.
9. Reduce the $+V$ and $-V$ supply voltages to 10 V.
10. If the output is not zero, readjust R_P for zero output.
11. Record V_{in} as V_{in_2} in the data log.
12. Turn off the power supply.
13. Calculate the change in V_{in} from the formula

$$V_{in} = V_{in_2} - V_{in_1}$$

14. Calculate the change in power supply voltage from the formulas

$$\Delta V_{ps} = 15\text{ V} - 12\text{ V} = \_\_\_\_\_\_\_\_\_\_$$

$$\Delta V_{ps} = 15\text{ V} - 10\text{ V} = \_\_\_\_\_\_\_\_\_\_$$

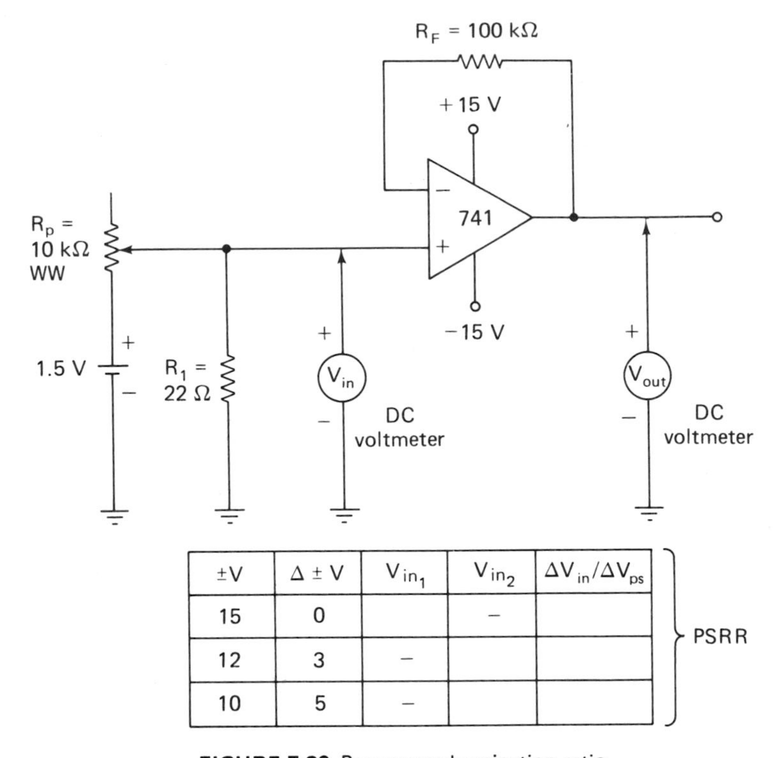

±V	Δ ± V	V_{in_1}	V_{in_2}	$\Delta V_{in}/\Delta V_{ps}$
15	0		—	
12	3	—		
10	5	—		

FIGURE 7-20 Power supply rejection ratio.

15. Calculate the power-supply rejection ratio (PSRR) from the formula

$$\text{PSRR} = \frac{\Delta V_{\text{in}}}{\Delta V_{\text{ps}}}$$

QUESTIONS FOR FIGURE 7-20

1. What is the PSRR of the circuit for a ΔV_{ps} of 3 V? ____________
2. What is the PSRR of the circuit for a ΔV_{ps} of 5 V? ____________

Experiment 7-21 Op-Amp Voltage-Level Detector

OBJECTIVE: To demonstrate how an op amp can detect a specific voltage level.

Required components:

1 1.2-kΩ resistor (R_4)

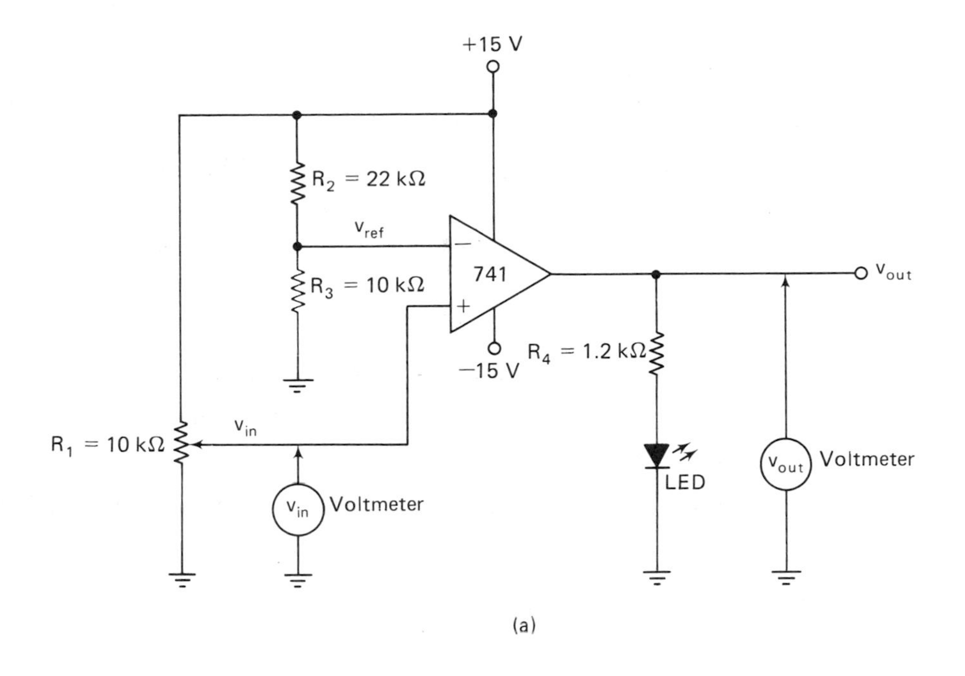

(a)

V_{ref} = ______

V_{in}	LED on/off	V_{out}
$< V_{ref}$		
$> V_{ref}$		

(b)

FIGURE 7-21 Voltage level detector.

1 10-kΩ resistor (R_3)

1 22-kΩ resistor (R_2)

1 10-kΩ potentiometer (R_1)

1 LED at $V_F \approx 2$ V

Test Procedure

1. With the power supply off, construct the circuit shown in Figure 7-21a.
2. Turn the power supply on.

3. With a voltmeter measure v_{ref} and record its value as indicated in Figure 7-21b.
4. Adjust R_1 (v_{in}) to ground or 0 V.
5. Indicate whether the LED is on or off in the data log of Figure 7-21b when v_{in} is less than v_{ref} ($v_{in} < v_{ref}$).
6. Measure v_{out} and record its value in the proper place of the data log.
7. Adjust R_1 until the LED turns on.
8. Measure v_{in} and compare its value to v_{ref}.
9. Indicate whether the LED is on or off in the data log when v_{in} is greater than v_{ref} ($v_{in} > v_{ref}$).
10. Measure v_{out} and record its value in the proper place of the data log.

QUESTIONS FOR FIGURE 7-21

1. When v_{in} is +3 V, the LED is ___________.
2. v_{ref} is determined by ___________ and ___________.
3. When v_{in} is +6.2 V, the LED is ___________ .

Experiment 7-22 Op-Amp Low-Pass Filter

OBJECTIVE: To show how an op-amp low-pass filter will pass frequencies below the cutoff frequency (f_c) and attenuate the frequencies above this point.

Required components:

2 10-kΩ resistors (R_1 and R_2)

1 22-kΩ resistor (R_3)

2 0.01 μF capacitors (C_1 and C_2)

Test Procedure

1. With the power supply off, construct the circuit shown in Figure 7-22a.
2. Turn on the power supply.

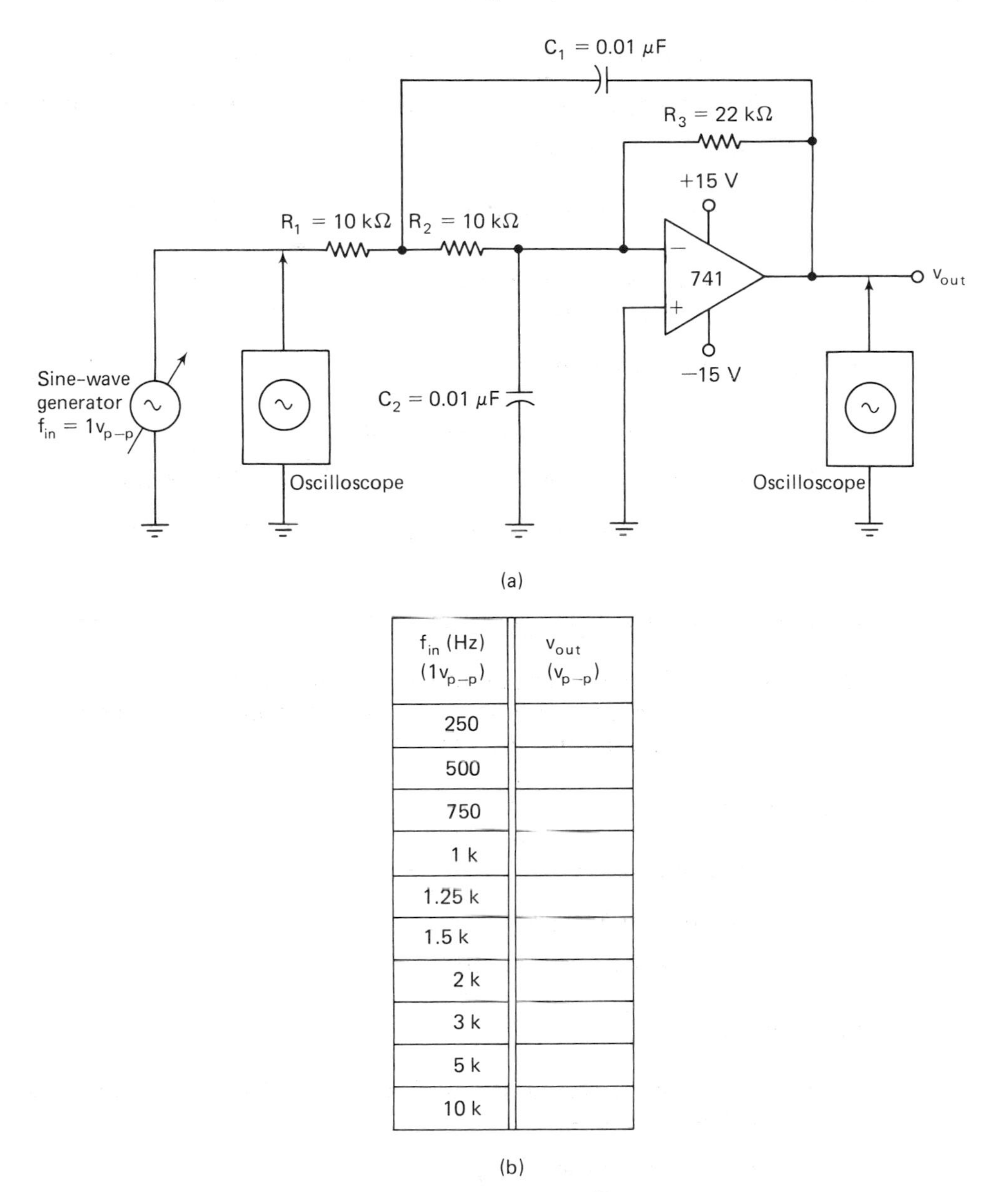

f_{in} (Hz) ($1v_{p-p}$)	v_{out} (v_{p-p})
250	
500	
750	
1 k	
1.25 k	
1.5 k	
2 k	
3 k	
5 k	
10 k	

(b)

FIGURE 7-22 Low-pass filter.

3. Adjust the sine-wave generator for the first frequency given in the data log shown in Figure 7-22b.

4. Set the sine-wave generator output amplitude for a 1 V p-p signal for each frequency given in the data log. Make sure this voltage level is maintained for each frequency setting.

5. Measure the V p-p of v_{out} and record its value in the data log.
6. Repeat steps 3, 4, and 5 for all frequencies given in the data log.
7. Turn off the power supply.
8. Calculate f_c from the formula: $f_c \approx \dfrac{1}{2\pi\sqrt{R_1 R_2 C_1 C_2}}$
9. Draw a frequency response curve from the results shown in the data log (try to use semilog graph paper).

QUESTIONS FOR FIGURE 7-22

1. The f_c for the low-pass filter is about __________ .
2. Frequencies below f_c are __________ by this circuit.
3. Frequencies above f_c are __________ by this circuit.

Experiment 7-23 Op-Amp High-Pass Filter

OBJECTIVE: To demonstrate how an op-amp high-pass filter will block or attenuate frequencies below the cutoff frequency (f_c) and pass frequencies above this point.

Required components:

1 10-kΩ resistor (R_1)

2 22-kΩ resistors (R_2 and R_3)

2 0.01 μF capacitors (C_1 and C_2)

Test Procedure

1. With the power supply off, construct the circuit shown in Figure 7-23a.
2. Turn on the power supply.
3. Adjust the sine-wave generator for the first frequency given in the data log shown in Figure 7-23b.
4. Set the sine-wave generator output amplitude for a 1 V p-p signal for each frequency given in the data log. Make sure this voltage level is maintained for each frequency setting.
5. Measure the V p-p of v_{out} and record its value in the data log.

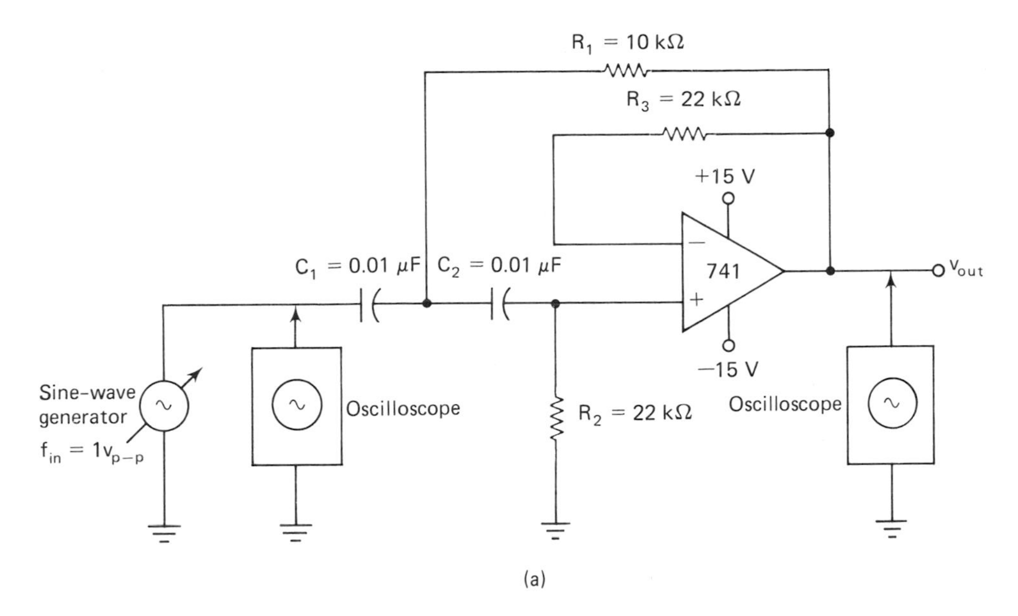

(a)

f_{in} (Hz) ($1v_{p-p}$)	V_{out} (v_{p-p})
250	
500	
750	
1 k	
1.25 k	
1.5 k	
2 k	
3 k	
5 k	
10 k	

(b)

FIGURE 7-23 High-pass filter.

6. Repeat steps 3, 4, and 5 for all frequencies given in the data log.
7. Turn off the power supply.
8. Calculate f_c from the formula: $f_c \approx \dfrac{1}{2\pi\sqrt{R_1 R_2 C_1 C_2}}$

9. Draw a frequency response curve from the results shown in the data log (try to use semilog graph paper).

QUESTIONS FOR FIGURE 7-23

1. The f_c for the high-pass filter is about ___________ .
2. Frequencies below f_c are ___________ by this circuit.
3. Frequencies above f_c are ___________ by this circuit.

Experiment 7-24 Op-Amp Bandpass Filter

OBJECTIVE: To show how an op-amp bandpass filter will pass a certain group of frequencies and reject those above and below these limits.

Required components:

2 10-kΩ resistors (R_1 and R_2)

1 100-kΩ resistor (R_3)

2 0.01 μF capacitors (C_1 and C_2)

Test Procedure

1. With the power supply off, construct the circuit shown in Figure 7-24a.
2. Turn on the power supply.
3. Adjust the sine-wave generator for the first frequency given in the data log shown in Figure 7-24b.
4. Set the sine-wave generator output amplitude for a 1 V p-p signal for each frequency given in the data log. Make sure this voltage level is maintained for each frequency setting.
5. Measure the V p-p of v_{out} and record its value in the data log.
6. Repeat steps 3, 4, and 5 for all frequencies given in the data log.
7. Turn off the power supply.

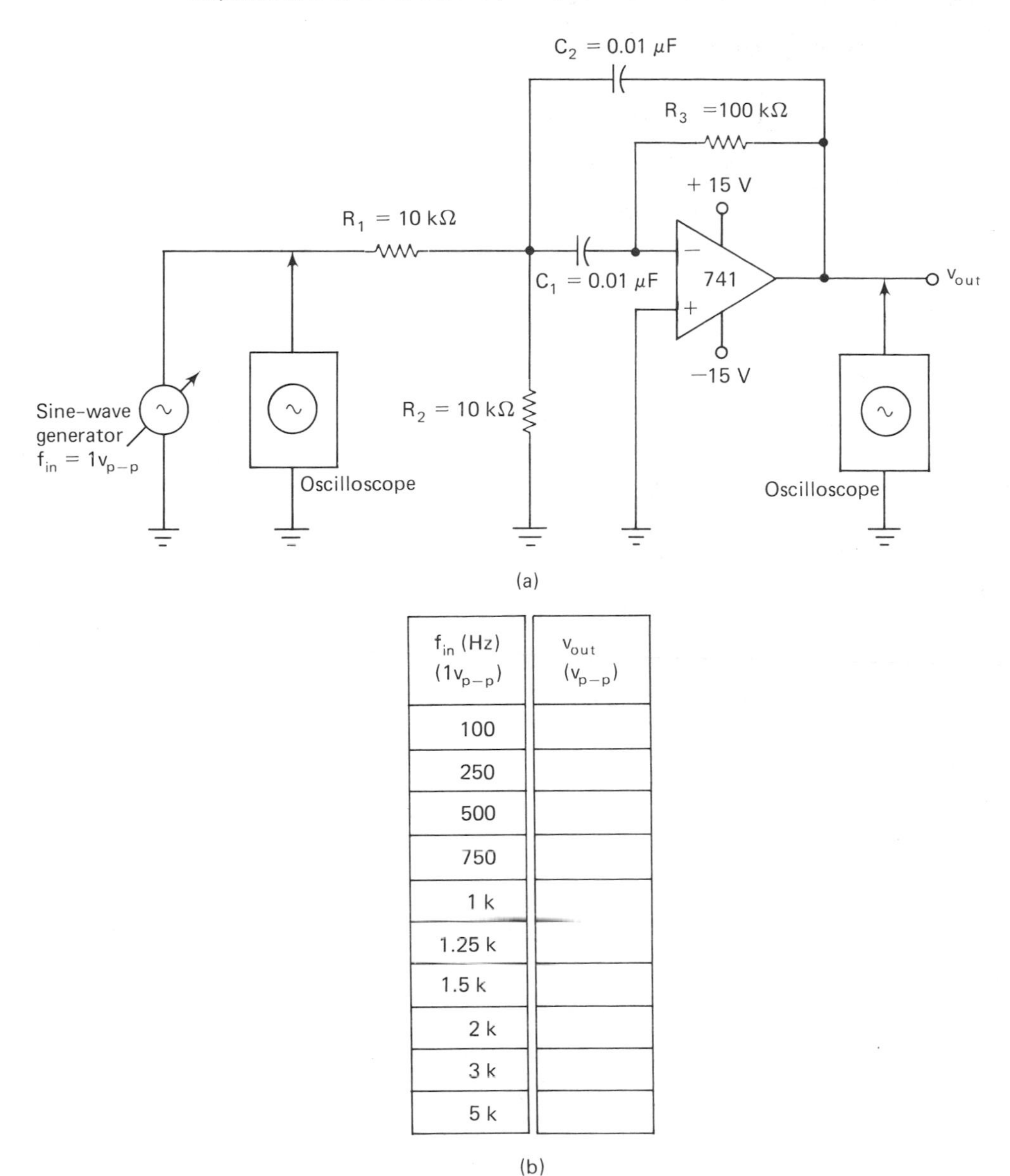

f_{in} (Hz) ($1v_{p-p}$)	V_{out} (v_{p-p})
100	
250	
500	
750	
1 k	
1.25 k	
1.5 k	
2 k	
3 k	
5 k	

(b)

FIGURE 7-24 Bandpass filter.

8. Calculate f_r from the formula: $f_r \approx \dfrac{1}{2\pi\sqrt{R_p R_3 C_1 C_2}}$, where

$$R_p = \frac{R_1 R_2}{R_1 + R_2}$$

9. Draw a frequency response curve from the results shown in the data log (try to use semilog graph paper).

QUESTIONS FOR FIGURE 7-24

1. The resonant frequency of the circuit is about ____________ hertz.
2. The approximate bandwidth of the circuit is ____________ hertz.
3. Frequencies below the bandpass are ______________________.
4. Frequencies above the bandpass are ______________________.

Experiment 7-25 Op-Amp Notch Filter

OBJECTIVE: To demonstrate how an op-amp notch filter will pass all frequencies above and below a specific bandwidth while rejecting the frequencies within the bandwidth.

Required components:

1 1-kΩ resistor (R_2)
1 10-kΩ resistor (R_1)
1 47-kΩ resistor (R_3)
1 1-mΩ resistor (R_4)
2 0.01 μF capacitors (C_1 and C_2)

Test Procedure

1. With the power supply off, construct the circuit shown in Figure 7-25a.
2. Turn on the power supply.
3. Adjust the sine-wave generator for the first frequency given in the data log shown in Figure 7-25b.
4. Set the sine-wave generator output amplitude for a 1 V p-p signal for each frequency given in the data log. Make sure this voltage level is maintained for each frequency setting.
5. Measure the V p-p of v_{out} and record its value in the data log.
6. Repeat steps 3, 4, and 5 for all frequencies given in the data log.
7. Turn off the power supply.
8. Calculate f_r from the formula: $f_r \approx \dfrac{1}{2\pi\sqrt{R_1 R_4 C_1 C_2}}$

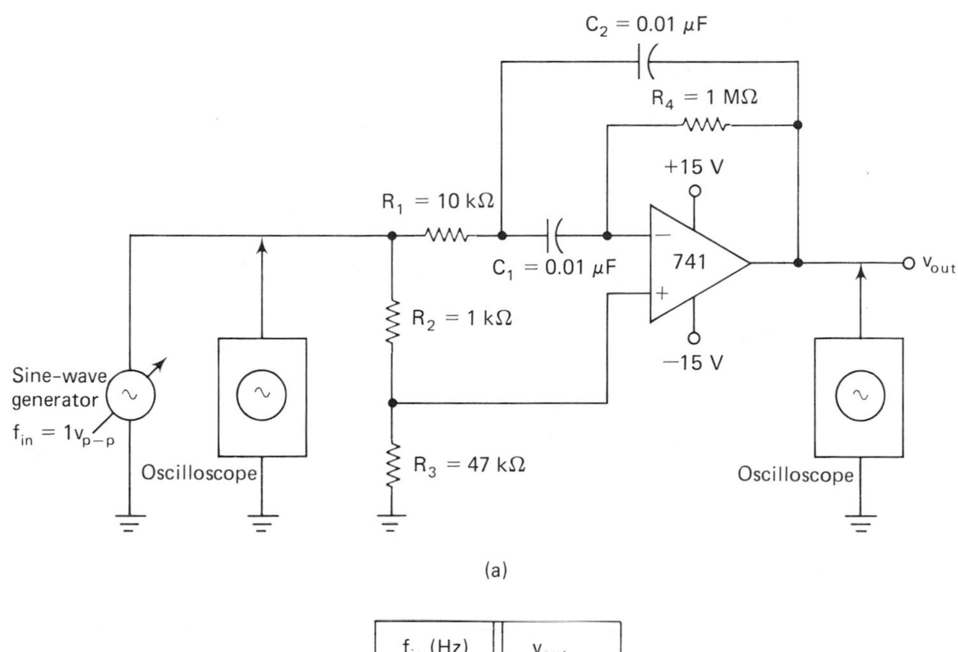

(a)

f_{in} (Hz) ($1v_{p-p}$)	V_{out} (v_{p-p})
50	
75	
100	
125	
150	
175	
200	
225	
250	

(b)

FIGURE 7-25 Notch filter.

9. Draw a frequency response curve from the results shown in the data log (try to use semilog graph paper).

QUESTIONS FOR FIGURE 7-25

1. The resonant frequency of the circuit is about ____________ hertz.

2. The approximate bandwidth of the circuit is ____________ hertz.
3. Frequencies below the notch are ______________________________ .
4. Frequencies above the notch are ______________________________ .

Experiment 7-26 Op-Amp Square-Wave Generator

OBJECTIVE: To show how an op amp can generate a square-wave voltage.

Required components:

1 10-kΩ resistor (R_1)
1 18-kΩ resistor (R_3)
2 22-kΩ resistors (R_1 and R_2)
1 0.01 μF capacitor (C_1)
1 0.05 μF capacitor (C_1)
1 0.1 μF capacitor (C_1)

Test Procedure

1. With the power supply off, construct the circuit shown in Figure 7-26a. Use the values given in the first row of the data log of Figure 7-26b for R_1 and C_1.
2. Turn the power supply on.
3. Using an oscilloscope, measure the voltage waveform across C_1 and record the peak-to-peak value in the space provided.
4. Using the oscilloscope, measure the voltage waveform at f_{out} and record the peak-to-peak value in the space provided.
5. Calculate the output frequency using the formula: $f_{out} \approx \frac{1}{2 R_1 C_1}$
6. Record the calculated f_{out} in the proper place of the data log.
7. Measure f_{out} and record in the proper place of the data log.
8. Substitute R_1 and C_1 with the remaining values given in the data log and repeat steps 5, 6, and 7.

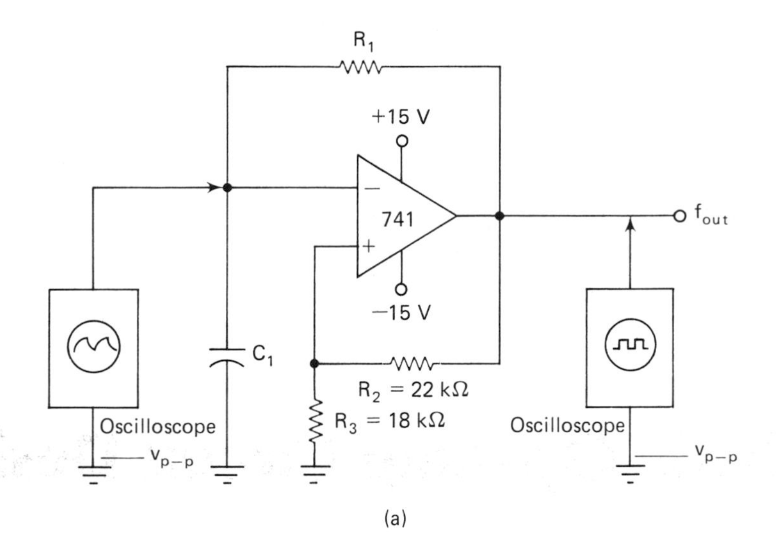

R_1 (Ω)	C_1 μF	f_{out} (Hz) Calculated	Measured
10 k	0.05		
10 k	0.1		
10 k	0.01		
22 k	0.01		

(b)

FIGURE 7-26 Square-wave generator.

QUESTIONS FOR FIGURE 7-26

1. The voltage waveform at the inverting input of the op amp is ____________.
2. The voltage waveform at the output of the op amp is ____________.
3. When R_1 increases, f_{out} ____________.
4. When C_1 decreases, f_{out} ____________.

Chapter 8

Basic Op-Amp Circuit Design

Many op-amp circuits evolve from simple basic design procedures. Most of the formulas needed for basic design have been covered in the preceding chapters. This chapter will provide practical, easy-to-follow, step-by-step design procedures for some basic op-amp circuits. The results of the design techniques can be proven with the test procedures given in Chapter 7 and other sections of the book.

A standard glossary of definitions, which will be used throughout each circuit design procedure, is given below.

V_{in}—DC input voltage

V_{out}—DC output voltage

I_{in}—DC input current
(current through source)

I_{out}—DC output current
(current through load)

v_{in}—AC peak-to-peak input signal voltage

v_{out}—AC peak-to-peak output signal voltage

i_{in}—AC peak-to-peak input signal current
(current through source)

i_{out}—AC peak-to-peak output signal current (current through load)

A_V—DC voltage gain

A_I—DC current gain

A_v—AC peak-to-peak voltage gain

A_i—AC peak-to-peak current gain

R_S—source resistance

R_L—load resistance

R_{in}—input resistance

R_F—feedback resistance

All other R's numbered as needed by circuit

C_{in}—input capacitance

C_{out}—output capacitance

Z_{in}—input impedance

Z_{out}—output impedance

$+V$—positive power-supply voltage

$-V$—negative power-supply voltage

Certain design procedures are common to all design problems and do not have to be repeated for each circuit. These procedures include:

1. *Selecting a ± power supply.* Make sure that the required V_{out} is about 2 V less than the $\pm V$ used. Choose the proper ± power supply required or desired for each application. (Example: $+V = +15$ V, $v_{out} = +13$ V; $-v = -15$ V, $v_{out} = -13$ V.)
2. *Using test equipment for design verification.* The design circuit problems in this chapter can be accurately tested with the same type of equipment listed in Chapter 7.
3. *Offset null adjusting.* With circuits requiring a high degree of accuracy, offset null adjustments should be made according to Experiment 7-12 or 7-14.
4. *Selecting the proper op amp.* Choose an op amp that has the required open-loop gain, slew rate, and frequency response for the designed circuit.

DESIGN 8-1 COMPARATOR CIRCUITS

Op-amp comparator circuits are used to clamp large-output DC voltage levels with the use of smaller DC control voltages. The input control voltages may be positive or negative, which clamp the output voltage at zero or some specific positive or negative level. Section 2-1 shows one method of utilizing comparators. These design examples will show another method using clamping diodes.

8-1-1 Basic Positive-Clamped Comparator

Referring to Figure 8-1, the diode in the feedback loop causes the circuit to respond in the open-loop mode for positive input levels and in the closed-loop mode for negative input levels. A reference voltage ($-V_{ref}$) is compared through R_1 with the variable V_{in} through R_2. The voltage waveforms show

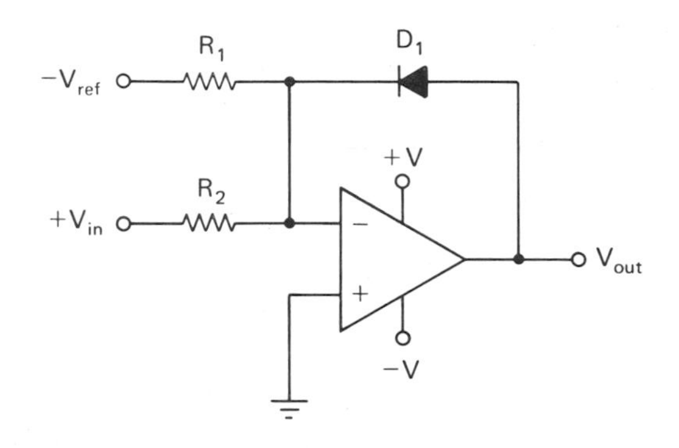

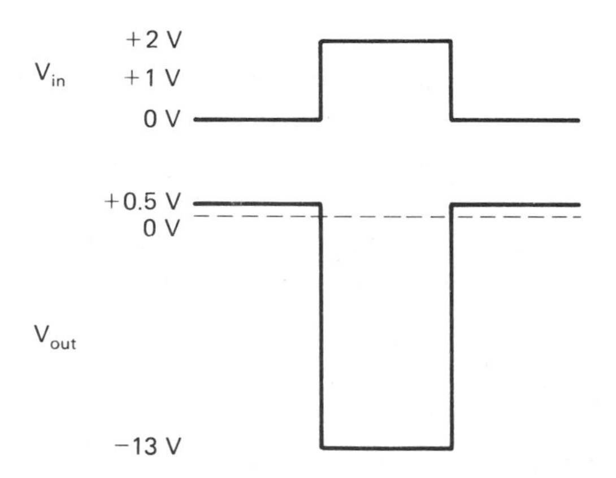

FIGURE 8-1 Positive-clamping level comparator (negative output).

the results for a desired clamping level of +2 V. When V_{in} is below +2 V, V_{out} is approximately +0.5 V (the forward voltage drop of the diode). When V_{in} reaches the desired clamping level, V_{out} changes states and swings to approximately -13 V ($-V_{sat}$). $-V_{ref}$ can be any voltage greater than V_{in} up to the $-V$ supply.

DESIGN PROCEDURES

1. Choose R_1 (typically 100 kΩ).
2. Select V_{in} (desired input clamping level).

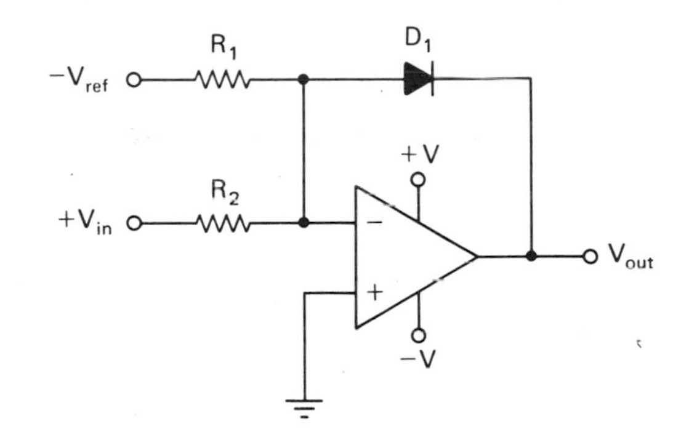

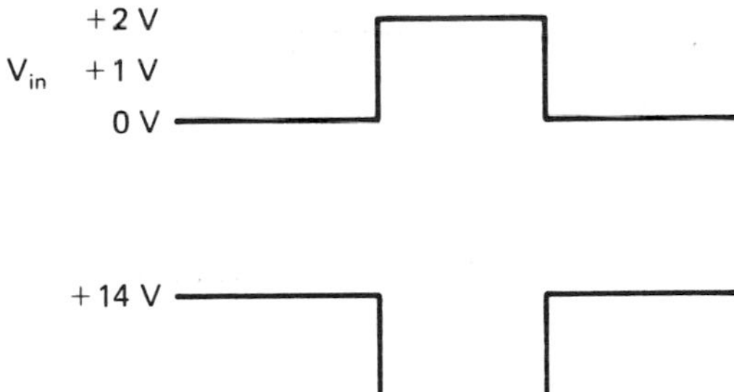

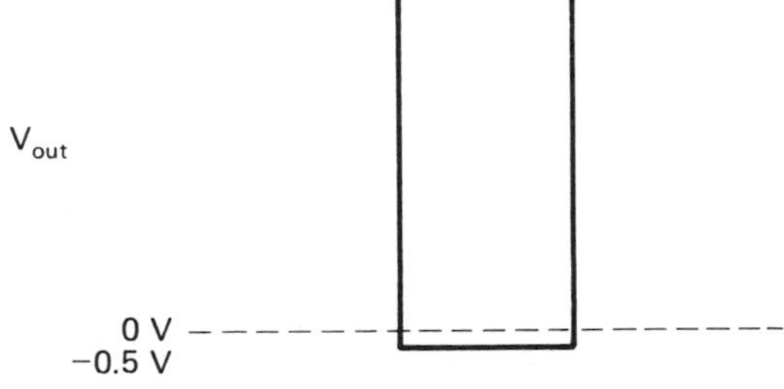

FIGURE 8.2 Positive-clamping level comparator (positive output).

3. Calculate

$$R_2 = \frac{R_1 V_{in}}{-V_{ref}}$$

4. Construct the circuit and verify the design results.

If the diode in the feedback loop is reversed as shown in Figure 8-2, V_{out} will be positive when V_{in} is less than the clamped input level, and about -0.5 V when V_{in} is greater than the clamped level.

8-1-2 Basic Negative-Clamped Comparator

The circuit shown in Figure 8-3 will respond to a negative input level. A reference voltage ($+V_{ref}$) is compared through R_1 with the variable V_{in}

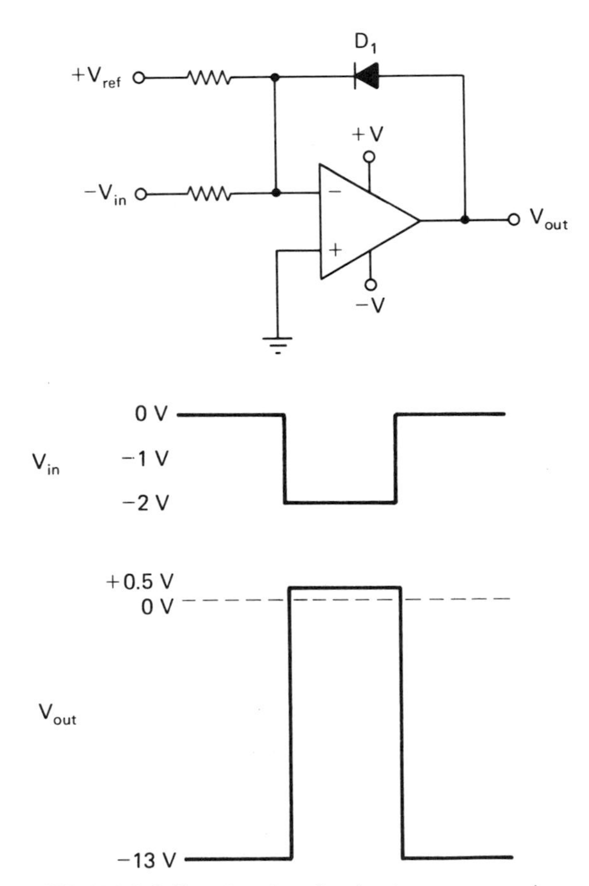

FIGURE 8-3 Negative-clamping level comparator (negative output).

through R_2. The voltage waveforms show the results for a desired clamping level of −2 V. When V_{in} is above −2 V, V_{out} is about −13 V ($-V_{sat}$). When V_{in} reaches the desired clamping level, V_{out} changes states and swings to about +0.5 V (the forward voltage drop of the diode). $+V_{ref}$ can be any voltage greater than V_{in} up to the $+V$ supply.

DESIGN PROCEDURES

1. Choose R_1 (typically 100 kΩ).
2. Select V_{in} (desired input clamping level).

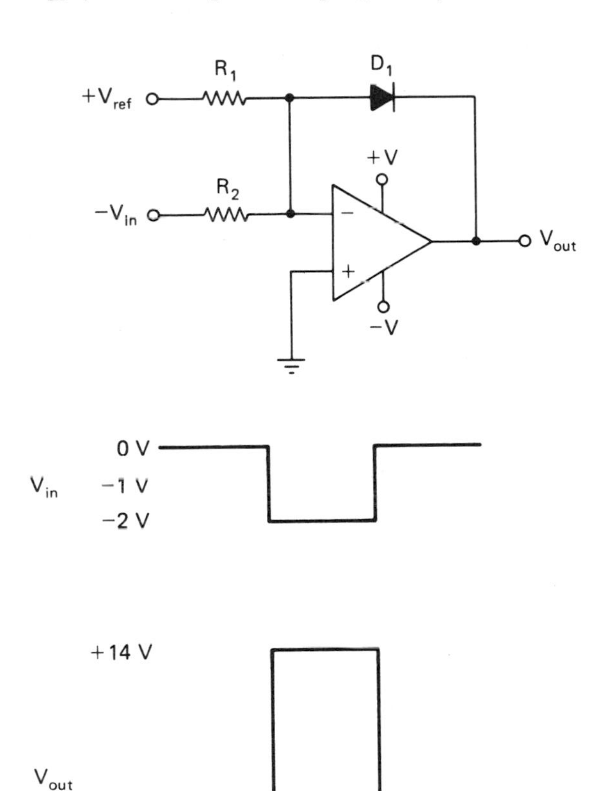

FIGURE 8-4 Negative-clamping level comparator (positive output).

3. Calculate

$$R_2 = \frac{R_1 V_{in}}{+V_{ref}}$$

4. Construct the circuit and verify the design results.

If the diode in the feedback loop is reversed as shown in Figure 8-4, V_{out} will be about -0.5 V when V_{in} is less than the clamped input level, and will be positive when V_{in} is greater than the clamped level.

8-1-3 Output Clamping to a Specific Level

The output of a comparator can be clamped to two specific levels with the use of a zener diode in the feedback loop, as shown in Figure 8-5. When V_{in} is below the input clamping level, zener diode is conducting in the zener region and V_{out} will equal $+V_Z$ (the zener breakdown voltage). When V_{in}

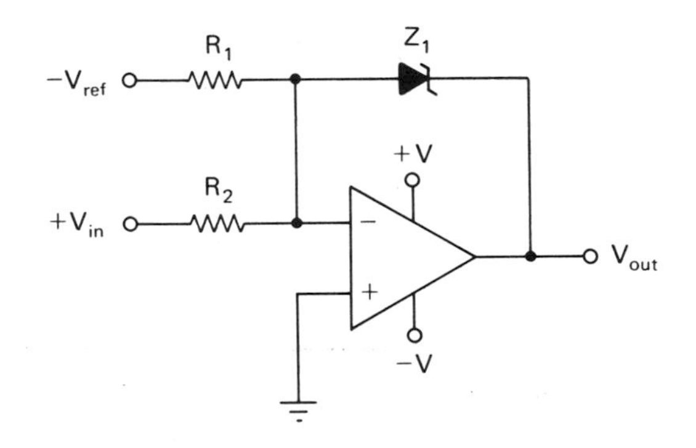

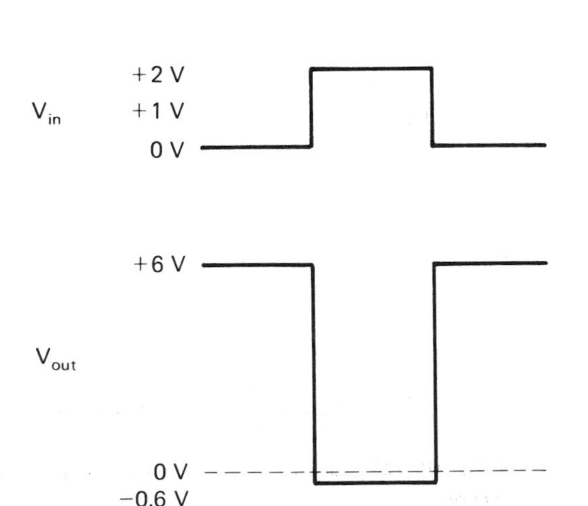

FIGURE 8-5 Specific level-clamping comparator.

goes above the input clamping level, the zener diode is forward-biased like a normal diode and the output will be clamped to -0.6 V.

If the zener diode in the feedback loop is reversed, V_{out} will be about +0.6 V when V_{in} is below the input clamping level. When V_{in} goes above the input clamping level, V_{out} will go to $-V_Z$.

DESIGN PROCEDURES

1. Choose R_1 (typically 100 kΩ).
2. Select V_{in} (desired input clamping level).
3. Calculate

$$R_2 = \frac{R_1 V_{in}}{-V_{ref}}$$

4. Choose a zener diode with the proper V_Z for the desired output clamping level.
5. Construct the circuit and verify the design results.

DESIGN 8-2 INVERTING AMPLIFIER CIRCUITS

Op-amp inverting amplifiers provide high voltage gain and high current gain. Basic inverting amplifiers are relatively easy to design and construct. These design examples will show how to determine voltage gain, current gain, and the values of related external components.

8-2-1 Inverting DC Amplifier

The input impedance (in this case R_{in}) of an inverting DC amplifier is usually chosen to be about 50 times greater than the source impedance, as shown in Figure 8-6. The output impedance is very low, typically 25 to 50 Ω, and is usually ignored in design.

DESIGN PROCEDURES

1. Determine or select V_{in}.
2. Determine source resistance R_s, either by direct ohmic measurement or using Ohm's law ($R_s = V_{in}/I_{in}$).

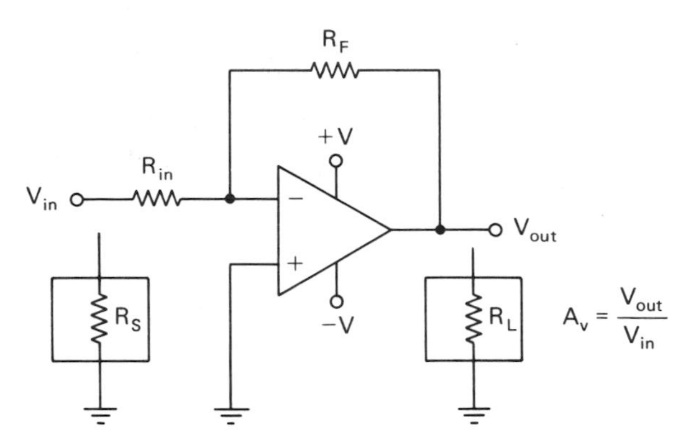

FIGURE 8-6 Inverting DC amplifier.

3. Choose R_{in} to be about 50 times greater than R_s.
4. Calculate the desired voltage gain by the formula $A_v = V_{out}/V_{in}$. (Remember that maximum V_{out} should be about 2 V less than the $\pm V$ supply.)
5. Calculate R_F from the formula $R_F = -A_v R_{in}$. (Disregard the minus sign in the calculations.)
6. Determine R_L by the following circuit to be driven, or arbitrarily select R_L for 2.2 kΩ.
7. Calculate V_{out} from the formula $V_{out} = (-R_F/R_{in})V_{in}$.
8. Calculate I_{in} from the formula $I_{in} = V_{in}/(R_s + R_{in})$.
9. Calculate I_{out} from the formula $I_{out} = V_{out}/R_L$.
10. Calculate A_I from the formula $A_I = I_{out}/I_{in}$.
11. Construct the circuit and verify the design results.

8-2-2 Inverting AC Amplifier

A capacitor is used at the input of an inverting amplifier when amplifying AC signals, as shown in Figure 8-7. This capacitor C_1 blocks any DC component from the source, thus minimizing any distortion of the AC signal at the output. Design procedures for an AC inverting amplifier are similar for those of the DC inverting amplifier, except that C_1 must be calculated for a specific break frequency.

Noise may be a problem, so remember that the larger R_F is made, the more susceptible the circuit is to noise.

Depending on the op amp chosen and the circuit gain selected, the output signal voltage will roll off rapidly beyond the bandwidth for a given circuit. (Refer to Section 1-3-10.)

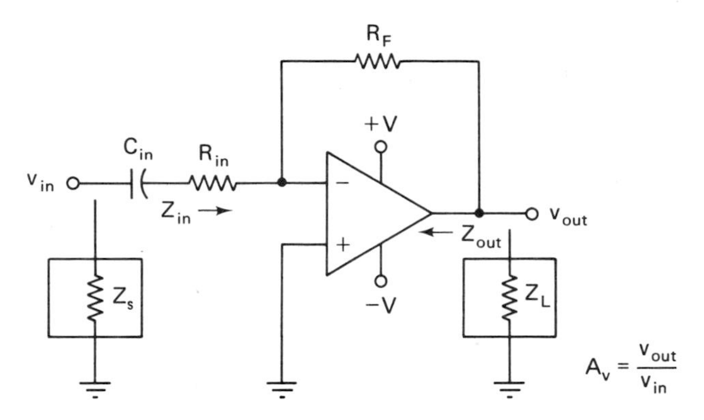

FIGURE 8-7 Inverting AC amplifier.

DESIGN PROCEDURES

1. Determine or select v_{in}.
2. Determine source impedance Z_s.
3. Choose R_{in} to be about 50 times greater than Z_s.
4. Calculate desired voltage gain by the formula $A_v = v_{out}/v_{in}$. (Remember that maximum v_{out} should be about 2 V less than the $\pm V$ supply.)
5. Calculate R_F from the formula $R_F = -A_v R_{in}$. (Disregard the minus sign in the calculations.)
6. Determine Z_L by the following circuit to be driven, or arbitrarily select Z_L for 2.2 kΩ.
7. Calculate v_{out} from the formula $v_{out} = -(R_F/R_{in})v_{in}$.
8. Calculate i_{in} from the formula $i_{in} = v_{in}/(Z_s + R_{in})$.
9. Calculate i_{out} from the formula $i_{out} = v_{out}/R_L$.
10. Calculate A_i from the formula $A_i = i_{out}/i_{in}$.
11. Calculate C_1 for a break frequency of 10 Hz from the formula
$$C_1 = \frac{1}{2\pi(Z_s + R_{in}) f_c}.$$
12. Construct the circuit and verify the design results.

8-2-3 Basic Two-Stage Cascaded Amplifier

Designing a basic two-stage cascaded inverting amplifier, as shown in Figure 8-8, is simply a matter of designing the gain of each stage. The important

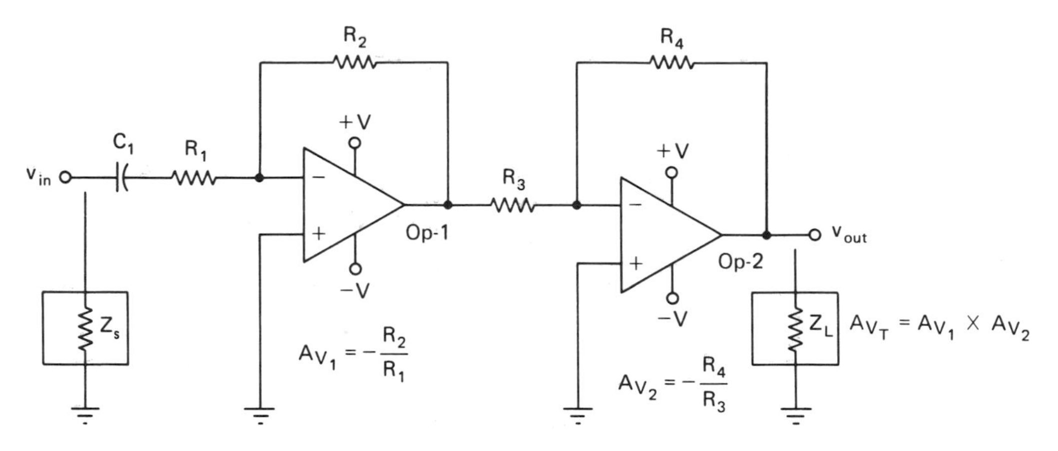

FIGURE 8-8 Two-stage inverting amplifier.

aspect to remember is not to have too much gain for each stage, since the total gain (A_{vT}) is equal to the first-stage gain (A_{v1}) times the second-stage gain (A_{v2}), or $A_{vT} = A_{v1} \times A_{v2}$.

Depending on the specific application, the first stage may be required to have more gain than the second stage. However, in this design problem the first stage could have the lower gain to reduce the input noise level. As an example, if $v_{in} = 10$ mV and $A_{v1} = 10$, v_{out} of OP-1 would be 100 mV. If $A_{v2} = 100$, then v_{out} of OP-2 would be 10 V. Total circuit gain is $A_v = v_{out}/v_{in} = 10\text{ V}/0.01\text{ V} = 1000$.

DESIGN PROCEDURES

1. Determine or select v_{in}.
2. Determine source impedance Z_s.
3. Choose R_1 to be about 50 times greater than Z_s.
4. Calculate C_1 for a break frequency of 10 Hz from the formula $C_1 = \dfrac{1}{2\pi(Z_s + R_1) f_c}$.
5. Choose the desired voltage gain for the first stage, A_{v1}.
6. Calculate R_2 from the formula $R_2 = -A_{v1}R_1$. (Disregard the minus sign in the calculations.)
7. Choose R_3 to be about 10 times greater than the output impedance of OP-1. (This impedance is about 25 to 50 Ω.)

8. Select the desired voltage gain for the second stage, A_{v2}. (Make sure that A_{vT} will not cause distortions of v_{out} for the v_{in} of step 1.)

9. Calculate R_4 from the formula $R_4 = -A_{v2}R_3$. (Disregard the minus sign in the calculations.)

10. Determine Z_L by the following circuit to be driven, or arbitrarily select Z_L for 2.2 kΩ.

11. Calculate the total circuit gain from the formula $A_{vT} = A_{v1} \times A_{v2}$.

12. Calculate v_{out} from the formula $v_{out} = A_{vT}v_{in}$. (Remember that v_{out} should be about 2 V less than the $\pm V$ supply.)

13. Calculate i_{in} from the formula $i_{in} = v_{in}/(Z_s + R_1)$.

14. Calculate i_{out} from the formula $i_{out} = v_{out}/Z_L$.

15. Calculate A_i from the formula $A_i = i_{out}/i_{in}$.

16. Construct the circuit and verify the design results.

DESIGN 8-3 NONINVERTING AMPLIFIER CIRCUITS

Op-amp noninverting amplifiers provide high voltage gain and high current gain together with high input impedance. Basic noninverting amplifiers are also relatively easy to design and construct. These design examples will show how to determine voltage gain, current gain, and the values of related external components.

8-3-1 Noninverting DC Amplifier

The input impedance of a noninverting DC amplifier is very large, usually a few megaohms. For this reason the circuit is capable of accommodating most source impedances without appreciable loading. The gain of the noninverting amplifier, as shown in Figure 8-9, is dependent on the same resistance ratio as the inverting amplifier.

DESIGN PROCEDURES

1. Determine source resistance R_s.

2. Choose R_{in} to equal R_s.

3. Determine or select V_{in}.

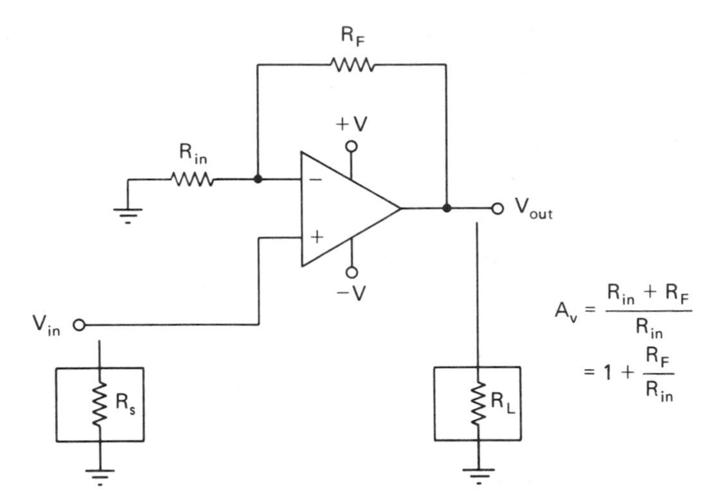

FIGURE 8-9 Noninverting DC amplifier.

4. Calculate desired voltage gain from the formula $A_v = V_{out}/V_{in}$. (Remember that maximum V_{out} should be about 2 V less than the $\pm V$ supply.)
5. Calculate R_F from the formula $R_F = A_v R_{in} - R_{in}$.
6. Determine R_L by the following circuit to be driven, or arbitrarily select R_L for 2.2 kΩ.
7. Calculate V_{out} from the formula $V_{out} = (1 + R_F/R_{in})V_{in}$.
8. Calculate I_{in} from the formula $I_{in} = V_{in}/(R_s + Z_{in})$. (Let $Z_{in} \approx$ 1 MΩ.)
9. Calculate I_{out} from the formula $I_{out} = V_{out}/R_L$.
10. Calculate A_I from the formula $A_I = I_{out}/I_{in}$.
11. Construct the circuit and verify the design results.

8-3-2 Noninverting AC Amplifier

The input impedance of a noninverting amplifier is usually between 5 and 50 kΩ and is determined by R_1, as shown in Figure 8-10. Capacitor C_1 blocks any DC component from the source, thus minimizing any distortion of the AC signal at the output. Design procedures for an AC noninverting amplifier are similar for those of the DC noninverting amplifier, except that C_1 must be calculated for a specific break frequency.

Depending on the op amp chosen and the circuit gain selected, the output signal voltage will roll off rapidly beyond the bandwidth for a given circuit. (Refer to Section 1-3-10.)

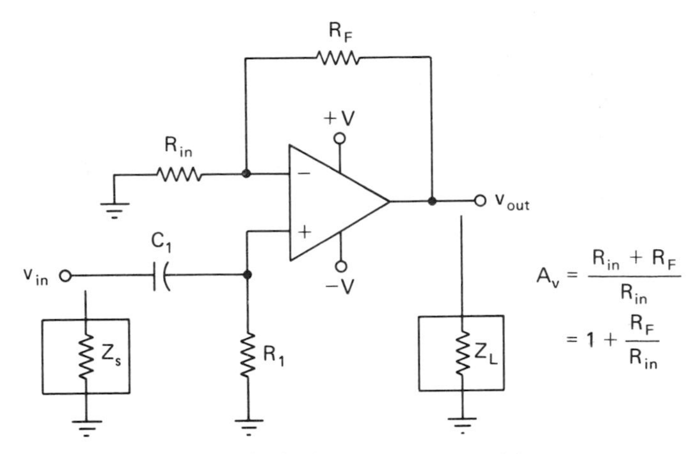

FIGURE 8-10 Noninverting AC amplifier.

DESIGN PROCEDURES

1. Choose R_1 between 5 and 50 kΩ.
2. Calculate C_1 from the formula $C_1 = \dfrac{1}{2\pi(Z_s + R_1)f_c}$.
3. Choose R_{in} equal to R_1.
4. Determine or select v_{in}.
5. Calculate desired voltage gain by the formula $A_v = v_{out}/v_{in}$. Remember that maximum v_{out} should be about 2 V less than the $\pm V$ supply.)
6. Calculate R_F from the formula $R_F = A_v R_{in} - R_{in}$.
7. Determine Z_L by the following circuit to be driven, or arbitrarily select Z_L for 2.2 kΩ.
8. Calculate v_{out} from the formula $v_{out} = (1 + R_F/R_{in})v_{in}$.
9. Calculate i_{in} from the formula $i_{in} = v_{in}/(Z_s + R_1)$.
10. Calculate i_{out} from the formula $i_{out} = v_{out}/Z_L$.
11. Calculate A_i from the formula $A_i = i_{out}/i_{in}$.
12. Construct the circuit and verify the design results.

8-3-3 Basic Two-State Cascaded Noninverting Amplifier

Designing a basic two-stage cascaded noninverting amplifier as shown in Figure 8-11 is accomplished by designing the gain of each stage. The total

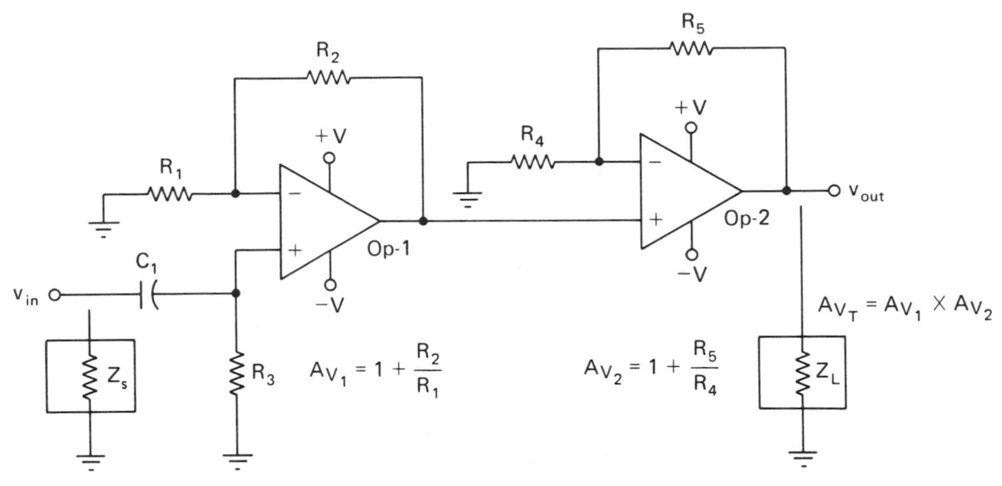

FIGURE 8-11 Two stage noninverting amplifier.

circuit gain is equal to the gain of the first stage times the gain of the second stage, $A_{vt} = A_{v1} \times A_{v2}$. As an example, if $A_{v1} = 11$ and $A_{v2} = 22$, $A_{vt} = 242$. If $v_{in} = 0.1$ V, the v_{out} of OP-1 would be 1.1 V and the v_{out} of OP-2 would be 24.2 V.

DESIGN PROCEDURES

1. Determine or select v_{in}.
2. Choose R_3 between 5 and 50 kΩ.
3. Calculate C_1 from the formula $C_1 = \dfrac{1}{2\pi(Z_s + R_3) f_c}$.
4. Choose R_1 equal to R_3.
5. Choose the desired voltage gain for the first stage, A_{v1}.
6. Calculate R_2 from the formula $R_2 = A_{v1}R_1 - R_1$.
7. Choose R_4 equal to R_1.
8. Select the desired voltage gain for the second stage, A_{v2}. (Make sure that A_{vt} will not cause distortion of v_{out} for the v_{in} of step 1.)
9. Calculate R_5 from the formula $R_5 = A_{v2}R_4 - R_4$.
10. Determine Z_L by the following circuit to be driven, or arbitrarily select Z_L for 2.2 kΩ.
11. Calculate total circuit gain from the formula $A_{vt} = A_{v1} \times A_{v2}$.
12. Calculate v_{out} from the formula $v_{out} = A_{vt}v_{in}$. (Remember that v_{out} should be about 2 V less than the $\pm V$ supply.)

13. Calculate i_{in} from the formula $i_{in} = v_{in}/(Z_s + R_3)$.
14. Calculate i_{out} from the formula $i_{out} = v_{out}/Z_L$.
15. Calculate A_i from the formula $A_i = i_{out}/i_{in}$.
16. Construct the circuit and verify the design results.

DESIGN 8-4 VOLTAGE-FOLLOWER CIRCUITS

Voltage followers are used to transfer signals from large impedances to small impedances while maintaining a gain of near unity. In other words, they are buffer or impedance-matching devices with controlled output voltage, but have the capability of current amplification.

8-4-1 Noninverting DC Voltage Follower

A basic noninverting DC voltage follower requiring no external components is shown in Figure 8-12. The input impedance is several megohms, and the output impedance is about 25 to 50 Ω. Voltage gain will be approximately 1, with current gains usually less than 1000, depending on the source resistance and the load resistance.

DESIGN PROCEDURES

1. Determine or select the required V_{in}.
2. Calculate A_v required from the formula $A_v = V_{out}/V_{in}$.

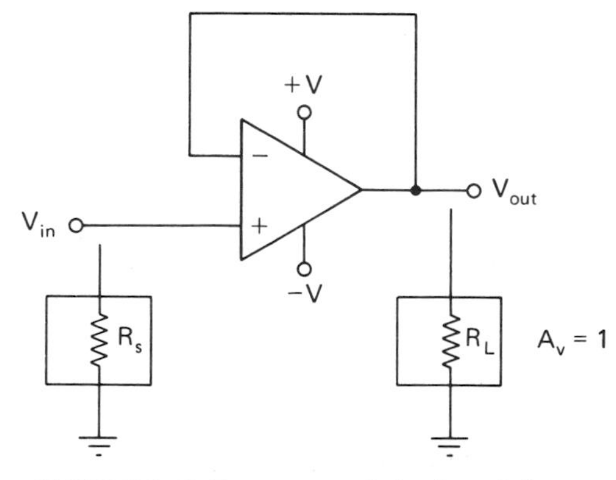

FIGURE 8-12 Noninverting DC voltage follower.

3. Calculate I_{in} from the formula $I_{in} = V_{in}/(R_s + Z_{in})$. (Let $Z_{in} \approx 2\ M\Omega$.)
4. Calculate I_{out} from the formula $I_{out} = V_{out}/R_L$.
5. Calculate A_I from the formula $A_I = I_{out}/I_{in}$.
6. Construct the circuit and verify the design results.

8-4-2 Noninverting AC Voltage Follower

The input impedance of a noninverting AC voltage follower, as shown in Figure 8-13, is determined by R_1. This resistor is typically selected to be 50 times greater than the source impedance (Z_s). Feedback resistor R_F is equal to R_1. Capacitor C_1 blocks any DC component from the source that may cause output distortion. The value of C_1 is determined by the desired frequency range.

DESIGN PROCEDURES

1. Determine or select the required v_{in}.
2. Calculate the A_v required from the formula $A_v = v_{out}/v_{in}$.
3. Calculate C_1 from the formula $C_1 = \dfrac{1}{2\pi(Z_s + R_1) f_c}$. (Choose for a break frequency of 10 Hz.)
4. Choose R_1 to be about 50 times greater than Z_s.

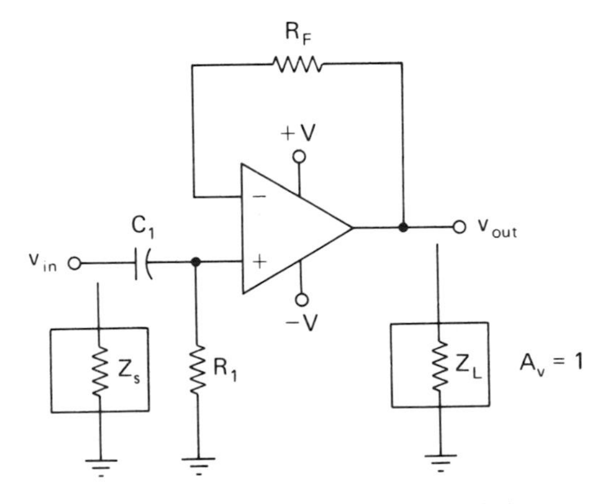

FIGURE 8-13 Noninverting AC voltage follower.

5. Choose R_F equal to R_1.
6. Calculate i_{in} from the formula $i_{in} = v_{in}/(Z_s + R_1)$.
7. Calculate i_{out} from the formula v_{out}/Z_L.
8. Calculate A_i from the formula $A_i = i_{out}/i_{in}$.
9. Construct the circuit and verify the design results.

8-4-3 Inverting DC Voltage Follower

An inverting DC voltage follower, as shown in Figure 8-14, is the same configuration as an inverting amplifier. Resistors R_{in} and R_F are equal, resulting in unity gain. The input impedance of this follower is much less than that of the noninverting follower and is dependent on R_{in}.

DESIGN PROCEDURES

1. Determine or select V_{in}.
2. Calculate A_v from the formula $A_v = V_{out}/V_{in}$.
3. Choose R_{in} to be 50 times greater than R_s.
4. Choose R_F to equal R_{in}.
5. Calculate I_{in} from the formula $I_{in}/(R_s + R_{in})$.
6. Calculate I_{out} from the formula $I_{out} = V_{out}/R_L$.

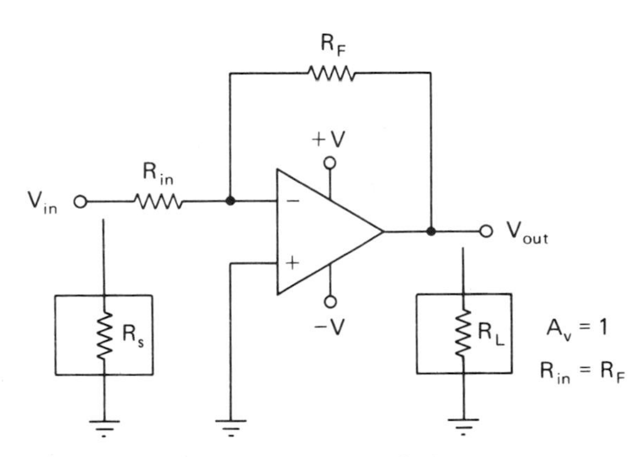

FIGURE 8-14 Inverting DC voltage follower.

7. Calculate A_I from the formula $A_I = I_{out}/I_{in}$.
8. Construct the circuit and verify the design results.

8-4-4 Inverting AC Voltage Follower

The inverting AC voltage follower shown in Fig. 8-15 is similar to the inverting DC voltage follower, except for C_1, which blocks the DC component from the source, thereby minimizing distortion at the output.

DESIGN PROCEDURES

1. Determine or select v_{in}.
2. Calculate A_v from the formula $A_v = v_{out}/v_{in}$.
3. Calculate C_1 from the formula $C_1 = \dfrac{1}{2\pi(Z_s + R_{in})\, f_c}$. (Choose for a break frequency of 10 Hz.)
4. Choose R_{in} to be 50 times greater than Z_s.
5. Choose R_F to equal R_{in}.
6. Calculate i_{in} from the formula $i_{in} = v_{in}/(Z_s + R_{in})$.
7. Calculate i_{out} from the formula $i_{out} = v_{out}/Z_L$.
8. Calculate A_i from the formula $A_i = i_{out}/i_{in}$.
9. Construct the circuit and verify the design results.

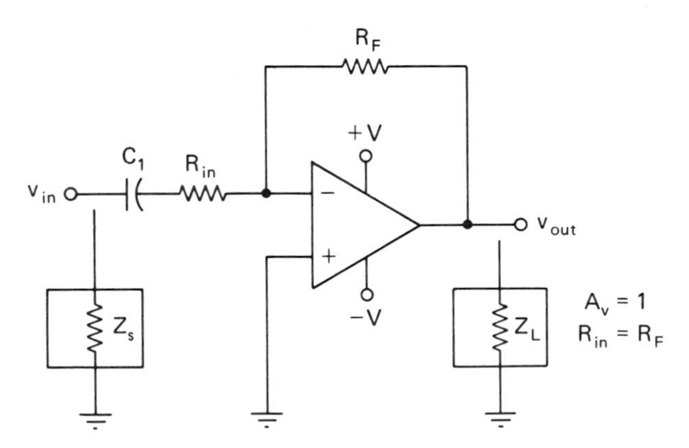

FIGURE 8-15 Inverting AC voltage follower.

DESIGN 8-5 VOLTAGE-SUMMING-AMPLIFIER CIRCUITS

Op amp voltage-summing amplifiers are used for summing several input signals to provide a single output signal. The output signal may be a direct mathematical summation of the input signals or may include a determined amount of gain. With a direct mathematical summation, all of the input resistors and the feedback resistors are of the same value. Where gain is desired, the feedback resistor is made larger. The summing amplifier may be a scaling adder, where the input resistors are selected to provide different gains for each input. Summing amplifiers may be inverting or noninverting; however, the inverting type is less complex to design and easier to construct. If an in-phase signal is needed from an inverted summing amplifier, it is usually easier to add an inverting follower stage.

8-5-1 Inverting DC Summing Amplifier

The inverting DC summing amplifier shown in Figure 8-16 is a basic circuit that can be used for direct mathematical summation, with summation at a specific gain, or with scaling adder inputs. The design procedures for this circuit will cover all three aspects of its use.

DESIGN PROCEDURES

1. Determine or select V_{in_1}, V_{in_2} and V_{in_3}.
2. Calculate the A_v required from the formula

$$A_v = \frac{V_{out}}{V_{in_1} + V_{in_2} + V_{in_3}}$$

3.* A. *For direct mathematical summation, select*

$$R_1 = R_2 = R_3 = R_F \quad \text{(typically 10 to 25 k}\Omega\text{)}$$

B. *For summation with gain, select*

$$R_1 = R_2 = R_3 \quad \text{(typically 10 to 25 k}\Omega\text{)}$$

* Select the proper step with respect to the type of summing amplifier you are designing. All other design steps apply to all three summing amplifiers.

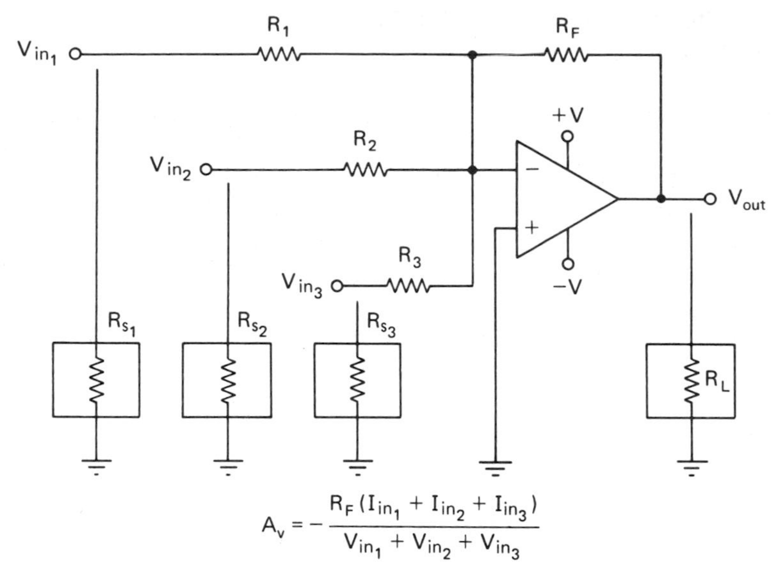

$$A_v = -\frac{R_F(I_{in_1} + I_{in_2} + I_{in_3})}{V_{in_1} + V_{in_2} + V_{in_3}}$$

FIGURE 8-16 Inverting DC summing amplifier.

C. *For scaling adder input resistors, calculate*

$$R_1 = \frac{V_{in_1}}{I_{in_1}} - R_{s_1}$$

$$R_2 = \frac{V_{in_2}}{I_{in_2}} - R_{s_2}$$

$$R_3 = \frac{V_{in_3}}{I_{in_3}} - R_{s_3}$$

Select $I_{in_1} = I_{in_2} = I_{in_3}$ (typically 0.1 mA).

4.* *Calculate I_{in} for direct mathematical summation or summation with gain from the formulas*

$$I_{in_1} = \frac{V_{in_1}}{R_{s_1} + R_1}$$

$$I_{in_2} = \frac{V_{in_2}}{R_{s_2} + R_2}$$

$$I_{in_3} = \frac{V_{in_3}}{R_{s_3} + R_3}$$

* Select the proper step with respect to the type of summing amplifier you are designing. All other design steps apply to all three summing amplifiers.

5. Calculate I_{out} from the formula $I_{out} = V_{out}/R_L$.
6. Calculate A_I from the formula $A_I = I_{out}/(I_{in_1} + I_{in_2} + I_{in_3})$.
7.* *Calculate R_F for summation with gain and scaling adder inputs from the formula*

$$R_F = \frac{-A_v(V_{in_1} + V_{in_2} + V_{in_3})}{I_{in_1} + I_{in_2} + I_{in_3}}$$

(Disregard minus signs in making the calculations.)

8. Construct the circuit and verify the design results.

8-5-2 Inverting AC Summing Amplifier

The inverting AC summing amplifier is similar to the inverting DC summing amplifier, except for the input capacitors, as shown in Figure 8-17. Follow the same design procedures given for the inverting DC summing amplifier and use the additional design procedures given below for the inverting AC summing amplifier.

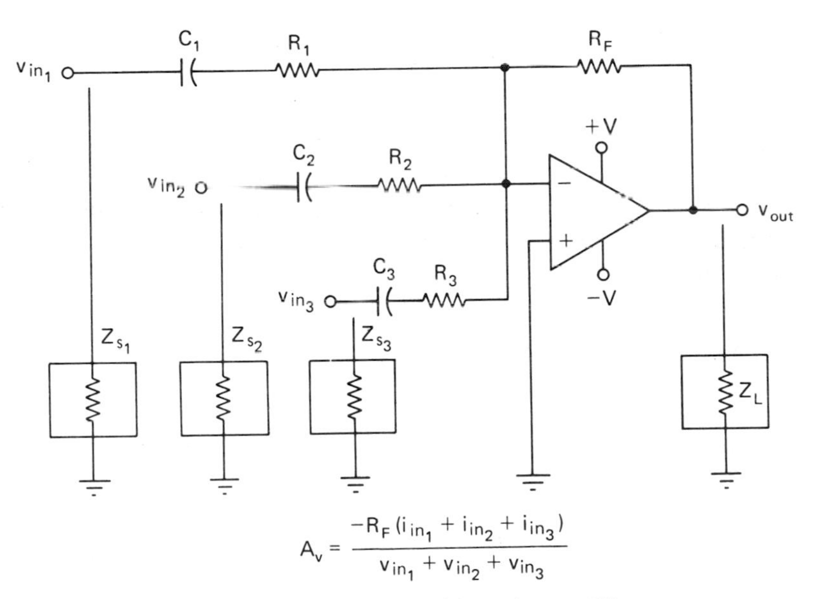

$$A_v = \frac{-R_F(i_{in_1} + i_{in_2} + i_{in_3})}{v_{in_1} + v_{in_2} + v_{in_3}}$$

FIGURE 8-17 Inverting AC summing amplifier.

* Select the proper step with respect to the type of summing amplifier you are designing. All other design steps apply to all three summing amplifiers.

ADDITIONAL DESIGN PROCEDURES

1. Use lowercase AC signal symbols, as shown in Figure 8-17.
2. The input resistors R_1, R_2, and R_3 for the direct mathematical summing and summing circuits with gain should be about 50 to 100 times greater than their source impedances.
3. Calculate C_1, C_2, and C_3 from the formulas

$$C_1 = \frac{1}{2\pi(R_{s_1} + R_1)\, f_c}$$

$$C_2 = \frac{1}{2\pi(R_{s_2} + R_2)\, f_c}$$

$$C_3 = \frac{1}{2\pi(R_{s_3} + R_3)\, f_c}$$

(f_c = 10 Hz, typically.)

4. Construct the circuit and verify the design results.

DESIGN 8-6 DIFFERENTIAL AMPLIFIER CIRCUITS

Differential amplifiers are used where input voltage differences must be amplified. A differential amplifier has low voltage gain but high current gain. Both inputs are used and the circuit operation resembles that of a voltage comparator, except with controlled gain. The output will be inverted depending on the polarity of the inverting input with respect to the noninverting input.

8-6-1 Differential DC Amplifier

The differential DC amplifier shown in Figure 8-18 has input impedances of about 1 MΩ. The output impedance, as with most op amps, is between 25 and 50 Ω and is usually considered zero for design purposes. Input voltages are not critical and may be up to 70 to 80% of the $\pm V$ supply. The design procedures given are for symmetrical gain; however, other gains may be found from the formula

$$V_{\text{out}} = \left[\left(\frac{R_1 + R_3}{R_2 + R_3}\right)\left(\frac{R_3}{R_1}\right)\right] \times \left[\left(\frac{V_{\text{in}_2} - R_F}{R_1}\right)(V_{\text{in}_1})\right]$$

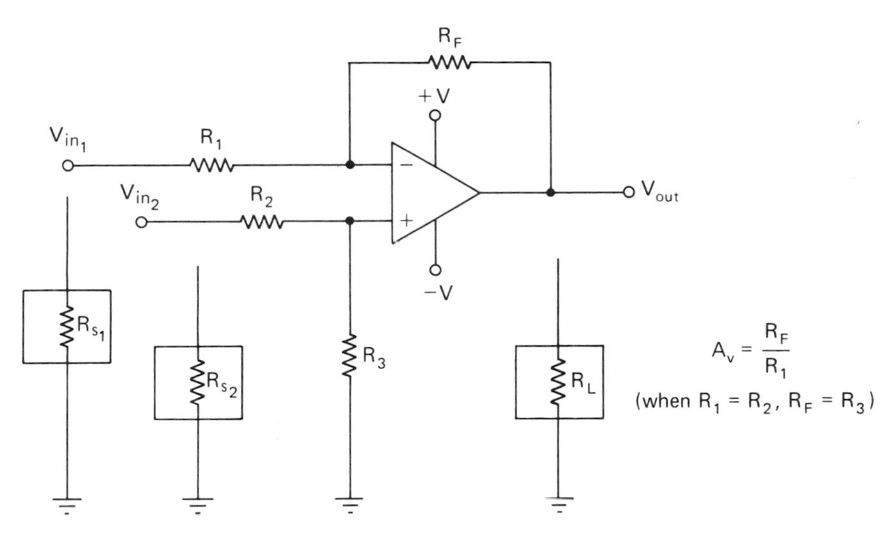

FIGURE 8-18 Differential DC amplifier.

DESIGN PROCEDURES

1. Determine or select V_{in_1} and V_{in_2}.
2. Select R_1 to be 50 times R_{s_1}.
3. Select R_2 to be equal to R_1.
4. Calculate A_v from the formula $A_v = V_{out}/(V_{in_2} - V_{in_1})$.
5. Calculate R_F from the formula $R_F = -A_v R_1$.
6. Select R_3 to equal R_F.
7. Calculate I_{in_1} from the formula

$$I_{in_1} = \frac{V_{in_1}}{R_{s_1} + R_1 + \dfrac{R_F Z_{in_1}}{R_F + Z_{in_1}}}$$

8. Calculate I_{in_2} from the formula

$$I_{in_2} = \frac{V_{in_2}}{R_{s_2} + R_2 + \dfrac{R_F Z_{in_2}}{R_F + Z_{in_2}}}$$

9. Calculate I_{out} from the formula $I_{out} = V_{out}/R_L$.
10. Calculate A_I from the formula $A_I = I_{out}/(I_{in_1} - I_{in_2})$.
11. Construct the circuit and verify the design results.

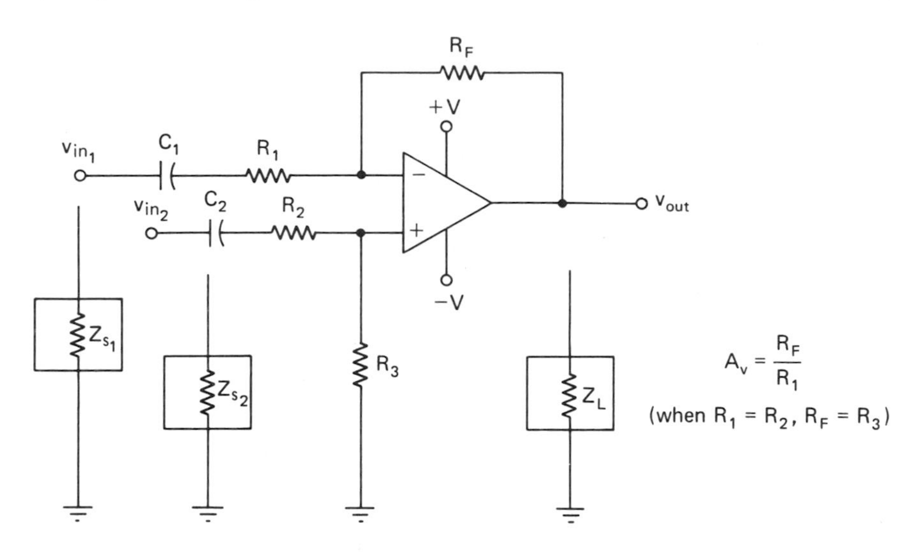

FIGURE 8-19 Differential AC amplifier.

8-6-2 Differential AC Amplifier

The differential AC amplifier is similar to the differential DC amplifier, except for the input capacitors, as shown in Figure 8-19. Follow the same design procedures given for the differential DC amplifier and use the additional design procedures given below for the differential AC amplifier.

ADDITIONAL DESIGN PROCEDURES

1. Calculate C_1 from the formula

$$C_1 = \frac{1}{2\pi[R_{s_1} + R_1 + (R_F Z_{\text{in}_1}/R_F + Z_{\text{in}_1})]\, f_c}.$$

2. Calculate C_2 from the formula

$$C_2 = \frac{1}{2\pi[R_{s_2} + R_2 + (R_3 Z_{\text{in}_2}/R_3 + Z_{\text{in}_2})]\, f_c}.$$

   (f = 10 Hz, typically.)

3. Construct the circuit and verify the design results.

DESIGN 8-7 SQUARE-WAVE-GENERATOR CIRCUIT

Using feedback capabilities, an op amp can produce a fairly stable oscillator circuit. The simplest oscillator circuit is the astable multivibrator, often

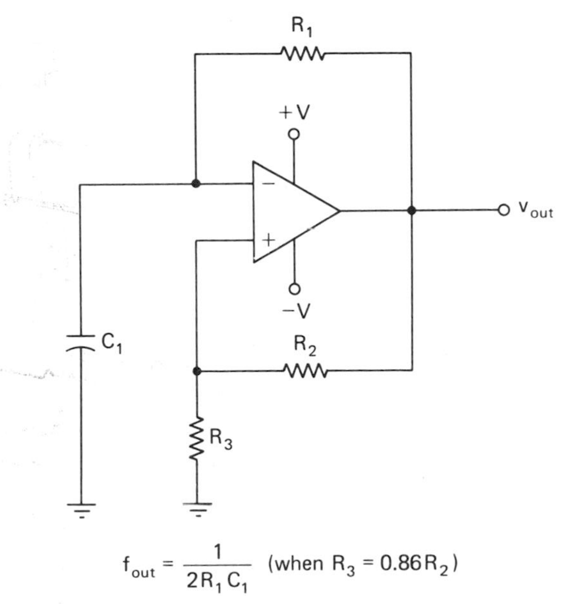

$$f_{out} = \frac{1}{2R_1C_1} \quad (\text{when } R_3 = 0.86R_2)$$

FIGURE 8-20 Square wave generator.

referred to as a square-wave generator, as shown in Figure 8-20. Exact frequencies can be obtained by the proper selection of R_1 and C_1, which provide an RC time constant that determines the desired frequency. Resistors R_2 and R_3 form a voltage divider whose ratio equals two time constants, which permits the frequency to be found by a simple formula.

DESIGN PROCEDURES

1. Select R_1 (usually 100 kΩ).
2. Select R_2 equal to R_1.
3. Calculate R_3 from the formula $R_3 = 0.86R_2$.
4. Choose the desired frequency.
5. Calculate C_1 from the formula $C_1 = \frac{1}{2}fR_1$.
6. Construct the circuit and verify the design results.

Chapter 9

Collection of Practical Op-Amp Circuits

This chapter presents a collection of practical op-amp circuits that will enable you to realize the versatility of the op amp. It is hoped that you may find some circuits in this chapter that will fit your particular application or that you will find satisfaction from the enjoyment of just constructing some of the circuits. Each circuit has a description of its use, theory of operation, and/or any related pertinent facts.

Special appreciation is given to Signetics Corporation, Fairchild Semiconductor Incorporated, and National Semiconductor Corporation for their contributions to this chapter.

9-1 POWER-SUPPLY APPLICATIONS

9-1-1 Voltage Regulator

The op-amp voltage regulator shown in Figure 9-1 is a noninverting amplifier. This circuit is a positive voltage regulator with R_{in} and R_F determining the gain. The zener diode provides the reference voltage. The regulated output voltage for the values given is then $3 \times (+5)$ V = +15 V. The output voltage to be regulated should be at least 2 V lower than the unregulated voltage from the power supply to keep the zener diode operating in its breakdown region.

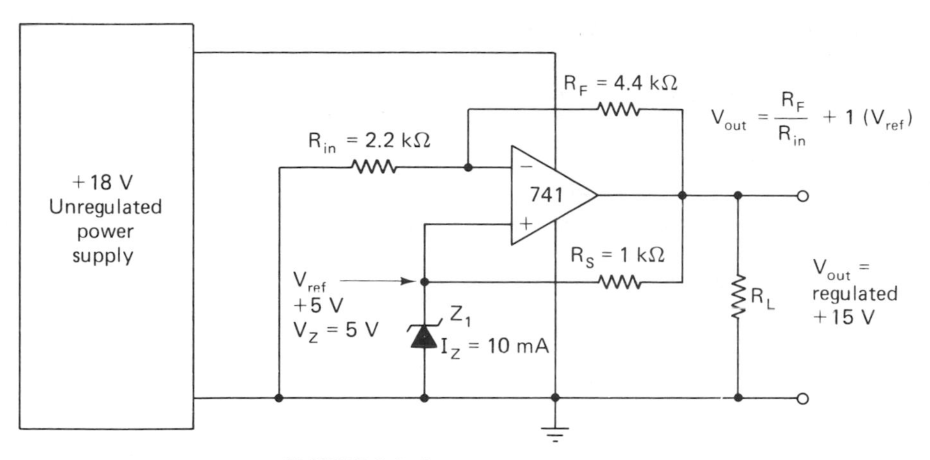

FIGURE 9-1 Op amp voltage regulator.

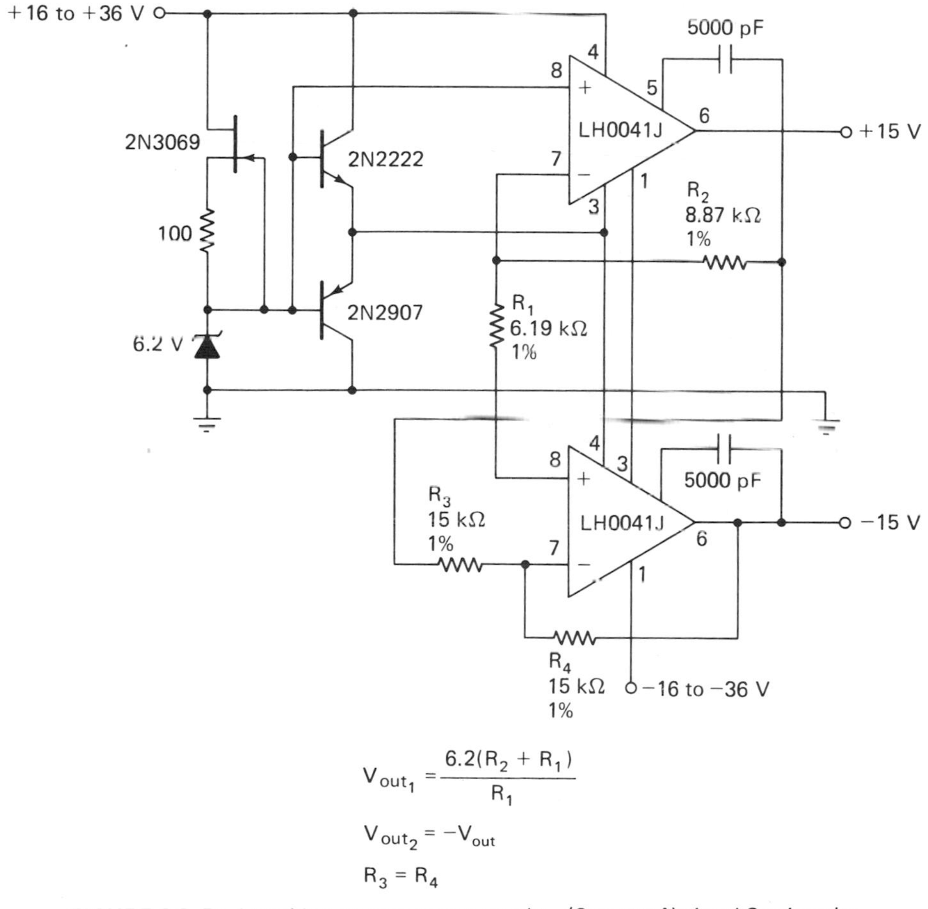

FIGURE 9-2 Dual tracking op-amp power supply. (Courtesy National Semiconductor Corp.)

9-1-2 Dual-Tracking Op-Amp Power Supply

A dual-tracking op-amp power supply is shown in Figure 9-2. The positive voltage regulator is similar to the circuit in Figure 9-1. The positive output is fed to an inverting follower to produce the negative output voltage.

9-1-3 Crowbar Overvoltage Protection

Some ICs cannot withstand overvoltage. The crowbar overvoltage protection circuit of Figure 9-3 is fast-acting to protect the load R_L against damaging effects of an overvoltage. The zener diode sets the inverting input at the reference +2.5 V. The trip adjust pot sets the noninverting input at +2.5 V. Differential input voltage is zero; therefore, the op-amp output is zero and the SCR is open. If the 5-V power supply increases, the voltage at the noninverting input increases. The output of the op amp increases, causing the SCR to fire. The SCR appears as a short around the load and activates the fuse, circuit breaker, or current limiter. When the trouble is cleared and the power supply is again at 5 V, the reset switch is used to disable the SCR, allowing full voltage to the load.

9-1-4 Crowbar Undervoltage Protection

In some applications it is desirable to shut down the power supply if the voltage level drops below a specific level. The circuit in Figure 9-4 is similar to the previous circuit except that the reference voltage is on the noninverting input. The trip adjust pot sets the inverting input at +2.5 V. Differential input voltage is zero, the op-amp output is zero, and the SCR is open. When the power-supply voltage drops below +5 V, the voltage at the invert-

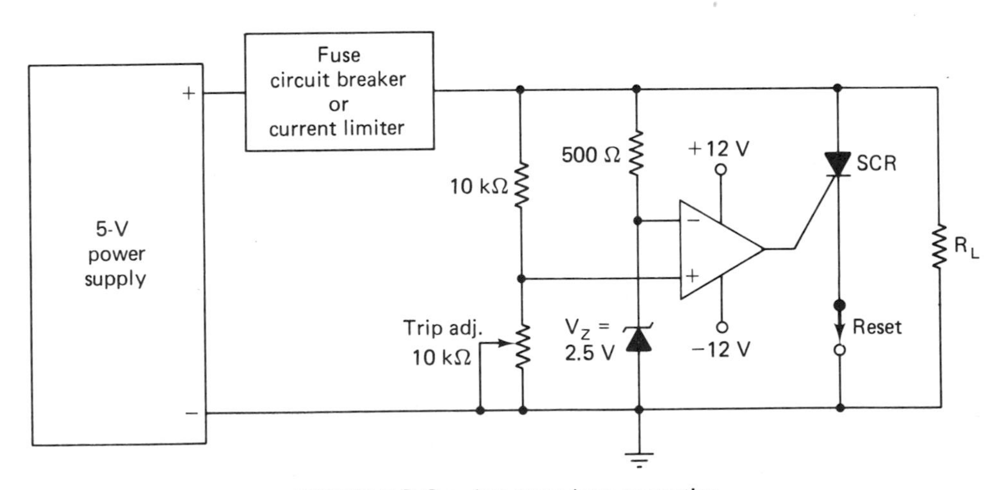

FIGURE 9-3 Crowbar overvoltage protection.

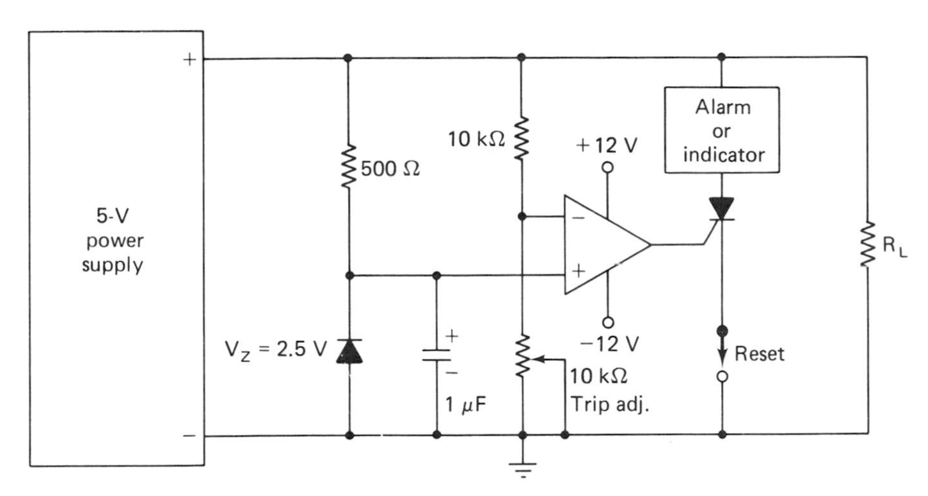

FIGURE 9-4 Crowbar undervoltage protection.

ing input decreases, the op-amp output voltage increases, and the SCR fires. Voltage across the load decreases and an alarm or indicator is activated. The capacitor across the zener diode temporarily holds the noninverting input at a lower voltage than the inverting input during initial turn-on. This allows the zener diode time to conduct and prevents false triggering the SCR.

9-2 AMPLIFIERS

9-2-1 Bridge Amplifier

A bridge amplifier is a balanced circuit where the input resistances are equal. The feedback resistor may be any value, depending on the desired output. All resistors in Figure 9-5 are considered equal, and with the bridge balanced, V_{out} is zero. A transducer is used as a sensing element and can be used in various parts of the circuit as shown. The transducer is a device that converts an environmental change to a resistive change and could be a thermistor, photodetector, or strain gauge. When the resistance of the transducer changes (ΔR_T), an output voltage will occur. The formulas give the V_{out} for each circuit.

9-2-2 Buffer Amplifier

A practical buffer amplifier for isolating circuits is shown in Figure 9-6. If the 1 MΩ causes any noise problems, the resistors could be reduced, but should be at least 10 times the impedance of the circuit feeding the buffer.

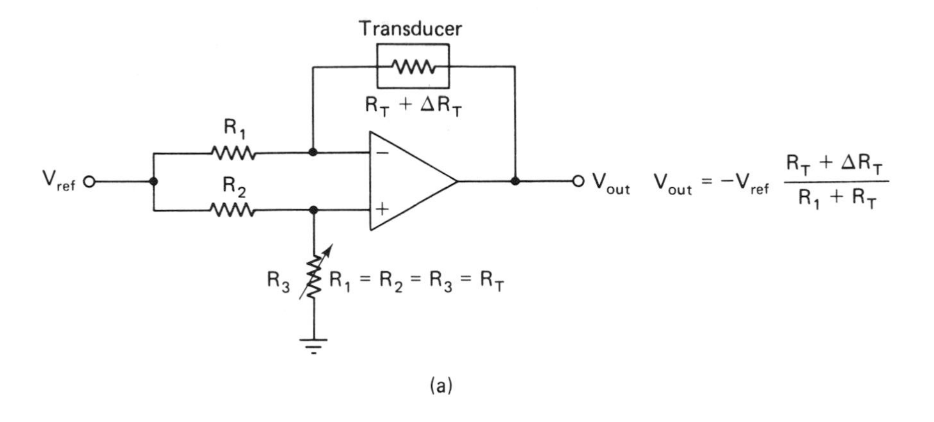

(a)

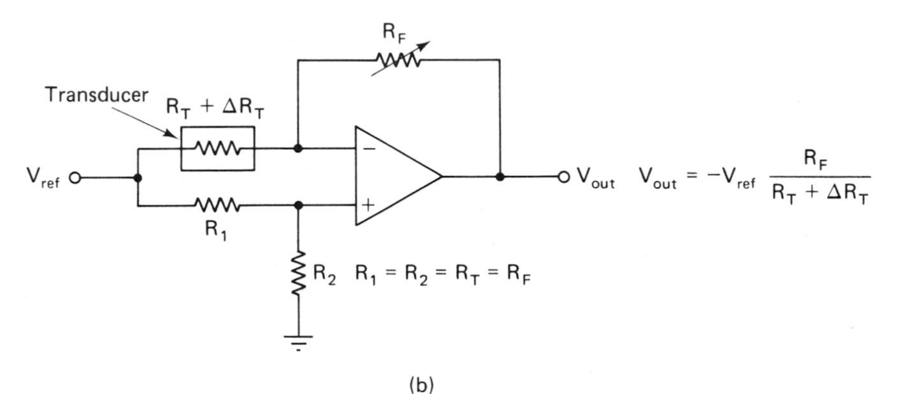

(b)

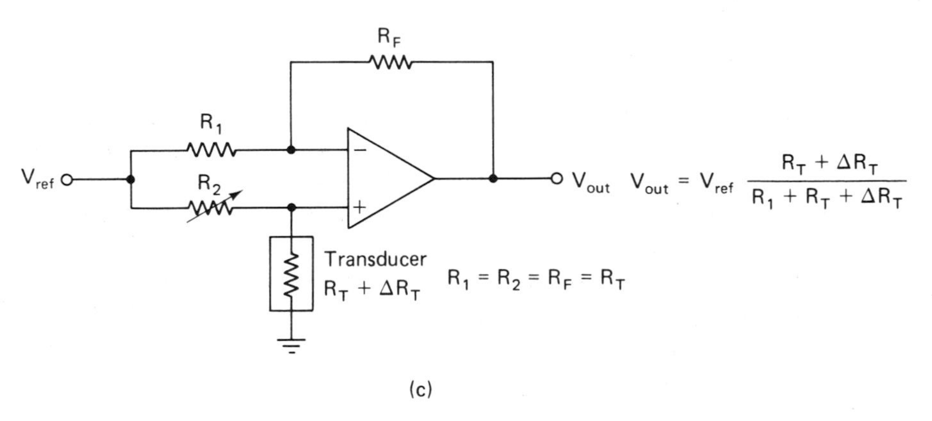

(c)

FIGURE 9-5 Bridge amplifier.

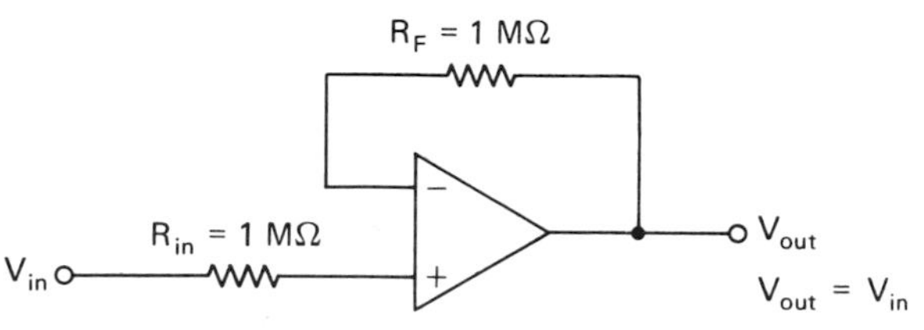

FIGURE 9-6 Buffer amplifier.

9-2-3 Current Amplifier

An op amp can detect a small current and amplify it up to its maximum output current, with the circuit shown in Figure 9-7. The amount of load current depends on the factor $F = R_2/R_1$ and the input current I_s. As an example, the small current from a solar cell can be amplified to cause the LED to give a visible indication. If $I_s = 0.1$ mA, then by the formula, $I_L =$ 10.1 mA. The LED might be part of an optical coupler that could drive a higher-voltage circuit.

9-2-4 Lamp Driver

A simple op-amp lamp driver is shown in Figure 9-8. The output of the op amp has to be positive-going for an NPN transistor to turn on and be of sufficient amplitude to drive the transistor into saturation. Resistor R_1 should be selected to limit the base current below the recommended maximum rating. The transistor I_c must have a maximum higher rating than the current that the lamp will draw.

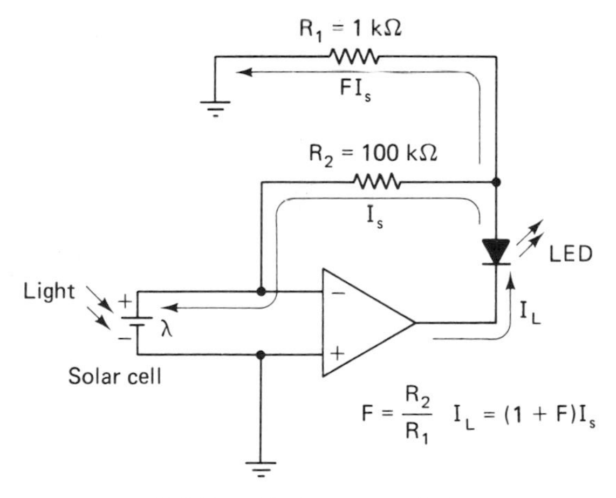

FIGURE 9-7 Current amplifier.

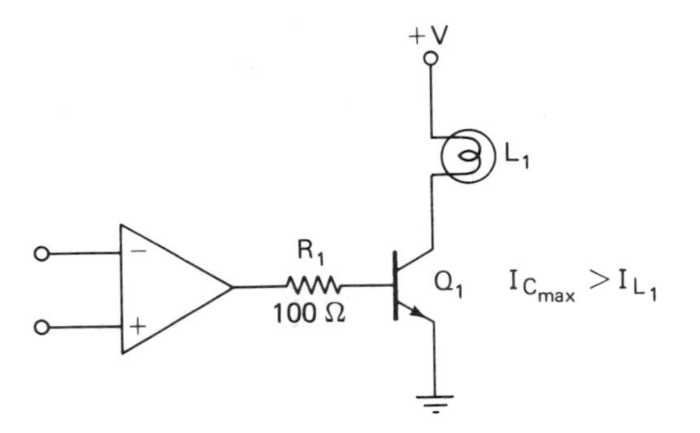

FIGURE 9-8 Lamp driver.

9-2-5 LED Driver

An op amp can drive an LED directly, provided that its maximum output current is not exceeded. A positive-going output will turn on the LED in Figure 9-9a, while a negative-going output will turn on the LED of Figure 9-9b. The value of R_1 is determined by the formulas given, where V_{FLED} is the forward voltage drop of the LED and I_{FLED} is the desired forward current through the LED.

9-2-6 Photodiode/Phototransistor Amplifier

For normal operation the photodiode must be reversed-biased, as shown in Figure 9-10a. Light striking the diode will decrease its resistance, causing

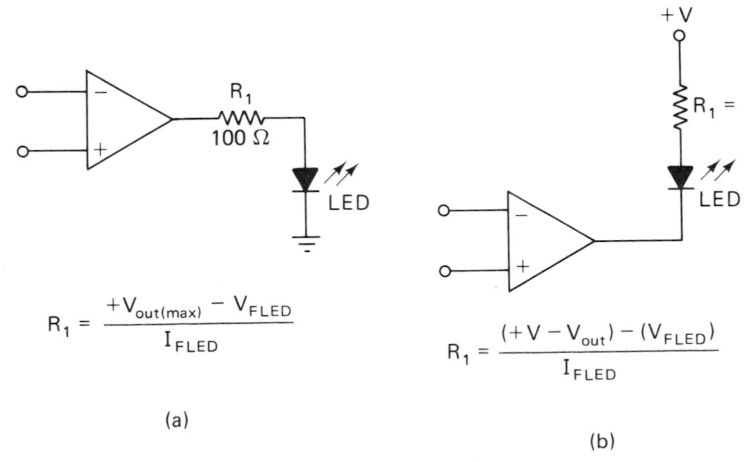

FIGURE 9-9 LED driver.

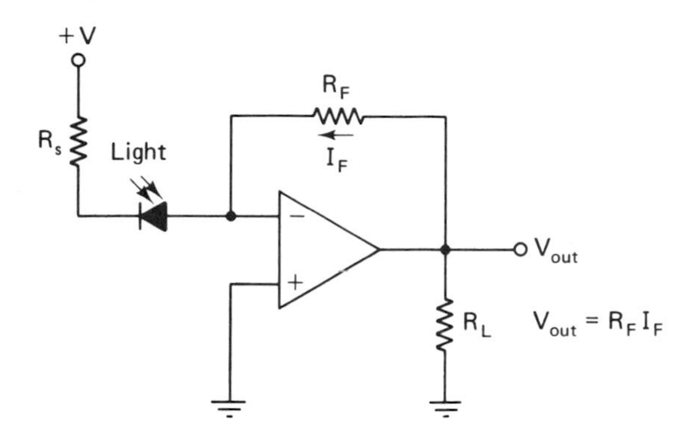

(a) Photodiode amplifier

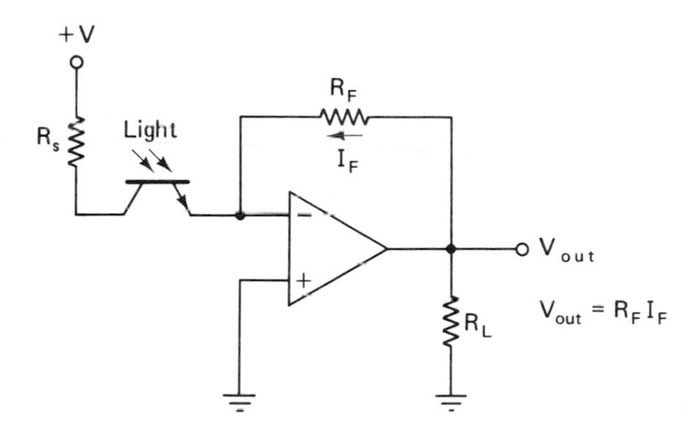

(b) Phototransistor amplifier

FIGURE 9-10 Photodiode/phototransistor amplifier: (a) photodiode amplifier; (b) phototransistor amplifier.

the output voltage to increase in a negative direction. The same action will occur for the phototransistor amplifier shown in Figure 9-10b. The output voltage depends on the current drawn through R_F. Considering the basic gain formula $A_v = -R_F/R_{in}$, when R_{in} decreases, A_v increases.

9-2-7 Photoresistor Amplifier

Similar to the two previous circuits, a photoresistor amplifier is shown in Figure 9-11. Again, the varying input resistance will cause a varying output.

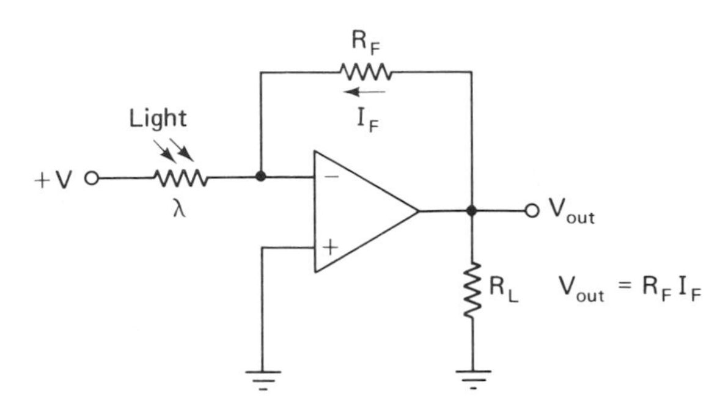

FIGURE 9-11 Photoresistor amplifier.

9-2-8 Solar-Cell Amplifier

The solar cell amplifier shown in Figure 9-12 is similar to the previously mentioned circuits, although its operation is somewhat different. The solar cell sees essentially a short circuit, since the inverting input is at virtual ground. The current generated by the solar cell is proportional to the light striking its surface. The current is converted to voltage by R_F as given by the formula. (Also see Section 9-2-3.)

9-2-9 Power-Booster Amplifier

Although the available power from op amps is usually sufficient, there are occasions when more power-handling capability is needed. The power-booster amplifier shown in Figure 9-13 is capable of driving moderate loads. The complementary transistor push-pull circuit will allow the output voltage to swing nearly to the maximum ± voltage supply and be able to handle more current. (See Figures 5-29 and 5-30.)

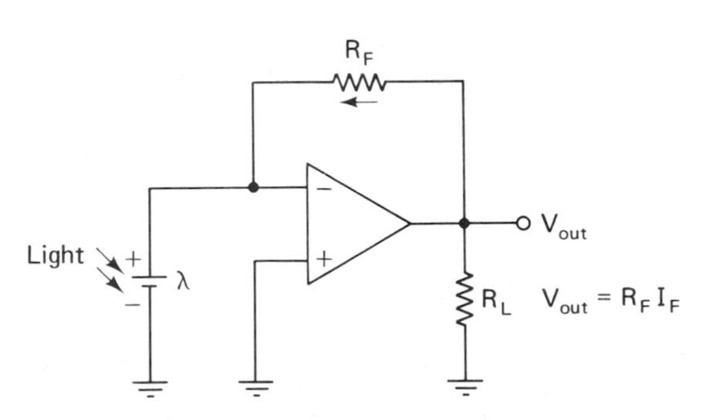

FIGURE 9-12 Solar cell amplifier.

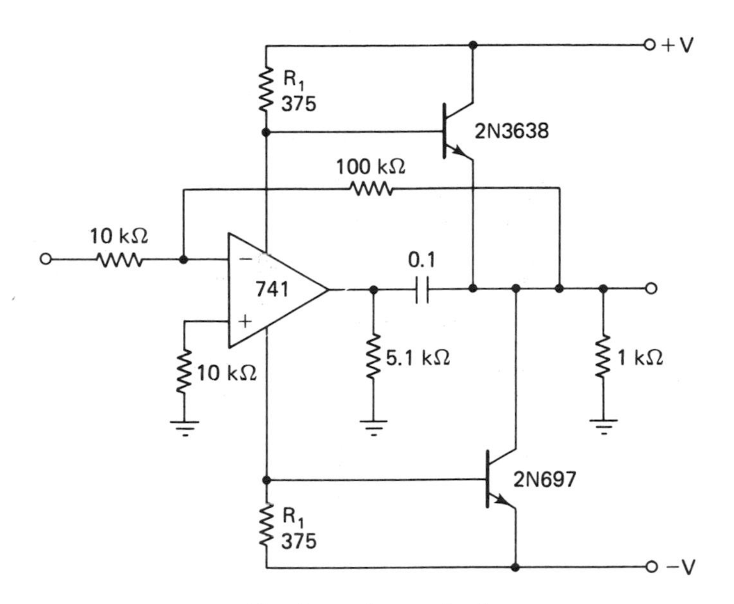

FIGURE 9-13 Power booster amplifier. (Permission to reprint granted by Signetics Corporation, a subsidiary of U.S. Philips Corp., 811 E. Arques Avenue, Sunnyvale, CA 94086.)

9-2-10 Phono Amplifier

The basic phono amplifier shown in Figure 9-14 uses an LM 380 power audio amplifier IC that will produce at least 2.5 W (rms) of power. A large signal output cartridge is required. This circuit uses a voltage-divider volume control and a high-frequency rolloff tone control. If the circuit tends to oscillate, a 2.7-Ω resistor in series with a 0.1-μF capacitor can be connected from pin 8 to ground.

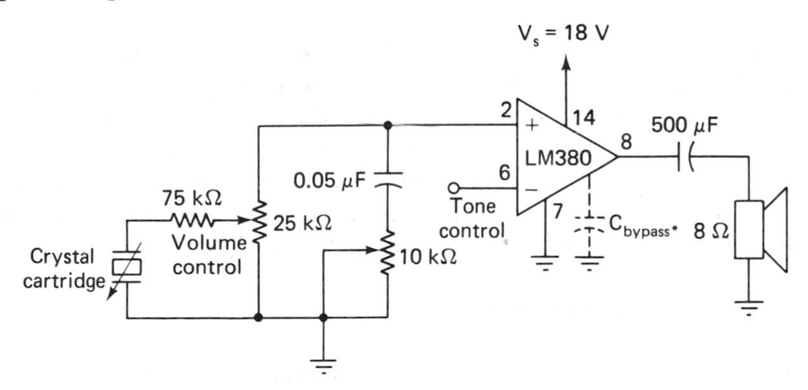

FIGURE 9-14 Phono amplifier. (Courtesy National Semiconductor Corp.)

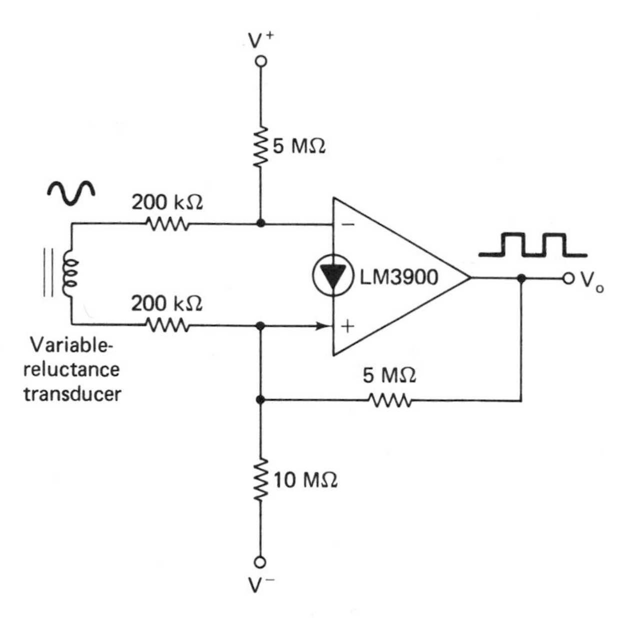

FIGURE 9-15 Squaring amplifier. (Courtesy National Semiconductor Corp.)

9-2-11 Squaring Amplifier

A squaring amplifier, as shown in Figure 9-15, is often needed to amplify low-level signals provided by variable-reluctance transducers. The output incorporates symmetrical hysteresis above and below the zero level, which improves noise immunity. The large input resistors provide a low-pass filter due to the "Miller effect" input capacitance of the amplifier. With the values shown, the output is approximately 0.3 V p-p.

9-2-12 Instrumentation Amplifier

A very popular circuit used for precision measurement and control is the instrumentation amplifier shown in Figure 9-16. This commonly used configuration consists of two input voltage followers feeding a differential amplifier. The followers exhibit extremely high input impedance with low error and allow the source (driving) resistances to be unbalanced by over 10 kΩ. The differential amplifier provides gain and high common-mode rejection. The gain is determined by R_6 to R_2 when $R_2 = R_5$ and $R_6 = R_7$.

9-2-13 Audio Bridge Amplifier

Twice the output power can be obtained from the circuit shown in Figure 9-17. The LM 377 amplifiers are useful in this configuration to drive float-

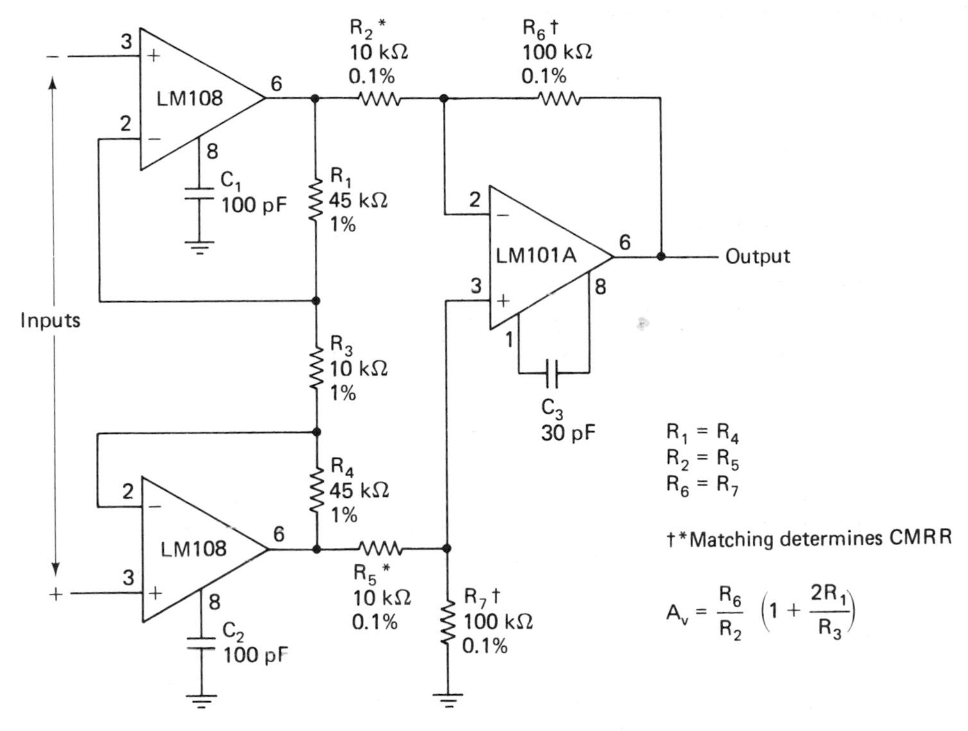

FIGURE 9-16 Differential input instrumentation amplifier with high common mode rejection. (Courtesy National Semiconductor Corp.)

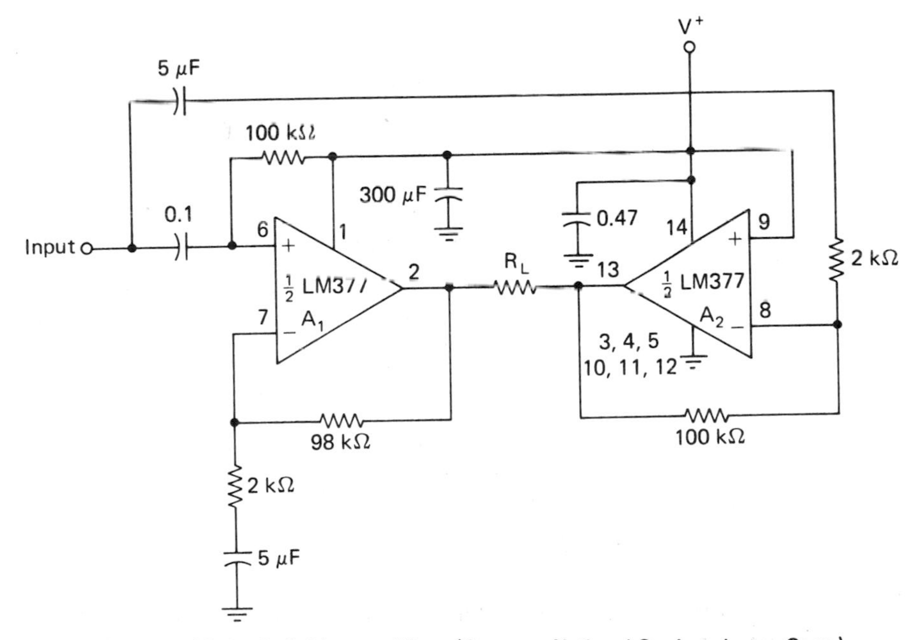

FIGURE 9-17 Audio bridge amplifier. (Courtesy National Semiconductor Corp.)

ing loads, such as loudspeakers or servo motors. The load impedance may be 8 or 16 Ω. Response of this circuit is 20 Hz to 160 kHz, with an output power up to 4 W.

9-2-14 DC Servo Amplifier

Op amps are extremely effective in controlling servo motors, as shown in Figure 9-18. The polarity and amplitude of the input voltage determine the speed and direction of the motor. This circuit has a gain of 10.

9-2-15 AC Servo Amplifier

An AC servo amplifier utilizes two op amps, as shown in Figure 9-19. The noninverting inputs are held at a DC reference voltage, while V_{in} controls the speed and direction of the motor.

9-2-16 Absolute-Value Amplifier

The circuit in Figure 9-20 generates a positive output voltage for either polarity of input. For positive signals, it acts as a noninverting amplifier, and for negative signals, as an inverting amplifier. For the best accuracy, input signals should be greater than 1 V.

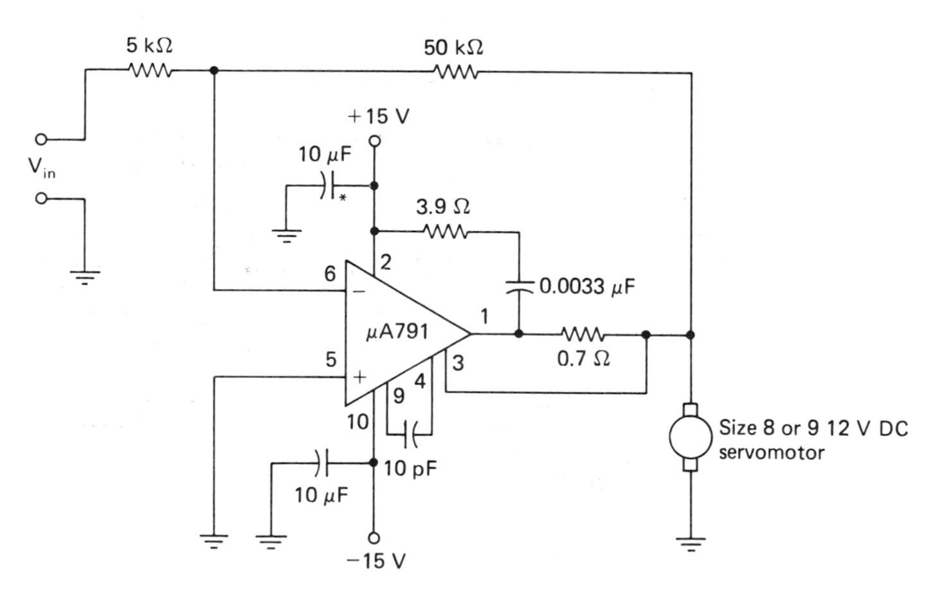

FIGURE 9-18 DC servo amplifier. (Courtesy Fairchild Camera & Instruments Corporation.)

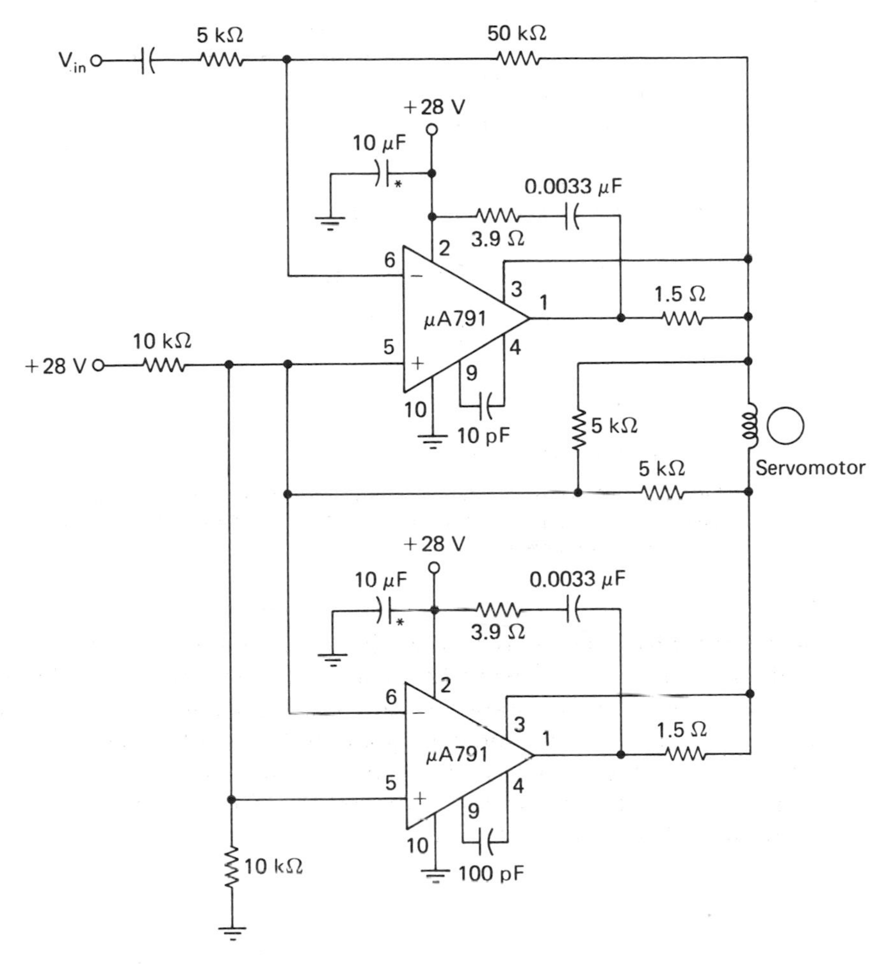

FIGURE 9-19 AC servo amplifier. (Courtesy Fairchild Camera & Instruments Corporation.)

9-3 OSCILLATORS AND WAVEFORM GENERATORS

9-3-1 Phase-Shift Oscillator

The phase-shift oscillator in Figure 9-21 operates on positive feedback applied to the inverting input. The output is shifted about 60° across each resistor–capacitor combination ($3 \times 60^\circ = 180^\circ$)—hence, oscillation.

9-3-2 Easily Tuned Sine-Wave Oscillator

The circuit shown in Figure 9-22 generates a sine wave by filtering a square wave. Two separate outputs are available. A voltage comparator produces

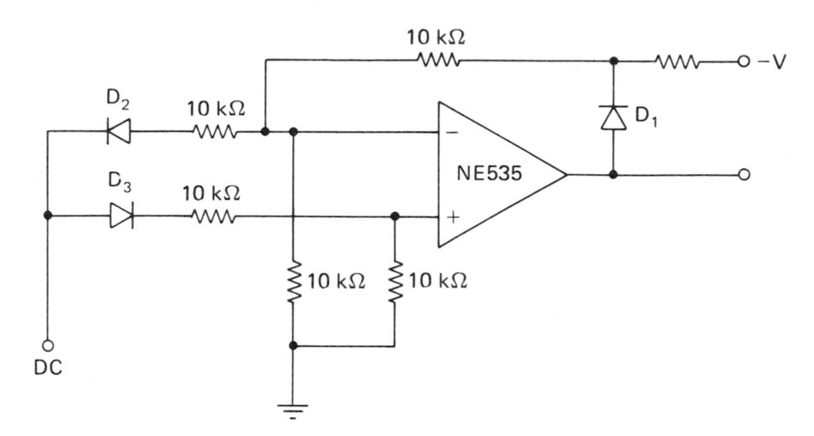

All resistor values are in ohms.

FIGURE 9-20 Absolute valve amplifier. (Permission to reprint granted by Signetics Corporation, a subsidiary of U.S. Philips Corp., 811 E. Arques Avenue, Sunnyvale, CA 94086.)

the square wave, which is fed to the tuned circuit of the filter amplifier, producing only the sine-wave fundamental. The sine wave is fed to the comparator to produce the square wave. A total frequency of below 20 Hz to above 20 kHz can be produced in graduated ranges by changing C_1 and C_2 as shown in the table. Resistor R_3 sets the desired frequency within the range. Resistor R_8 is the amplitude adjustment. The zener diode stabilizes the square wave being fed to the filter.

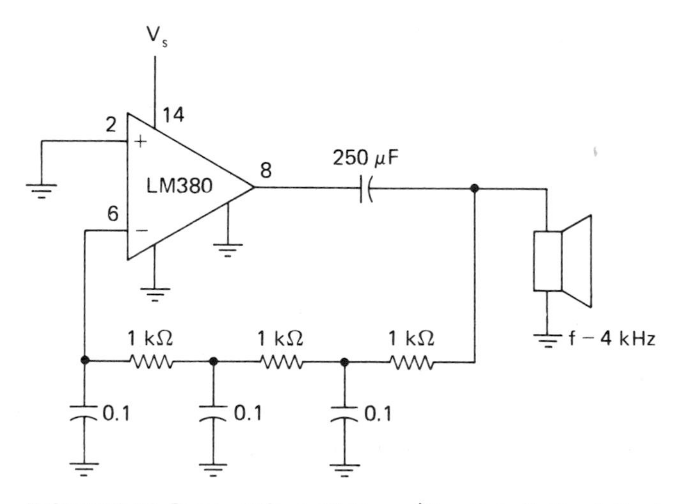

FIGURE 9-21 Phase shift oscillator. (Courtesy National Semiconductor Corp.)

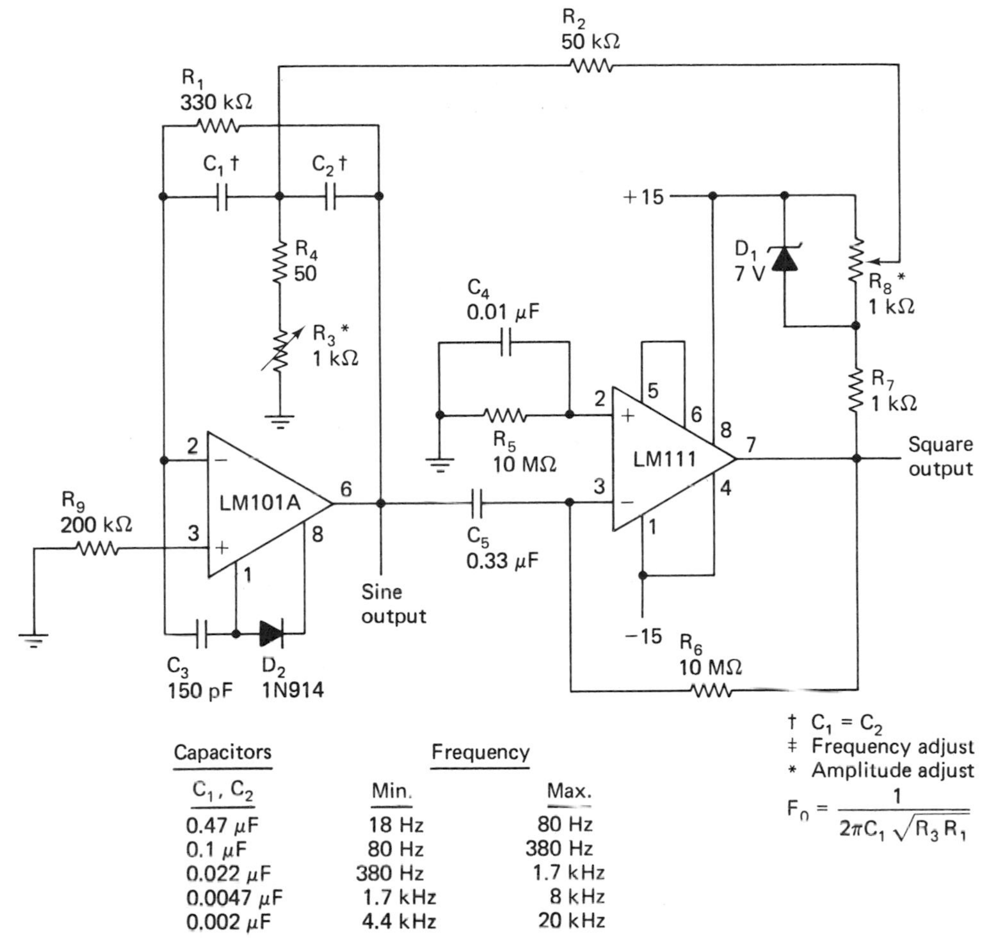

$$F_0 = \frac{1}{2\pi C_1 \sqrt{R_3 R_1}}$$

Capacitors C_1, C_2	Frequency Min.	Frequency Max.
0.47 μF	18 Hz	80 Hz
0.1 μF	80 Hz	380 Hz
0.022 μF	380 Hz	1.7 kHz
0.0047 μF	1.7 kHz	8 kHz
0.002 μF	4.4 kHz	20 kHz

FIGURE 9-22 Easily tuned sine wave oscillator. (Courtesy National Semiconductor Corp.)

9-3-3 Crystal Oscillator

The circuit in Figure 9-23 shows how a crystal oscillator is constructed using an LF111 voltage comparator. Similar to a standard square-wave generator, this circuit exhibits greater stability with positive feedback via the crystal.

9-3-4 Simple Staircase Generator

This circuit in Figure 9-24 is a basic integrator. The incoming pulses charge up the capacitor, which causes the output to go in a negative direction. When the charge on the capacitor reaches the firing potential of the UJT

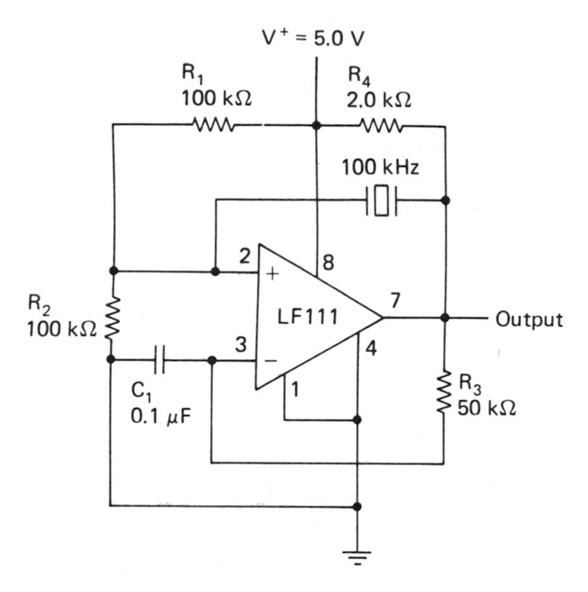

FIGURE 9-23 Crystal oscillator. (Courtesy National Semiconductor Corp.)

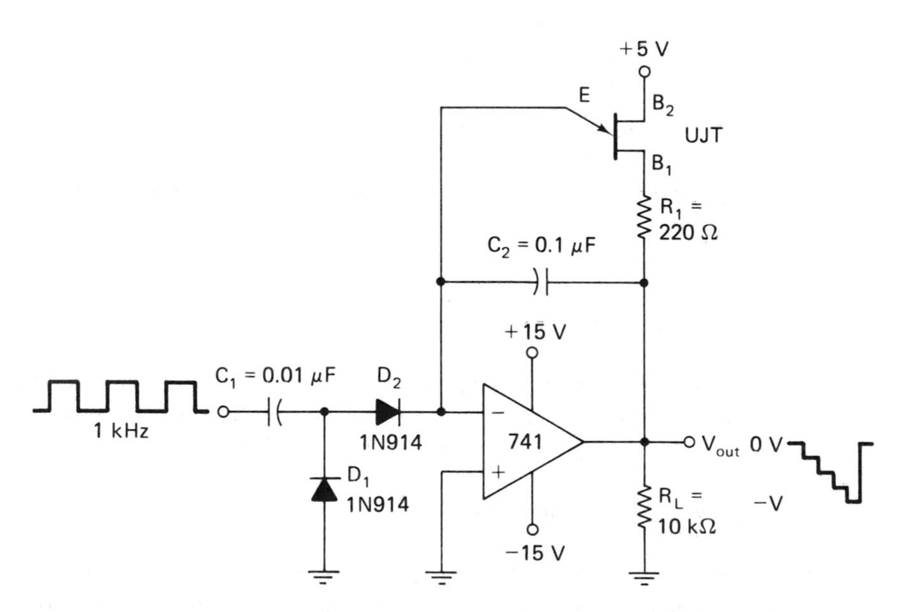

FIGURE 9-24 Simple staircase generator.

emitter, the UJT conducts and discharges the capacitor. The output returns to zero and with the UJT now open, the process starts over.

9-3-5 Free-Running Staircase Generator/Pulse Counter

An improved staircase generator using a quad op amp LM3900 IC is shown in Figure 9-25. This circuit is self-generating and uses one of the op amps as a pulse generator which feeds the noninverting input of a difference inte-

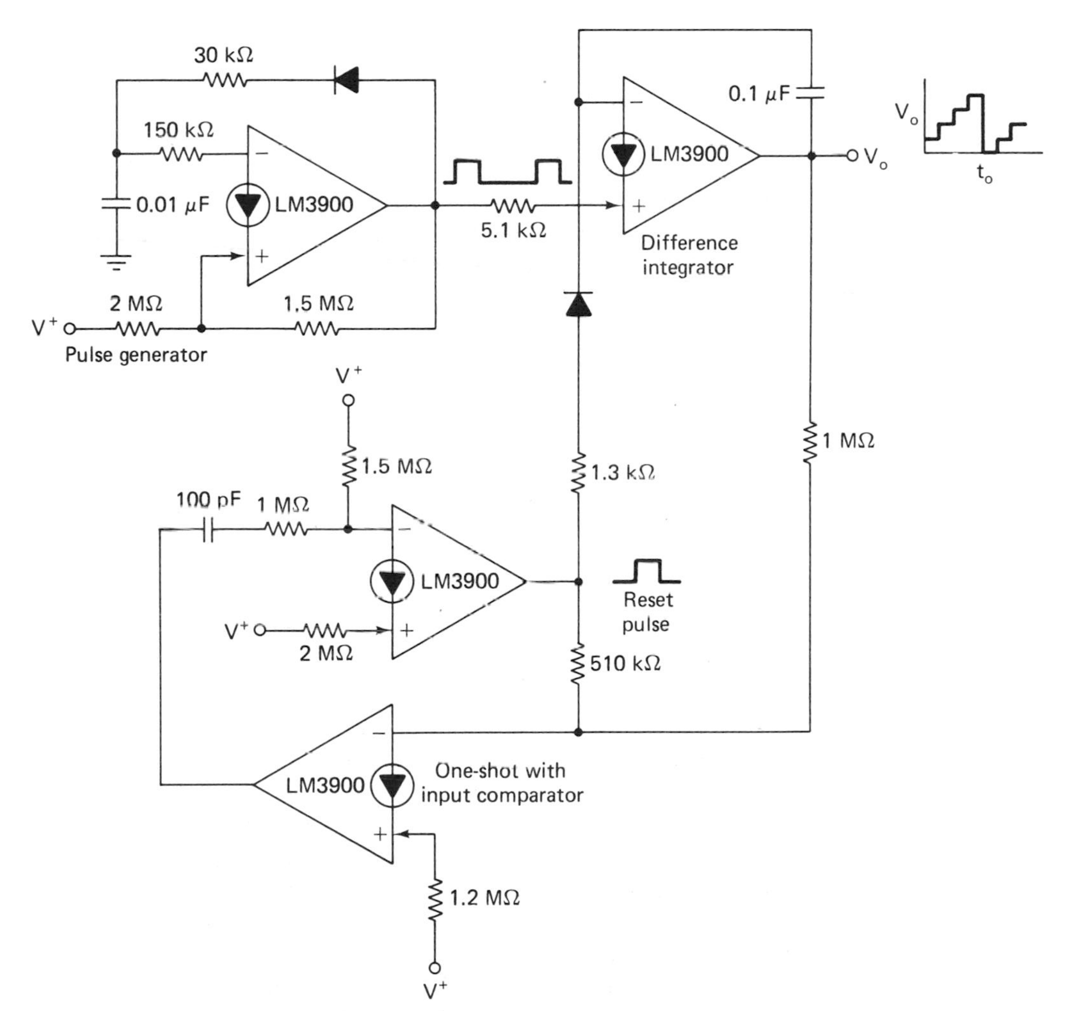

FIGURE 9-25 Free running staircase generator/pulse counter. (Courtesy National Semiconductor Corp.)

grator. The output is positive-going, which feeds a voltage comparator. When the charge on the capacitor fires the comparator, a one-shot multivibrator resets the difference integrator.

9-3-6 Digital-to-Analog Staircase Generator

Figure 9-26 shows a digital-to-analog converter circuit. A 7495 4-bit shift register is wired to produce a sequential counter whose outputs feed into an op-amp summing amplifier. The negative-going staircase output from the summing amplifier is fed to an inverting amplifier, producing a positive-going output with adjustable gain.

9-3-7 Monostable (One-Shot) Multivibrator

A positive-going-output one-shot multivibrator is shown in Figure 9-27a. Resistor R_2 keeps the output at zero in the quiescent state. A differentiated positive input pulse causes the output to switch to the positive voltage state where it is latched by R_5. When capacitor C_1 charges up to about $\frac{1}{4}$ of the $+V$ supply, the circuit latches back to the quiescent state. The diode allows rapid retriggering.

With slight modification the previous circuit can be used as a negative-going-output one-shot multivibrator, as shown in Figure 9-27b. Resistors R_2 and R_3 keep the inverting input essentially at ground, which forces V_{out} to the positive level. A differentiated negative input pulse causes the output to switch to zero. The output will remain low until C_1 discharges to about $\frac{1}{10}$ of the $+V$ supply.

Moving the R_4C_2 network to the inverting input will allow the circuit to operate with a differentiated positive input pulse.

9-3-8 Schmitt Trigger

Op amps wired as voltage-level detectors perform exceptionally well as Schmitt triggers, shown in Figure 9-28. The inverting circuit in Figure 9-28a has a normally high output. When the voltage on the inverting input reaches the high trip point, the output goes to zero. When the input voltage is reduced to the low trip point, the output again goes high. The difference between these two input voltages is the circuit hysteresis. Resistors R_F and R_B, together with the $+V$ supply, determine the high and low trip-point voltages.

The noninverting circuit shown in Figure 9-28b is similar except that the output is normally low. The input voltage is fed to the noninverting input and the hysteresis is somewhat greater.

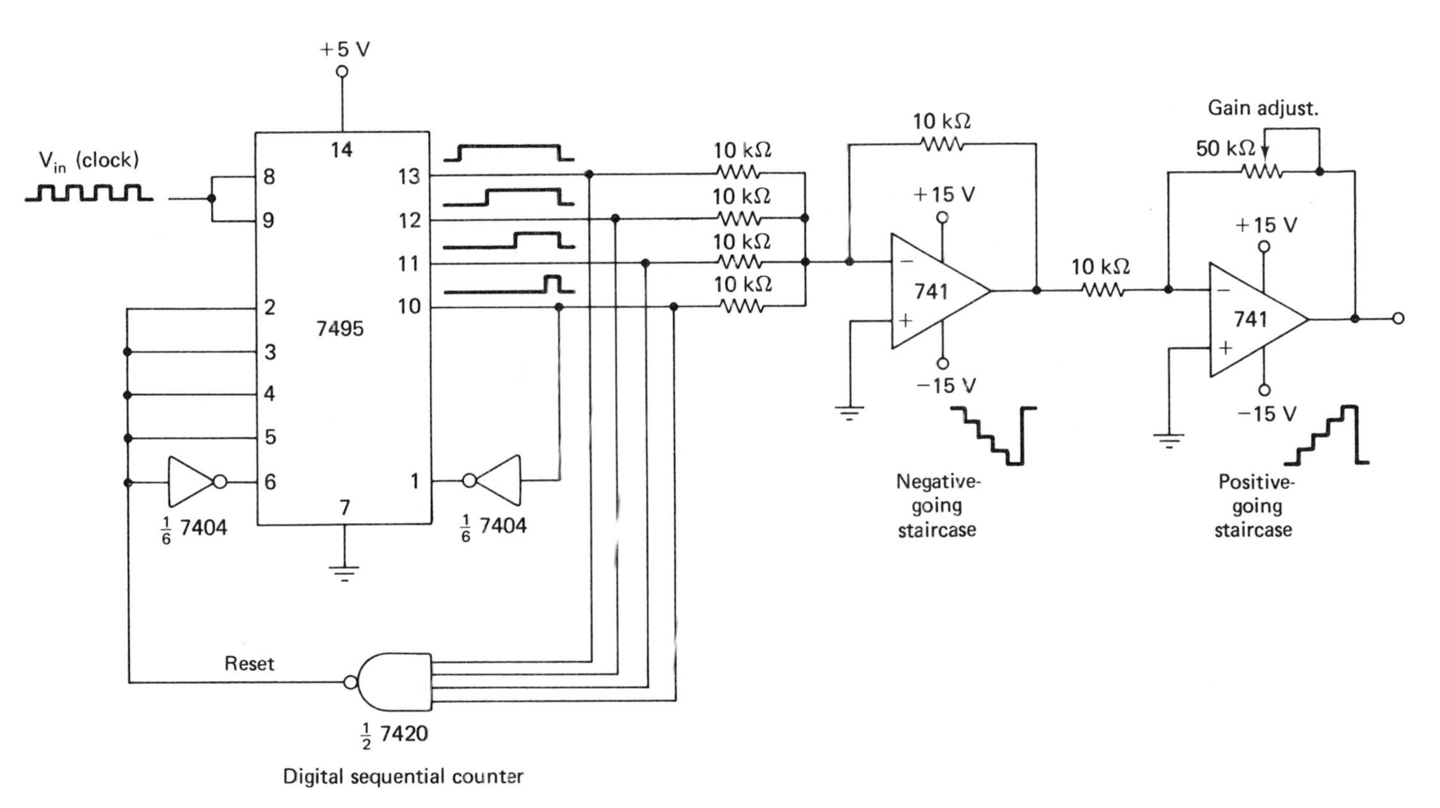

FIGURE 9-26 Digital-to-analog staircase generator.

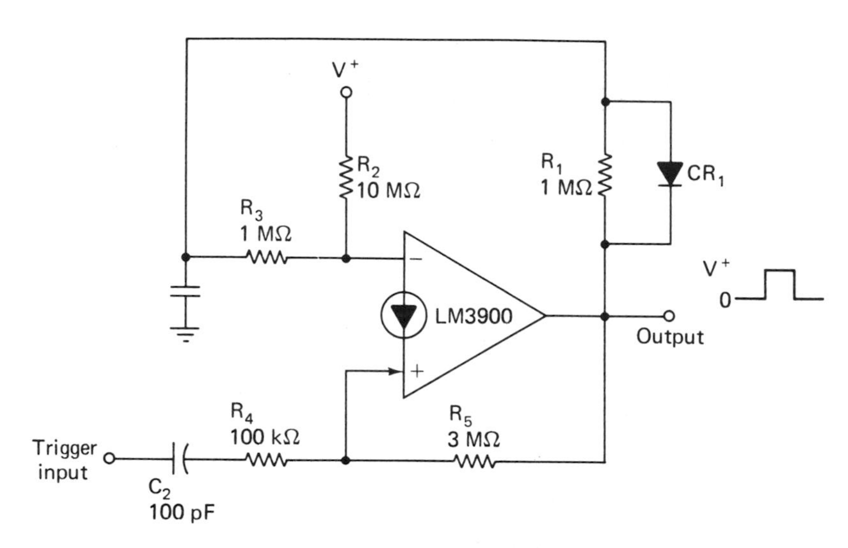

(a) Positive output pulse

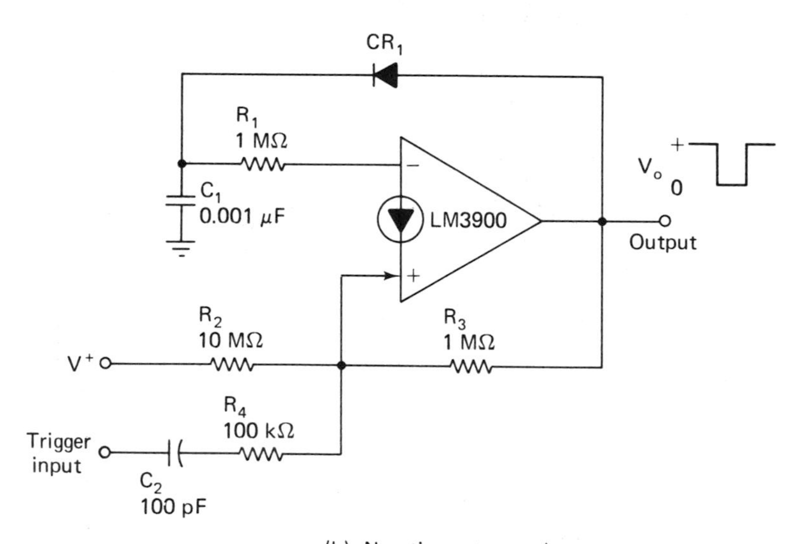

(b) Negative output pulse

FIGURE 9-27 Monostable (one-shot) multivibrator. (Courtesy National Semiconductor Corp.)

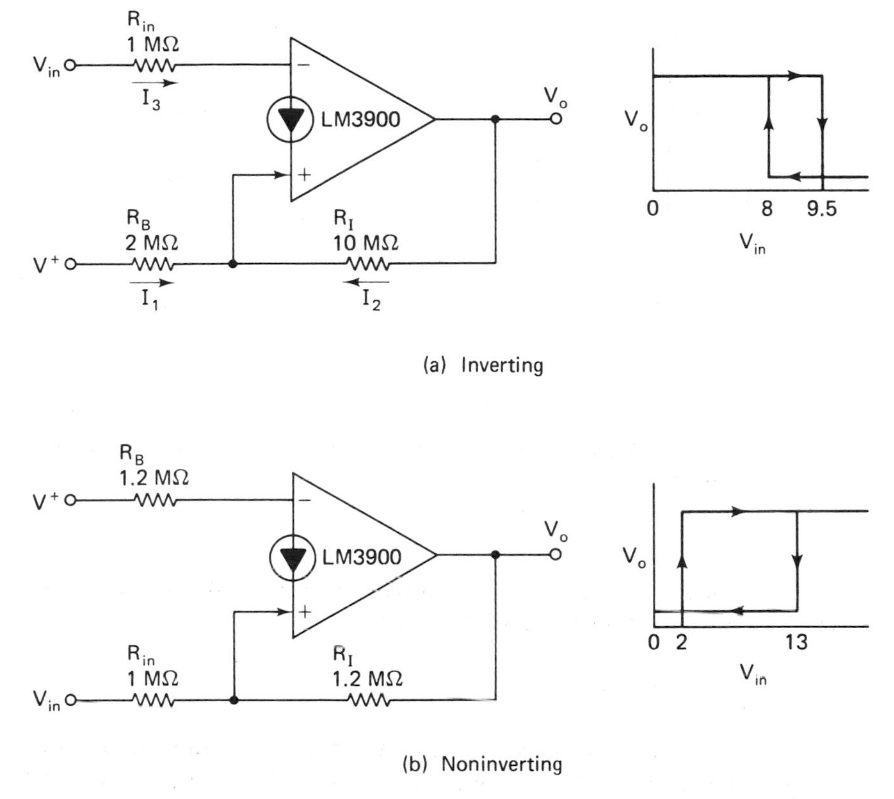

FIGURE 9-28 Schmitt trigger. (Courtesy National Semiconductor Corp.)

9-3-9 Programmable Unijunction Oscillator

If a diode and an *RC* charging circuit are added to the Schmitt trigger, a programmable unijunction oscillator can be produced, as shown in Figure 9-29. The output is normally high, and when the input voltage reaches the high trip point, the output falls to about zero and capacitor *C* discharges through the diode. The low trip point must be larger than +1 V to guarantee that the V_F of the diode plus the output voltage is less than the low trip-point voltage. Resistor R_2 can be made smaller to increase the discharge current.

9-3-10 Frequency Doubler

The simple frequency doubler shown in Figure 9-30 is very similar to the absolute-value amplifier of Figure 9-20. A sine-wave signal input results in a full-wave rectified output at twice the frequency of the input signal. Other

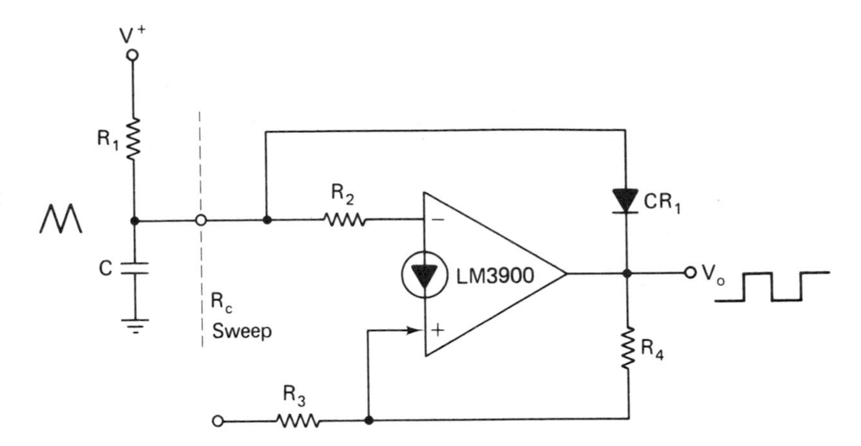

FIGURE 9-29 Programmable unijunction oscillator. (Courtesy National Semiconductor Corp.)

waveform-shaping circuits can follow this circuit to restore the pure sine wave or create other desired waveforms.

9-3-11 Pulse Generator

The pulse generator shown in Figure 9-31 is similar to a square-wave generator except that the output goes from zero to a positive level instead of from positive to negative. Frequency of the pulse generator is primarily determined by the capacitor, R_1, and the V_{ref}.

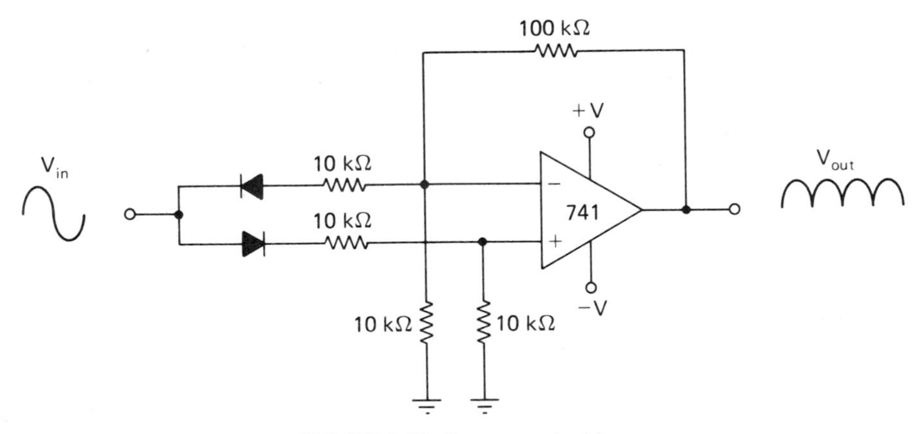

FIGURE 9-30 Frequency doubler.

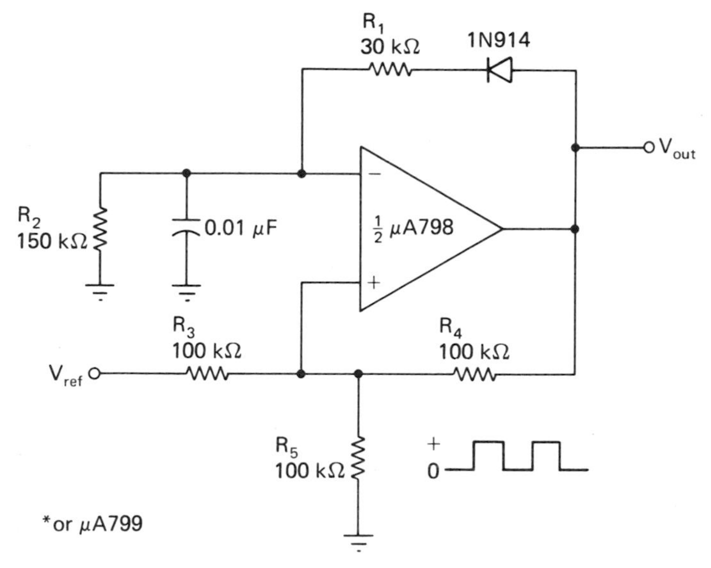

FIGURE 9-31 Pulse generator. (Courtesy Fairchild Camera & Instruments Corporation.)

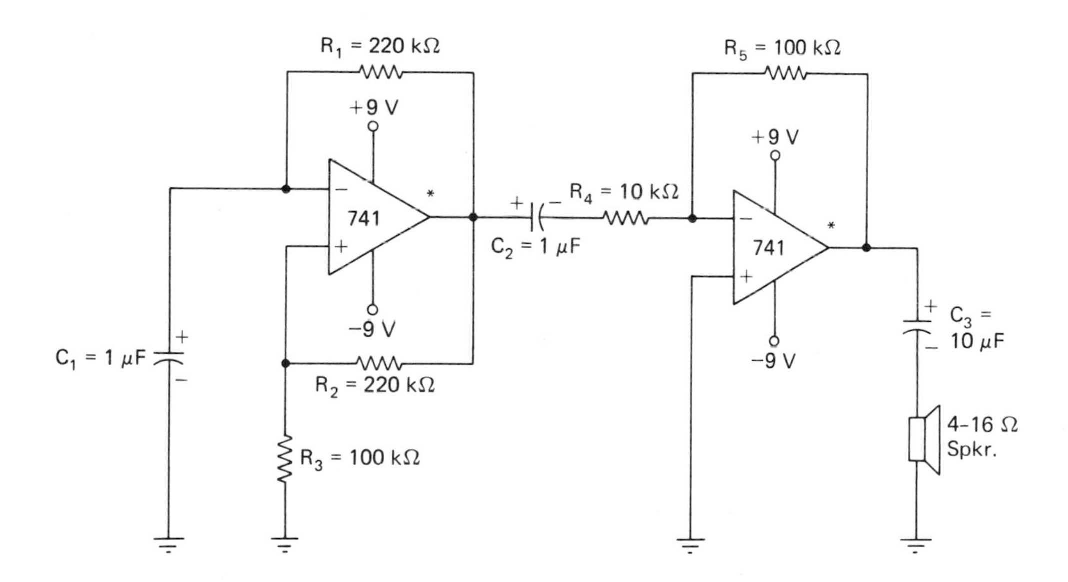

FIGURE 9-32 Two-tone alarm circuit.

9-3-12 Two-Tone Alarm Circuit

Given in Figure 9-32 is a novel but very effective alarm circuit. Two tones are heard alternately at the speaker. Varying C_2 changes the pitch of the two tones, while changing C_1 will change the switching time. Musically speaking, the two tones approximate tonic and minor third notes.

9-4 SIMPLE TEST INSTRUMENTS

9-4-1 Sensitive Low-Cost DC Voltmeter

The voltmeter shown in Figure 9-33 has extremely high input impedance. Using the LF536 op amp as a noninverting amplifier, the input impedance for ranges up to 10 V is 5000 MΩ. The 30-V range has an input impedance of 30 MΩ, while the 100-V range has 100 MΩ. The diodes protect the input against overvoltage; however, the meter cannot withstand more than a 50% overload.

9-4-2 Wide-Band AC Voltmeter

The circuit shown in Figure 9-34 is a wide-band ac voltmeter capable of measuring AC signals as low as 15 mV up to 5 V at frequencies from 100 Hz to 500 kHz. Altering the values of resistors R_1 through R_6 changes the full-scale sensitivity of the meter ($R \cong V_{in}/100\ \mu A$).

9-4-3 Triple-Range Ohmmeter

The ohmmeter shown in Figure 9-35 uses a linear scale, needs no calibration, and is insensitive to power-supply voltage. Resistances can be measured from 0 to 100 kΩ in three ranges. Like a standard voltmeter, resistance values are measured from a full-scale standpoint. A germanium diode protects the meter from overcurrent when the test points are left open.

9-4-4 Audio Circuit Tester

Shown in Figure 9-36 is a simple audio signal injector and signal tracer circuit that can be constructed in a single unit. The signal injector is a square-wave generator with a frequency selection ranging from 50 Hz to 20 kHz in four steps. An output adjustment can be set for the desired amplitude of the injected signal. The frequency ranges may be used to check the frequency response of a circuit. The signal tracer is a standard inverting amplifier with adjustable gain.

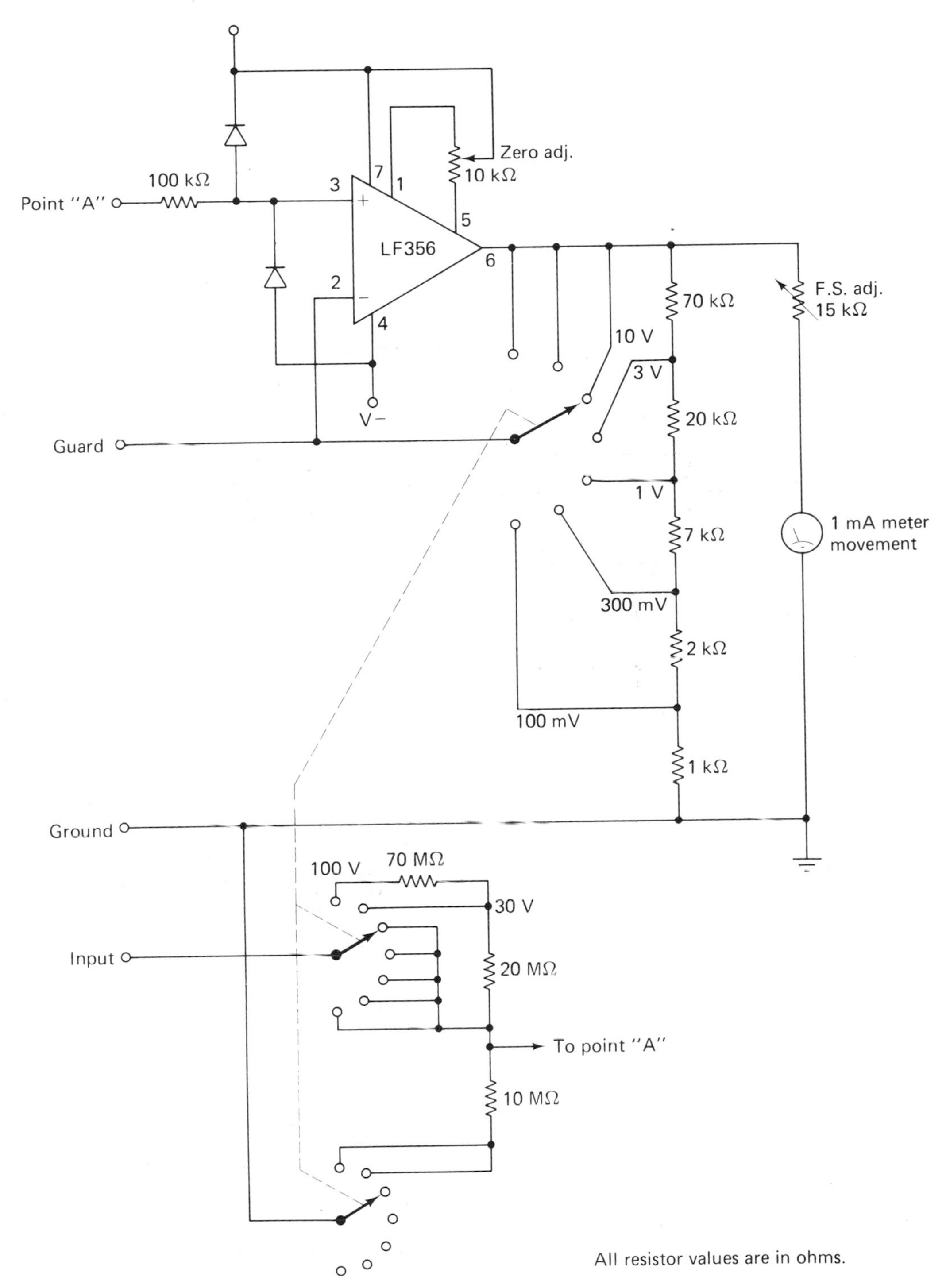

FIGURE 9-33 Sensitive low cost DC voltmeter. (Permission to reprint granted by Signetics Corporation, a subsidiary of U.S. Philips Corp., 811 E. Arques Avenue, Sunnyvale, CA 94086.)

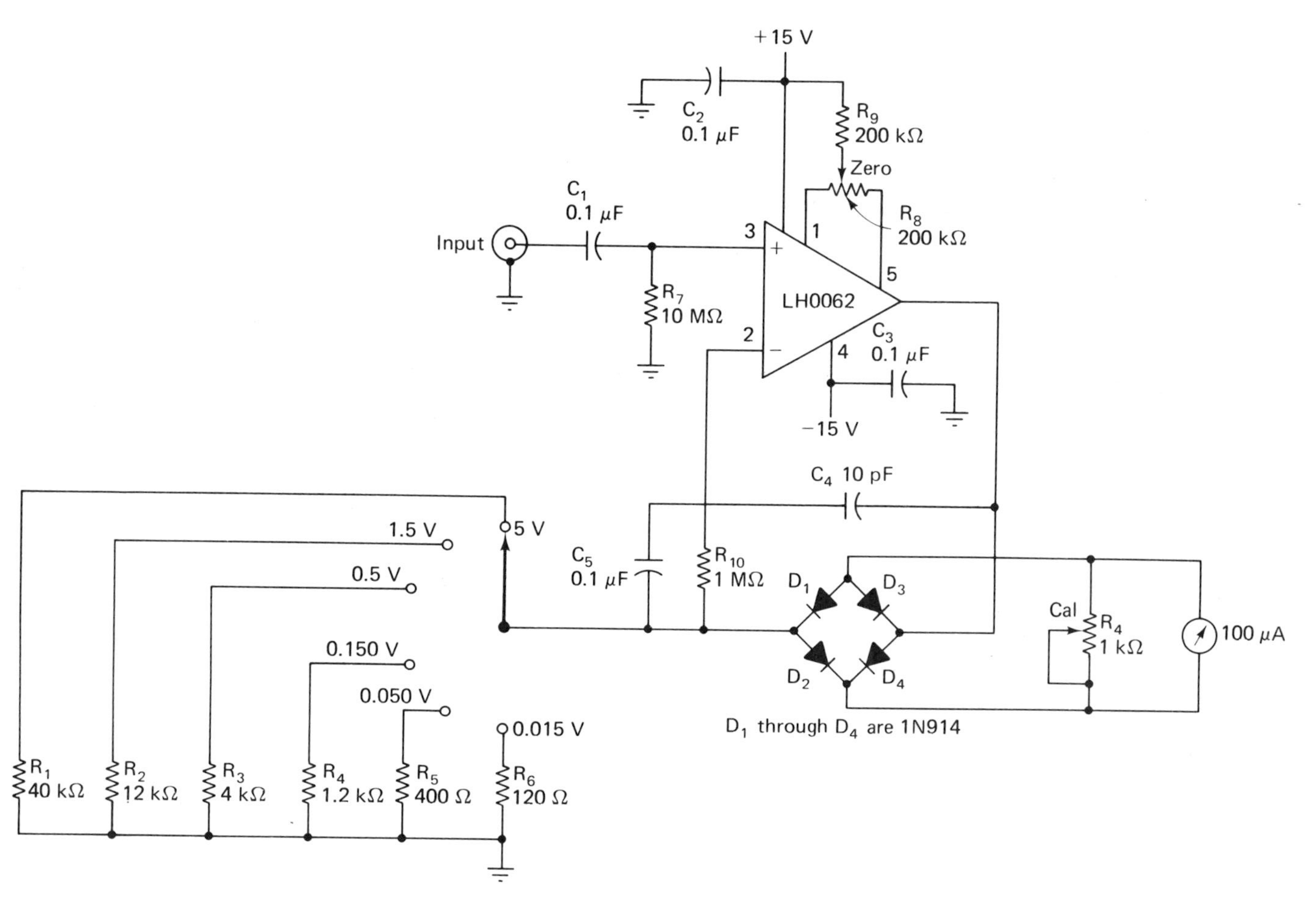

FIGURE 9-34 Wide band AC voltmeter. (Courtesy National Semiconductor Corp.)

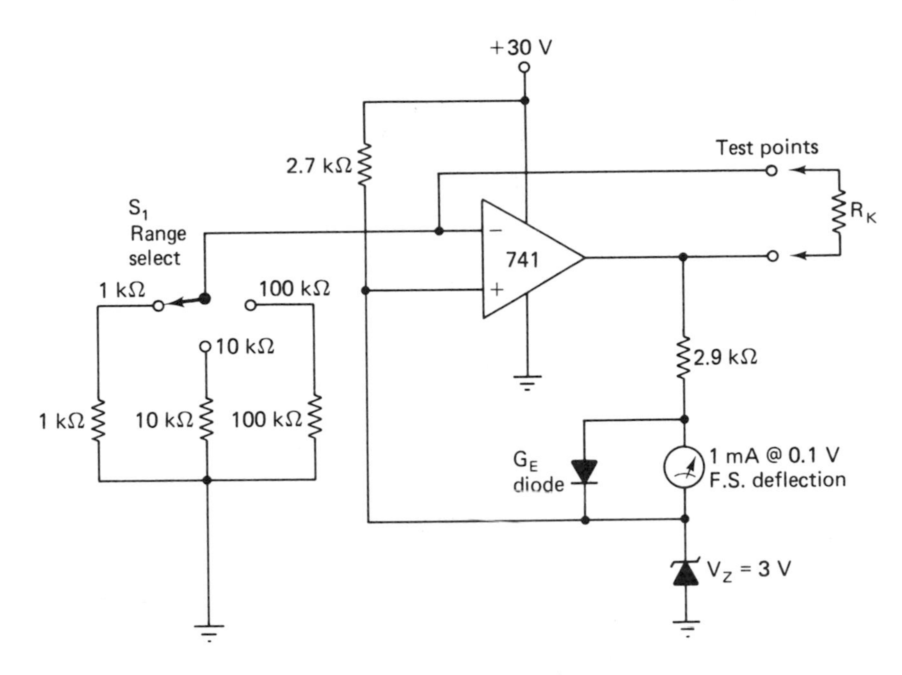

FIGURE 9-35 Triple range ohmmeter.

The injector is set for the desired frequency and amplitude, and the signal is then entered into the circuit via the signal injector probe. The signal tracer probe is then used to trace the signal through the circuit under test at various points to locate the faulty section or stage.

9-5 LOGIC CIRCUITS

9-5-1 AND Gate

The three-input AND gate shown in Figure 9-37 will give a high or 1 output ($\approx$+15 V) when all of the inputs are high or 1 (+15 V). Resistors R_1 and R_2 set the +375-mV reference at the inverting input. In the "off state" this voltage forces the output to zero. When all inputs are high, sufficient current is drawn through R_6 to develop a voltage drop at the noninverting input that is more positive than the reference voltage. The output will now be in the "on state" or high potential.

9-5-2 OR Gate

The OR-gate circuit in Figure 9-38 is identical to the AND gate in Figure 9-37 except that R_1 has been increased to set the reference voltage at the

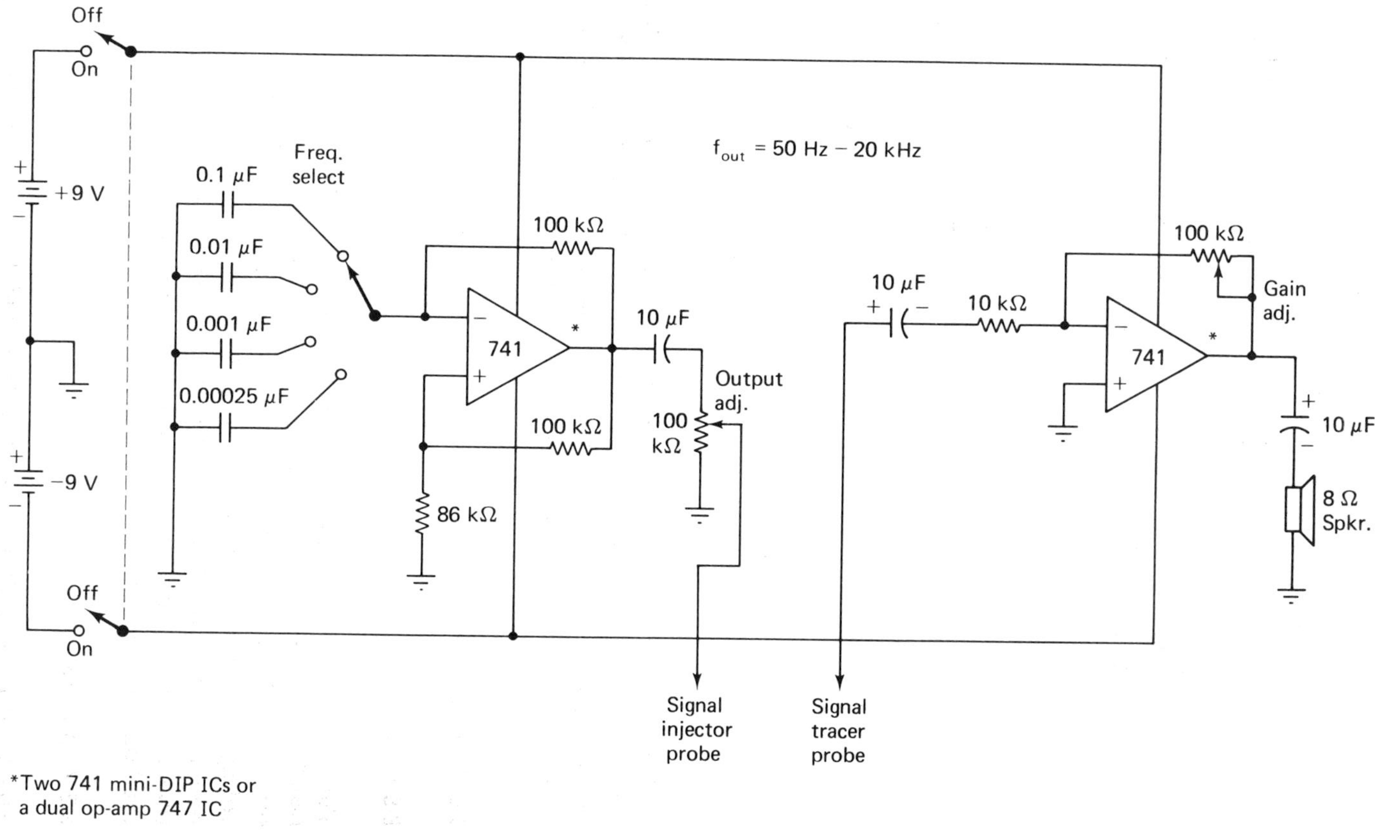

*Two 741 mini-DIP ICs or
a dual op-amp 747 IC

FIGURE 9-36 Audio circuit tester.

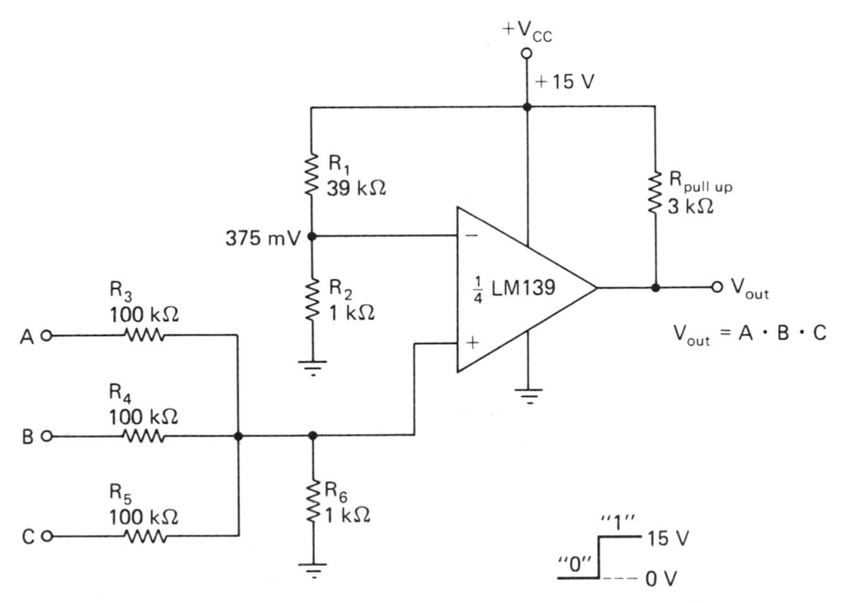

FIGURE 9-37 AND gate. (Courtesy National Semiconductor Corp.)

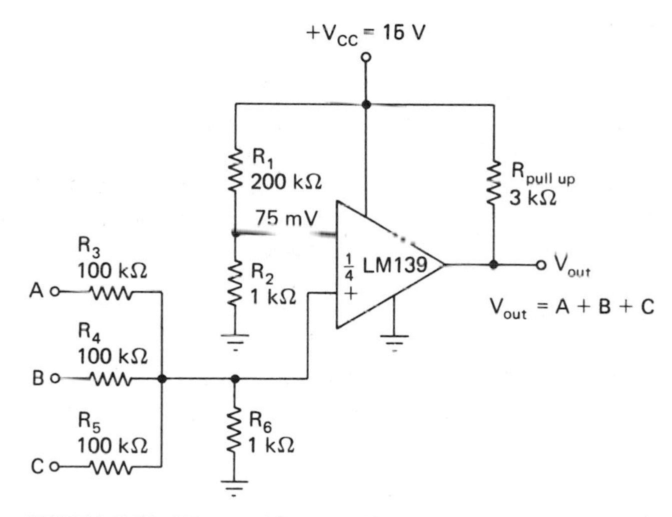

FIGURE 9-38 OR gate. (Courtesy National Semiconductor Corp.)

inverting input to 75 mV. The output will go high ($\approx$+15 V) when any one of the inputs goes high (+15 V). A 1 at any input will develop enough voltage drop across R_6 ($\approx$150 MV) to cause the op amp to switch states.

9-5-3 NAND Gate

Unlike the op-amp AND gate, which uses an active noninverting input, the NAND gate uses the inverting input, as shown in Figure 9-39. Any low

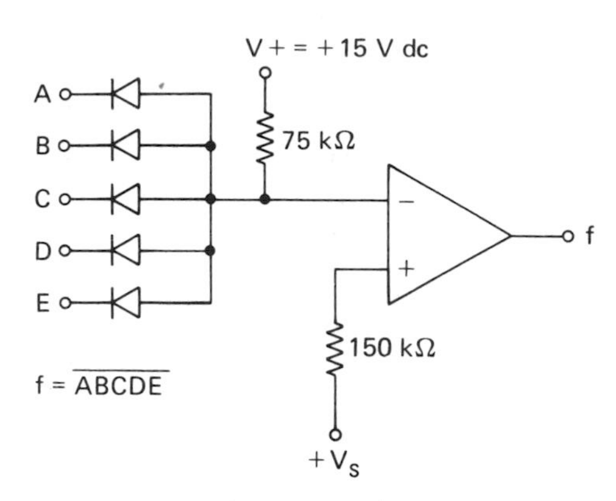

FIGURE 9-39 NAND gate. (Courtesy Fairchild Camera & Instruments Corporation.)

input (0 V) will cause the output to go high (≈+15 V). When all diode inputs are high, the inverting input is pulled high and the output is low. When any one diode input is low, the inverting input is pulled low and the op-amp output goes high.

9-5-4 NOR Gate

Like the NAND gate in Figure 9-39, the NOR gate shown in Figure 9-40 uses an active inverting input. The output will go high (≈+15 V) only if all inputs are low (0 V). If any input is high (+15 V), there will be sufficient positive voltage at the inverting input to cause the output to remain low. When all resistor inputs are low, the inverting input will be low, causing the op amp to switch to the high state.

9-5-5 RS Flip-Flop

The RS flip-flop shown in Figure 9-41 is considered "off" when Q output is low (0 V) and its complementary $\overline{Q}$ output is high (+15 V). The low out-

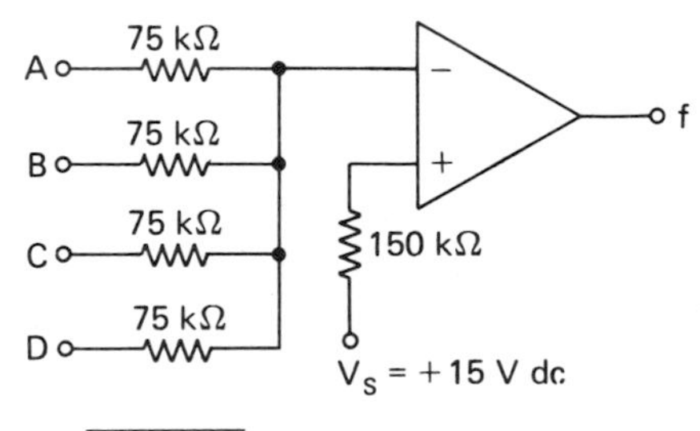

FIGURE 9-40 NOR gate. (Courtesy Fairchild Camera & Instruments Corporation.)

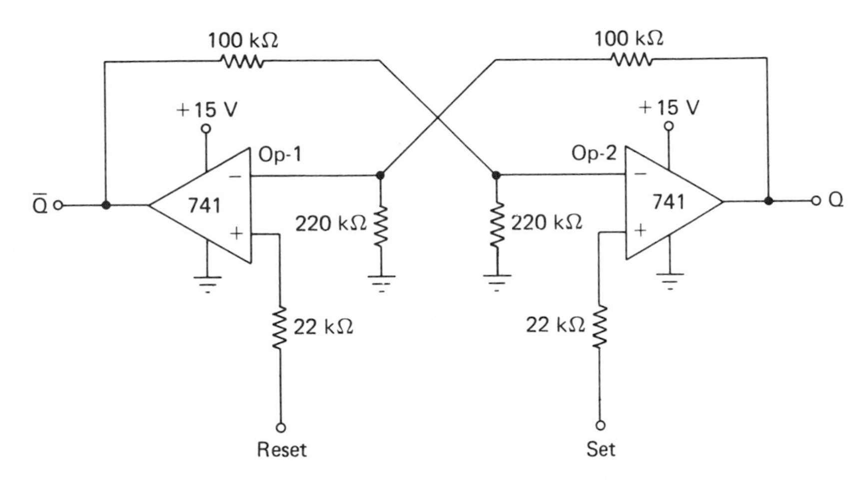

FIGURE 9-41 RS flip-flop.

put of Q is fed back to the inverting input of OP-1, which keeps $\overline{Q}$ output high. Similarly, the high output of $\overline{Q}$ is fed back to the inverting input of OP-2, which keeps Q output low. When a positive-going pulse is placed at the set input, Q output goes high and $\overline{Q}$ output goes low. The circuit will remain in this "on" state due to the latching action previously mentioned. A positive-going pulse on the reset input will force the flip-flop to the original "off" state. This logic circuit can be used as a temporary memory device.

9-6 MISCELLANEOUS CIRCUITS

9-6-1 Feedforward Frequency Compensation

Op amps without internal frequency compensation can be modified with a simple feed forward network, as shown in Figure 9-42, which will increase the slew rate and bandwidth of a particular circuit. Feedforward frequency compensation is achieved by connecting C_1 from the input to one of the compensation terminals. High frequencies are bypassed around the initial stages of the op amp, thereby increasing the bandwidth. Capacitor C_2 is used for stability in the feedback loop. The diode may be added to increase slew rate with fast-rising inputs. This added high-frequency gain will also amplify high-frequency noise, and frequency-compensation techniques should only be applied to the extent required by the circuit.

9-6-2 One-IC Intercom

The LM 380N-8 (mini-DIP) IC can be used to construct a simple but effective intercom system, as shown in Figure 9-43. Both speakers are used as

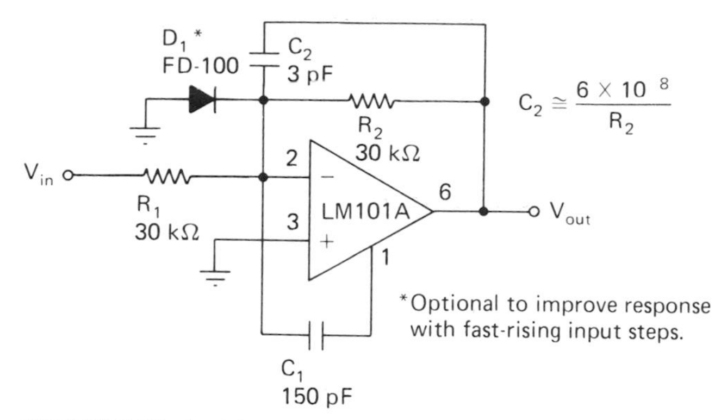

FIGURE 9-42 Feed-forward frequency compensation. (Courtesy National Semiconductor Corp.)

input (microphone) and output (speaker) transducers. The listen/talk switches are momentary (spring-return) types wired in the normal position, as shown. Operate only one switch at a time to prevent feedback oscillations. The audio output transformer steps up the speaker signal (when in the talk mode) to provide sufficient voltage to drive the amplifier. Resistor R_1 serves as a volume control. Capacitive filtering may be needed to prevent oscillation from occurring, depending on the distance of the remote speaker.

9-6-3 Simulated Inductor (Gyrator)

An op amp can be used to simulate an inductor (sometimes referred to as a gyrator), as shown in Figure 9-44. This type of circuit may be used to eliminate inductors from filters and tuned circuits. Inductance is characterized by an increase in output when frequency increases, with the effective inductance being equal to $L \approx R_1 R_2 C_1$.

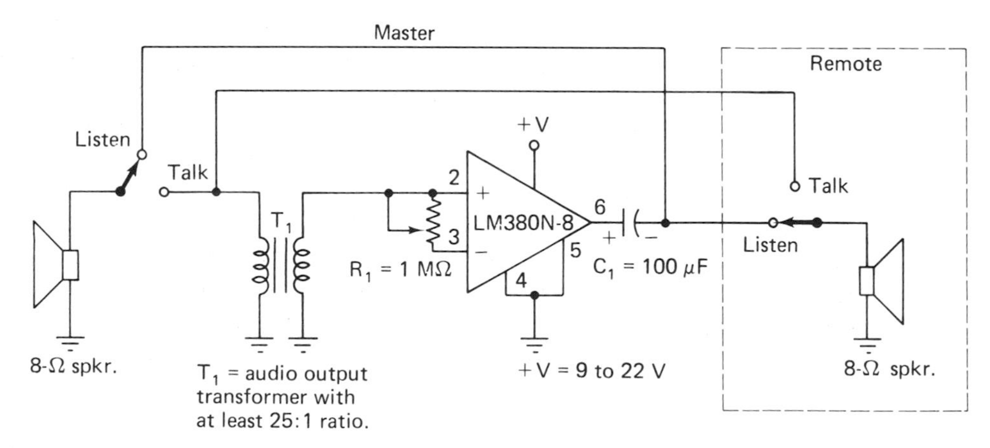

FIGURE 9-43 One IC intercom.

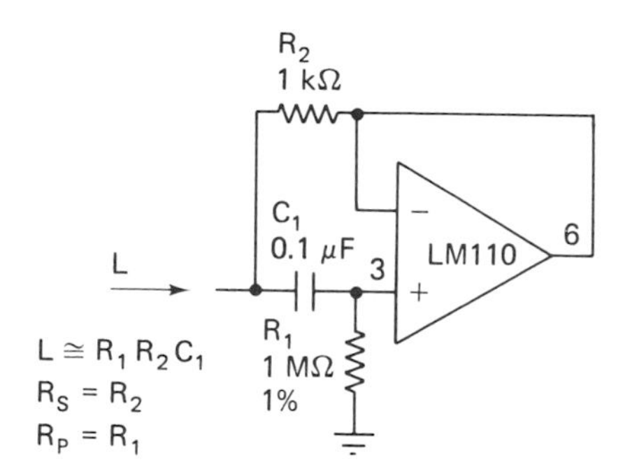

FIGURE 9-44 Simulated inductor (gyrator). (Courtesy National Semiconductor Corp.)

9-6-4 Op-Amp Tachometer

The basic op-amp tachometer shown in Figure 9-45a converts the input pulses to an average DC output via the *RC* averaging network. The output increases at a linear rate with an increase of input frequency. The resistor provides a discharge path for the capacitor to limit the integrating action; therefore, the output voltage is directly proportional to the input frequency. A meter calibrated for RPM may be connected to the op-amp output.

A frequency-doubling tachometer, as shown in Figure 9-45b, reduces the ripple on the DC output voltage, which would result in a more precise meter indication. The operation of the circuit is to average the charge and discharge transient currents of the input capacitor, C_{in}. Resistor R_{in} converts the voltage pulses to current pulses and limits surge currents. Two current pulses are drawn from the *RC* averaging network for each cycle of the input frequency.

9-6-5 Low-Frequency Mixer

A frequency mixer that allows two input frequencies to produce a sum and a difference frequency is shown in Figure 9-46. Because a diode exists at the noninverting input of the LM3900, this op amp can be used for nonlinear signal processing. Filtering is accomplished by the 1-MΩ and 150-pF feedback elements. The circuit has a gain of 10 with a corner frequency of 1 kHz. A signal with a larger amplitude can be placed at V_1 to serve as the local oscillator. The input diode is gated at this frequency (f_1). A smaller signal of a different frequency (f_2) is placed at V_2. The difference frequency ($f_2 - f_1$) is filtered from the resulting composite waveform and is present at the output. Relatively high frequencies can be applied at the inputs as long as the desired difference frequency is within the bandwidth capabilities of the amplifier and the *RC* low-pass filter.

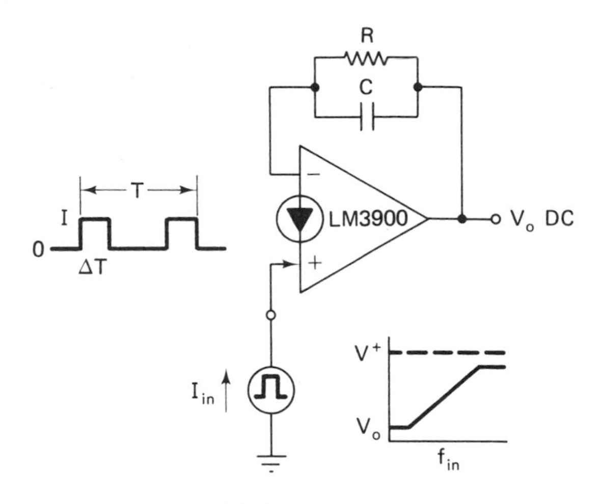

(a) Basic tachometer

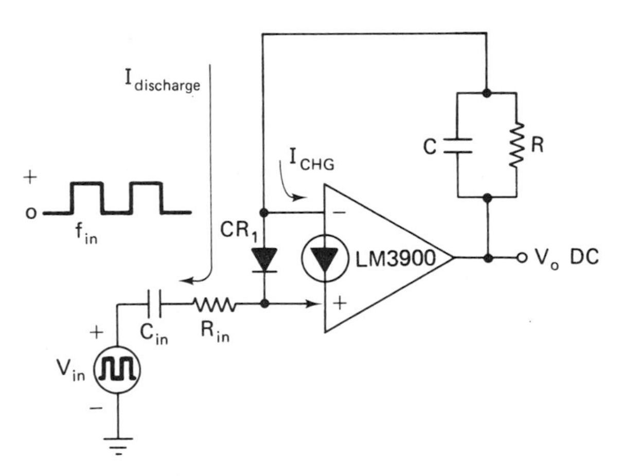

(b) Frequency-doubling tachometer

FIGURE 9-45 Op amp tachometer: (a) basic tachometer; (b) frequency doubling tachometer. (Courtesy National Semiconductor Corp.)

9-6-6 Window Voltage Detector

The window voltage detector shown in Figure 9-47 monitors an input voltage and indicates when this voltage goes either above or below the desired limits. The upper limit voltage (V_{UL}) is +5 V, and the lower limit voltage (V_{LL}) is +4 V. The input voltage (V_{in}) should be looking through a window whose limits are +4 V and +5 V. If V_{in} exceeds V_{UL}, the upper op amp's output swings negative and the LED lights. If V_{in} drops below V_{LL}, the

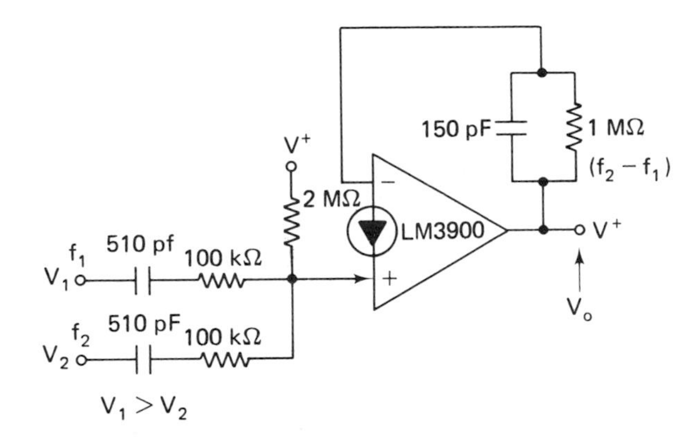

FIGURE 9-46 Low frequency mixer. (Courtesy National Semiconductor Corp.)

lower op amp's output swings negative and the LED lights. Window voltage detectors can be connected together to give multiple limit indications.

9-6-7 Sample-and-Hold Circuit

A basic sample-and-hold circuit is shown in Figure 9-48a. The op amp is in the noninverting follower configuration. When switch S_1 is closed, capacitor

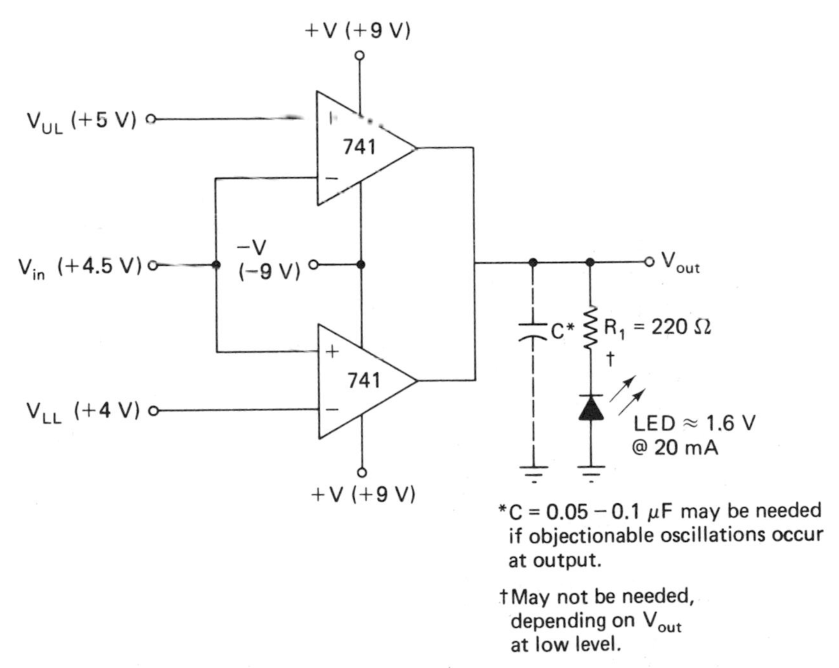

FIGURE 9-47 Window voltage detector.

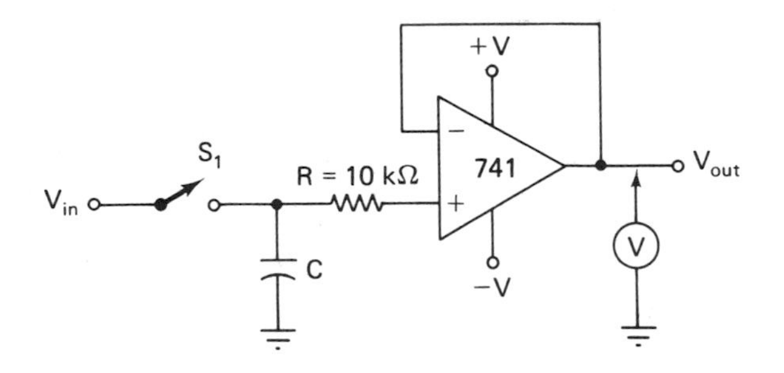

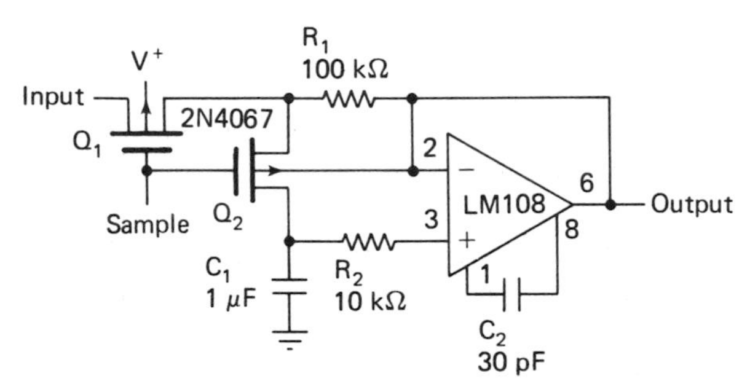

†Teflon, polyethylene, or polycarbonate dielectric capacitor

Worst-case drift less than 3 mV/s

FIGURE 9-48 Sample and hold circuit: (a) basic circuit; (b) improved circuit that eliminates leakages. (Courtesy National Semiconductor Corp.)

C charges to V_{in} max. After S_1 is opened, C remains charged and the output will be at the same potential; therefore, the circuit has sampled a voltage and is temporarily holding it.

However, leakages do occur, causing errors at the output, and this basic circuit is unable to sample rapidly changing transient voltages. An improved circuit using FET switches, which overcome the problems of the basic circuit, is shown in Figure 9-48b. Certain types of capacitors, such as paper and Mylar, exhibit a polarization phenomenon which causes the sampled voltage to drop off by about 50 mV and then stabilize when exercised over a 5-V range during the sample interval. Using the types of capacitors listed reduces this problem.

9-6-8 Bi-Quad Active Bandpass Filter

The bi-quad bandpass filter shown in Figure 9-49 is highly selective, with a center frequency of 1 kHz, a Q of 50, and a voltage gain of 100. However,

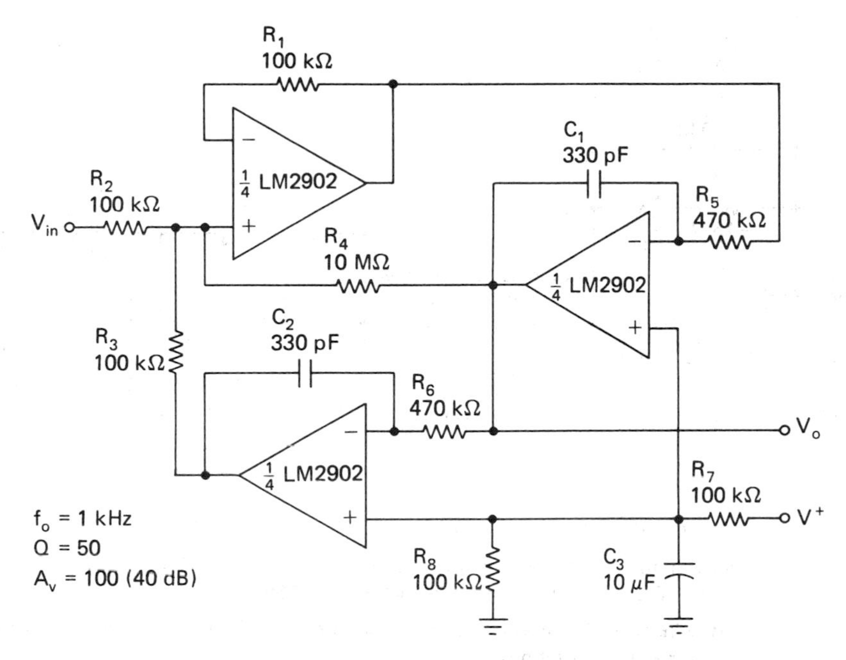

FIGURE 9-49 Bi-quad active bandpass filter. (Courtesy National Semiconductor Corp.)

this circuit is also called a state-variable filter and can simultaneously provide high-pass, low-pass, and bandpass outputs. The high-pass output can be taken from the output of the top op amp, the low-pass output from the output of the bottom op amp, and the bandpass output from the middle op amp, as shown. This circuit is easily tunable when $R_1 = R_3$, $R_5 = R_6$, and $C_1 = C_2$. Its center frequency is then determined as

$$f_c = \frac{1}{2\pi R_5 C_1}$$

9-6-9 Compressor/Expander Amplifiers

The circuit shown in Figure 9-50 has two functions. The compressor amplifier compresses high-amplitude input signals to prevent following circuits from clipping and creating other types of distortion. The resistor/diode networks, in parallel with the feedback resistor R_2, conduct on high peaks of the input signal, thereby reducing the gain of the amplifier.

The expander, being the counterpart of the compressor, receives the compressed signal (perhaps from a transmission line) and extends the amplitude to its full-amplitude range. In this case the resistor diode networks are placed in parallel with the input resistor. When the input signal reaches the

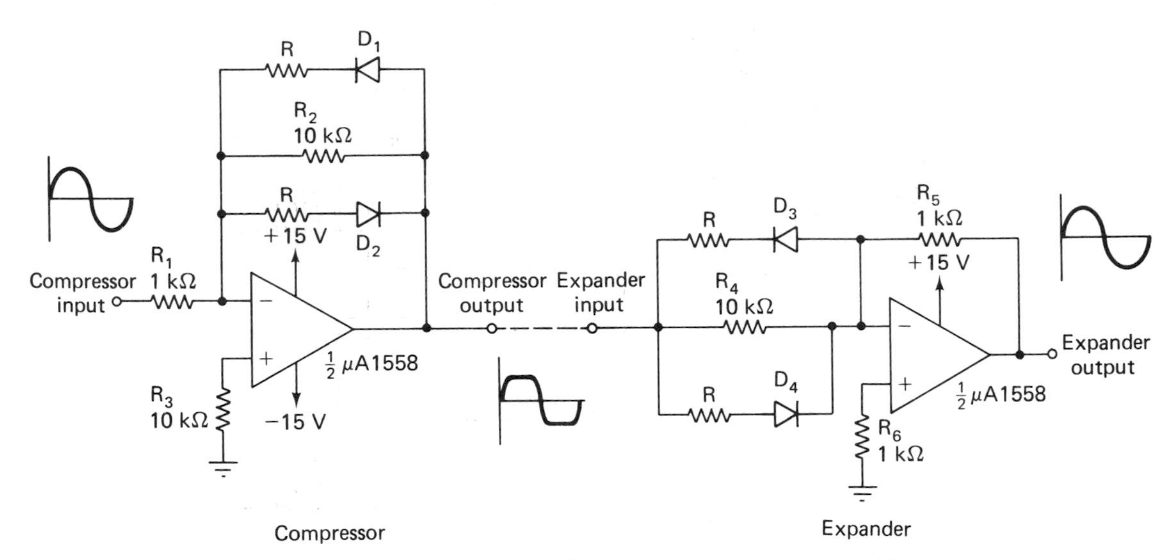

FIGURE 9-50 Compressor/expander amplifiers. (Courtesy Fairchild Camera & Instruments Corporation.)

compressed level, these networks conduct, thereby increasing the gain of the amplifier.

9-6-10 Log Generator

By combining the characteristics of bipolar transistors and op amps, circuits can be constructed to generate logarithmic voltage outputs. The circuit shown in Figure 9-51 generates a logarithmic output voltage for a linear input current. This log (logarithmic) generator is a low-level circuit capable of handling input currents from 10 nA to 1.0 mA, with a dynamic range of 5 decades or 100 dB at an accuracy of 3.0%. With the values given, the scale factor is 1 V/decade and

$$E_{out} = -\left[\log_{10}\left(\frac{E_{in}}{R_{in}}\right) + 5\right]$$

9-6-11 Antilog Generator

Slight modification of the log generator shown in Figure 9-51 will produce the antilog generator shown in Figure 9-52. A linear input current to this circuit produces an exponential (antilog) voltage output. With the values given

$$E_{out} = 10^{-(E_{in})}$$

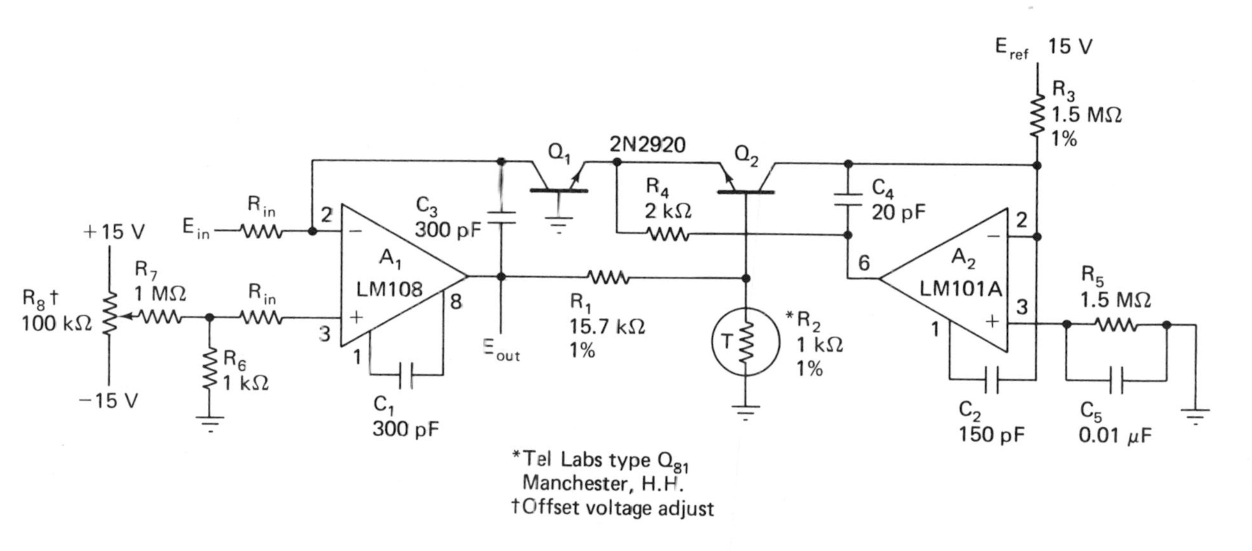

FIGURE 9-51 Log generator. (Courtesy National Semiconductor Corp.)

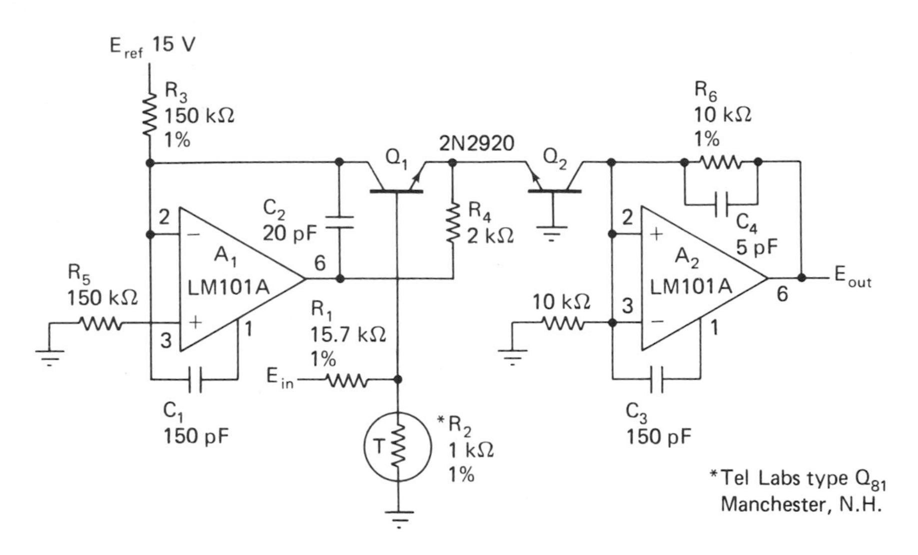

FIGURE 9-52 Antilog generator. (Courtesy National Semiconductor Corp.)

9-6-12 Multiplier/Divider

Using logarithms for multiplication is a simple process of addition. Adding two numbers in log form produces a product in log form. Taking the antilog gives the product in conventional form. Similarly, division is a process of subtracting logs and then finding the antilog of the difference to obtain the conventional quotient.

The circuit shown in Figure 9-53 is a multiplier/divider. Basically, it consists of the log generator shown in Figure 9-51 and the antilog generator of Figure 9-52.

For multiplication E_2 is set at a reference voltage and $E_{out} = 10(E_1 \cdot E_3)$, while for a multiplication/division operation, $E_{out} = E_1E_3/10E_2$. If a single number is desired in the numerator, let E_1 or E_3 equal 1.

9-6-13 Cube Generator

A cube (X^2) generator is shown in Figure 9-54. It is similar to the multiplier/divider circuit of Figure 9-53 except for the addition of a few components, and inputs E_2 and E_3 are at a reference voltage. Actually, the circuit is capable of any power function by changing the values of R_9 and R_{10} as given by the expression

$$E_{out} = E_{in}\, 16.7R_9/R_9 + R_{10}$$

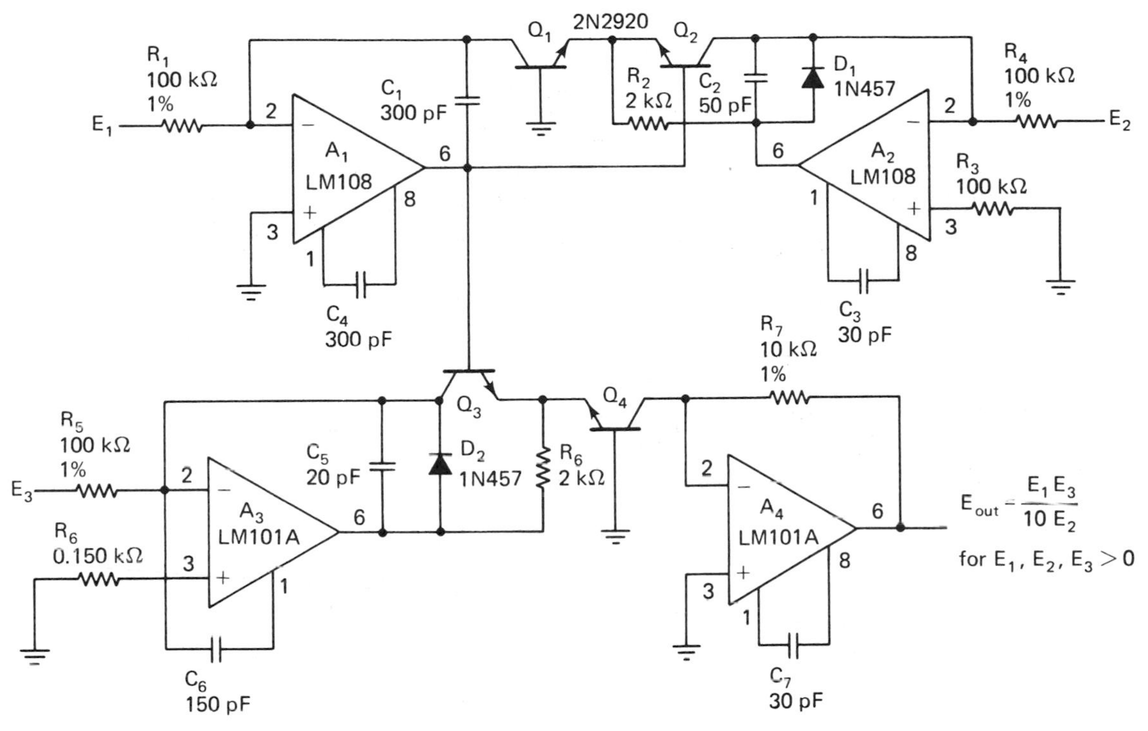

FIGURE 9-53 Multiplier/divider. (Courtesy National Semiconductor Corp.)

9-6-14 Root Extractor

Finding the root of a number using logs involves taking the log of the number, dividing it by, say, $\frac{1}{2}$ for the square root, and finding the antilog to convert it to conventional form. Such a root extractor is shown in Figure 9-55. Voltage divider R_4-R_5 determines the ratio of the log voltage for the desired root.

9-6-15 High-Speed Warning Device

A high-speed warning device to be used for automotive applications is shown in Figure 9-56. The input signal uses the engine speed available at the primary of the spark coil, thus eliminating electromechanical transducers at the transmission or speedometer cable. A switch in the transmission closes in top gear and enables the display and audible alarm switch. The display shows the desired speed limit and is set by the 100-kΩ pot in accordance with the gear/axle ratios, number of cylinders, wheel/tire size, and so on. An

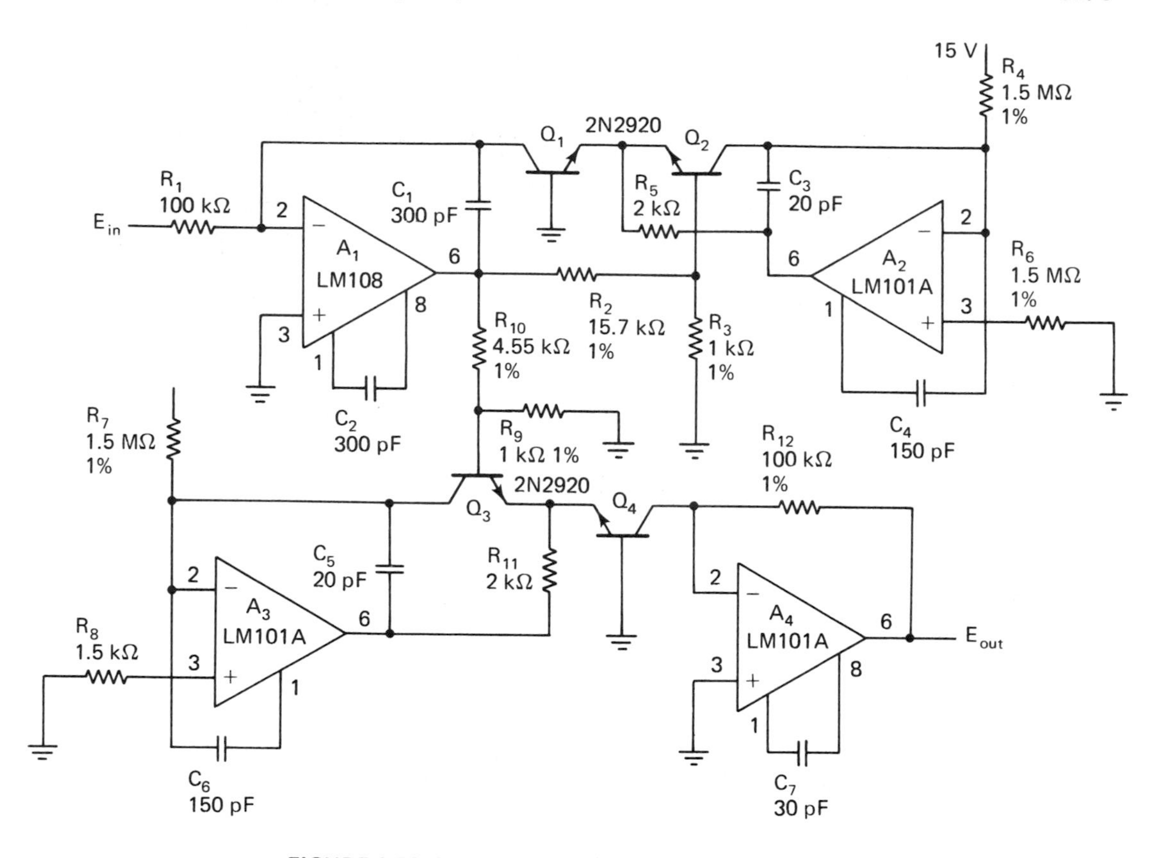

FIGURE 9-54 Cube generator. (Courtesy National Semiconductor Corp.)

LM2900 quad Norton op amp can be used to perform all the functions. Initially, A_1 amplifies and regulates the signal from the spark coil. A_2 converts frequency to voltage with the output proportional to engine RPM. A_3 compares this voltage with the reference voltage and turns on the output transistor at the set speed. When the vehicle speed exceeds the set limit, the tone generator will be energized. To extinguish these warnings, the driver will have to slow the vehicle to below the value set by the hysteresis.

9-6-16 Pulse-Width Modulator

The pulse-width modulator shown in Figure 9-57 is basically a square-wave generator except that the pulse width can be modified via the input. The duty cycle of the output can be altered because the addition of the control voltage at the input alters the trip points of the generator.

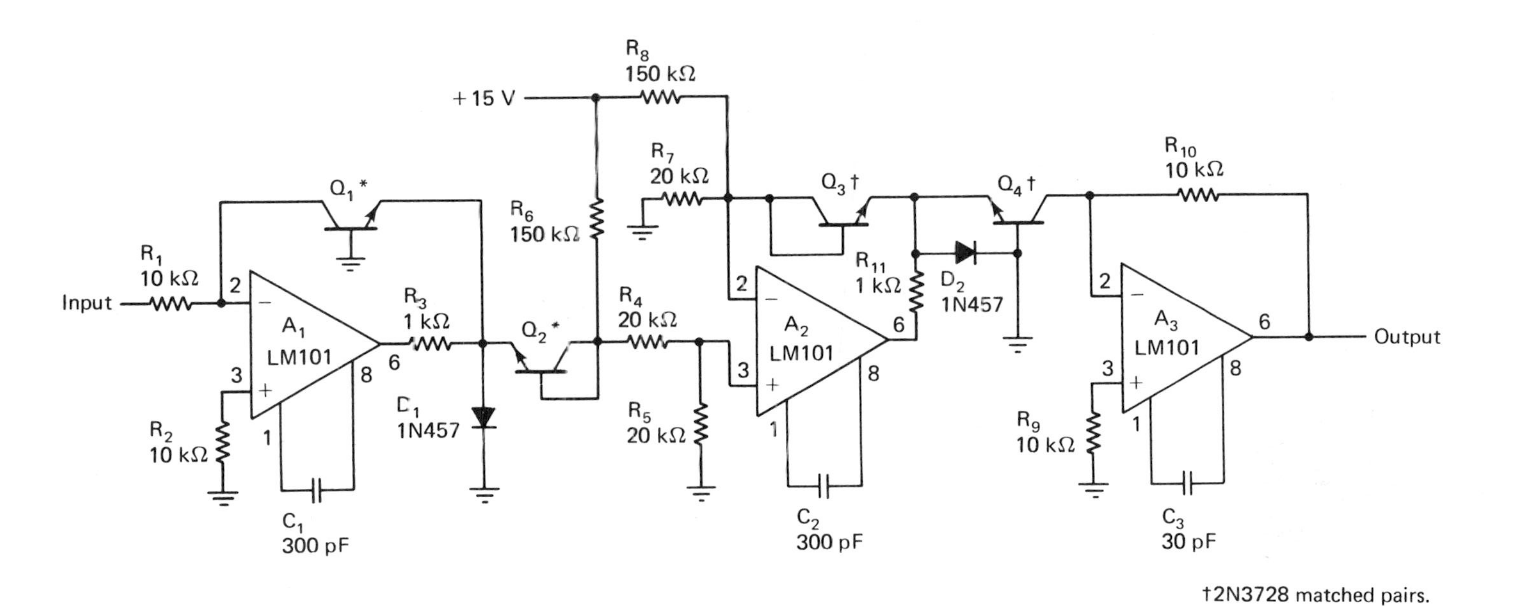

†2N3728 matched pairs.

FIGURE 9-55 Root extractor. (Courtesy National Semiconductor Corp.)

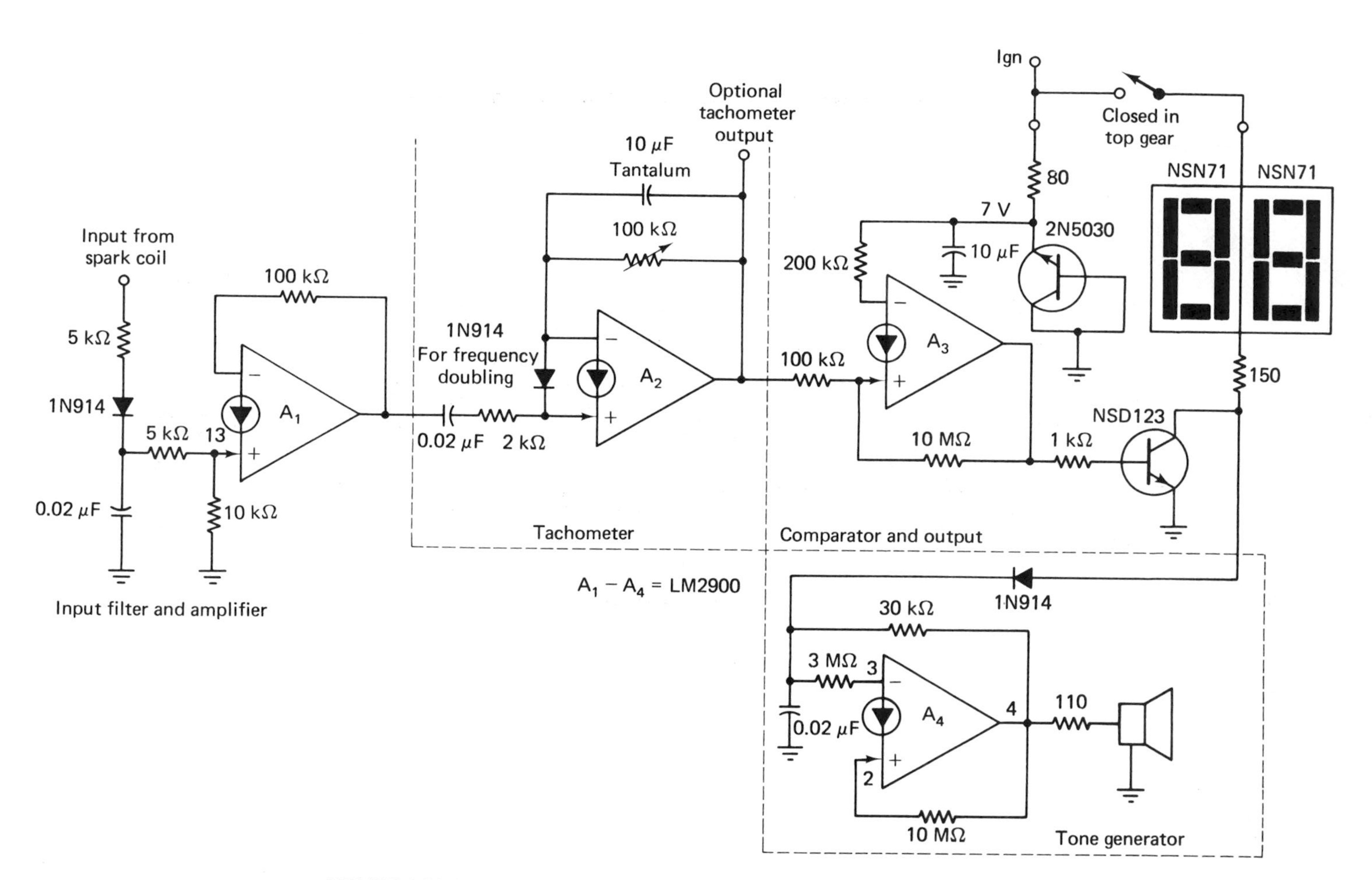

FIGURE 9-56 High speed warning device. (Courtesy National Semiconductor Corp.)

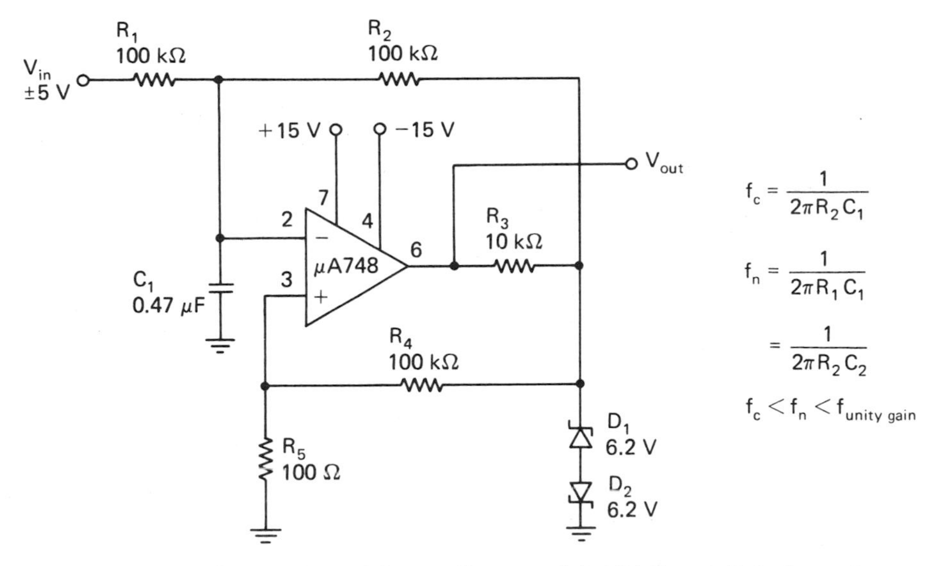

FIGURE 9-57 Pulse with modulator. (Courtesy Fairchild Camera & Instruments Corporation.)

9-6-17 Capacitance Multiplier

In low-impedance systems where large capacitances are required, the circuit shown in Figure 9-58 might be used. Not intended for tuned circuits or filters because of low Q, the capacitance multiplier can be used in timing circuits on servo compensation networks. Capacitance C is determined by the formula shown.

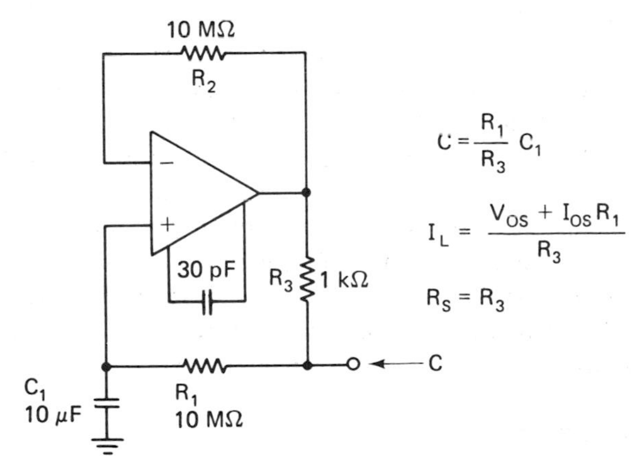

FIGURE 9-58 Capacitance multiplier. (Courtesy Fairchild Camera & Instruments Corporation.)

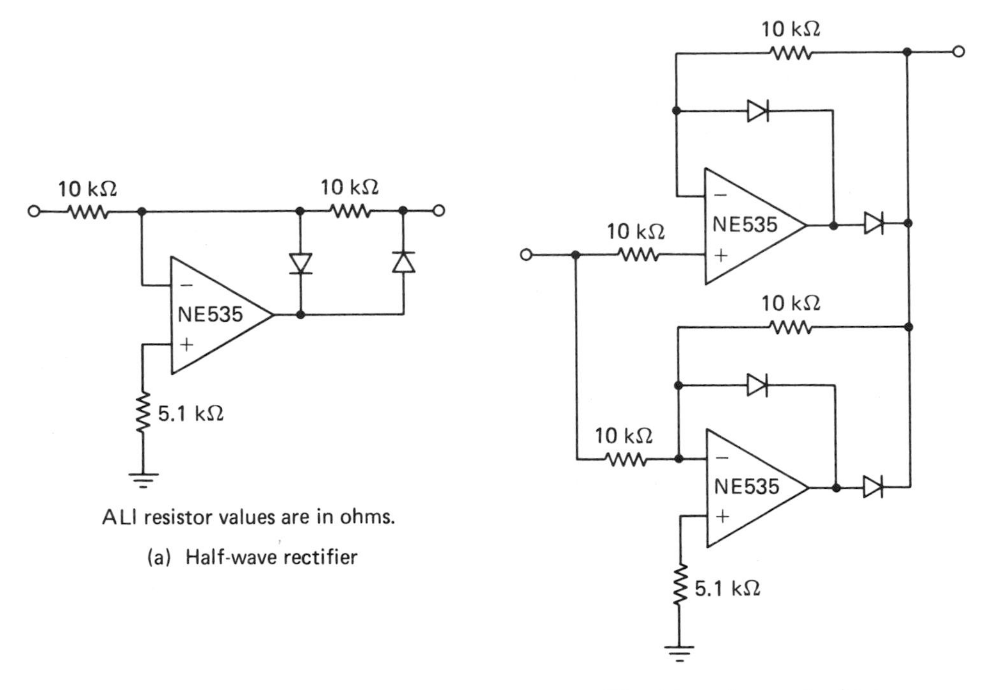

FIGURE 9-59 Precision rectifiers: (a) half wave rectifier; (b) full wave rectifier. (Permission to reprint granted by Signetics Corporation, a subsidiary of U.S. Philips Corp., 811 E. Arques Avenue, Sunnyvale, CA 94086.)

9-6-18 Precision Rectifiers

Normal diodes conduct only when the forward voltage across them exceeds 0.5 V, which leads to large errors when rectifying small signals. The precision rectifiers shown in Figure 9-59 provide accurate rectification. The half-wave rectifier of Figure 9-59a has a gain of 0 for positive signals and a gain of −1 for negative signals. Output impedance differs for the two input polarities, and buffering may be needed. The output impedance for the full rectifier of Figure 9-59b is low for both input polarities, and the errors are small at all signal levels. Reversing the diodes in each circuit will invert the output polarity.

9-6-19 Phase Shifter

The phase shifter shown in Figure 9-60 can shift an input signal up to approximately 180°. The output V_{out} has the same frequency and amplitude

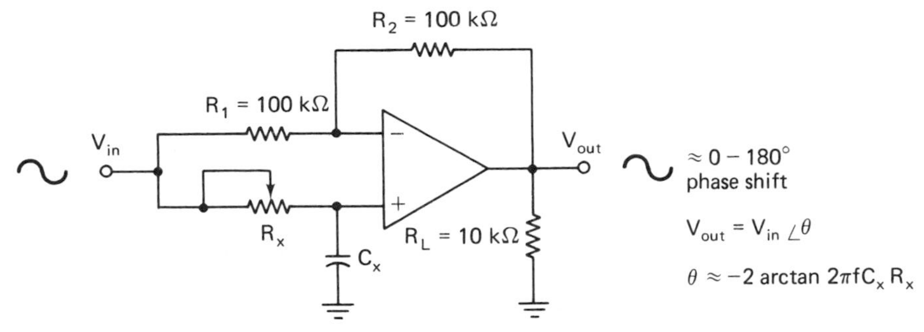

FIGURE 9-60 Phase shifter.

as V_{in} except that it lags by so many degrees depending on the values of C_x and R_x. Making R_x variable allows for fine adjustment.

9-6-20 Phase-Locked Loop

The LM3900 quad op amp can be used in a phase-locked loop, as shown in Figure 9-61. Output signal phases from the input op amp are compared by

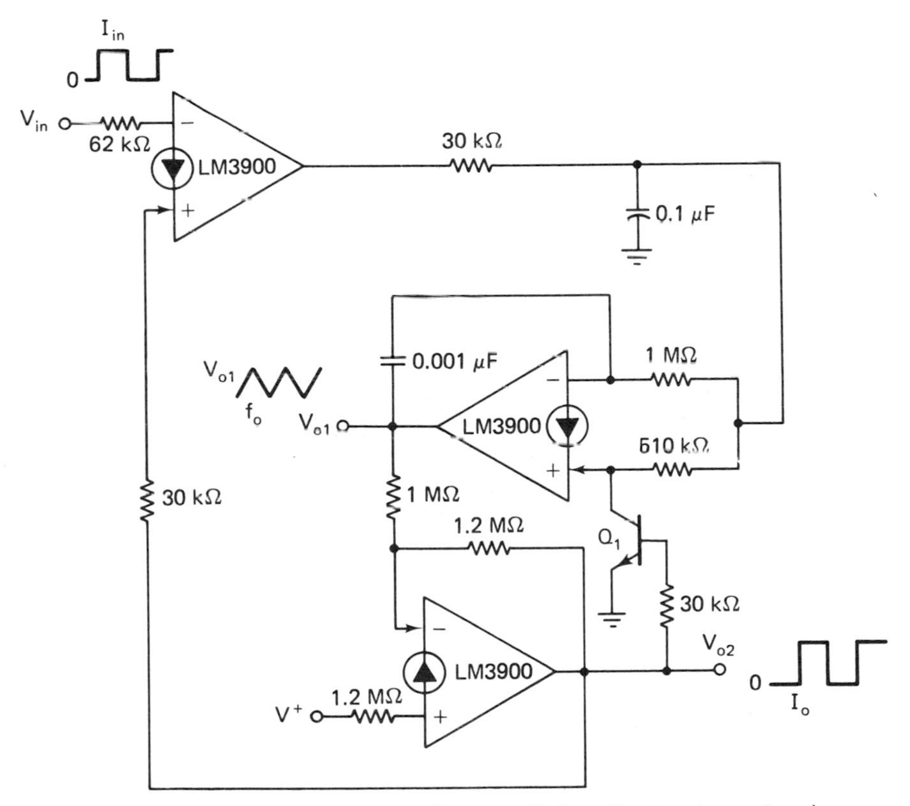

FIGURE 9-61 Phase-locked loop. (Courtesy National Semiconductor Corp.)

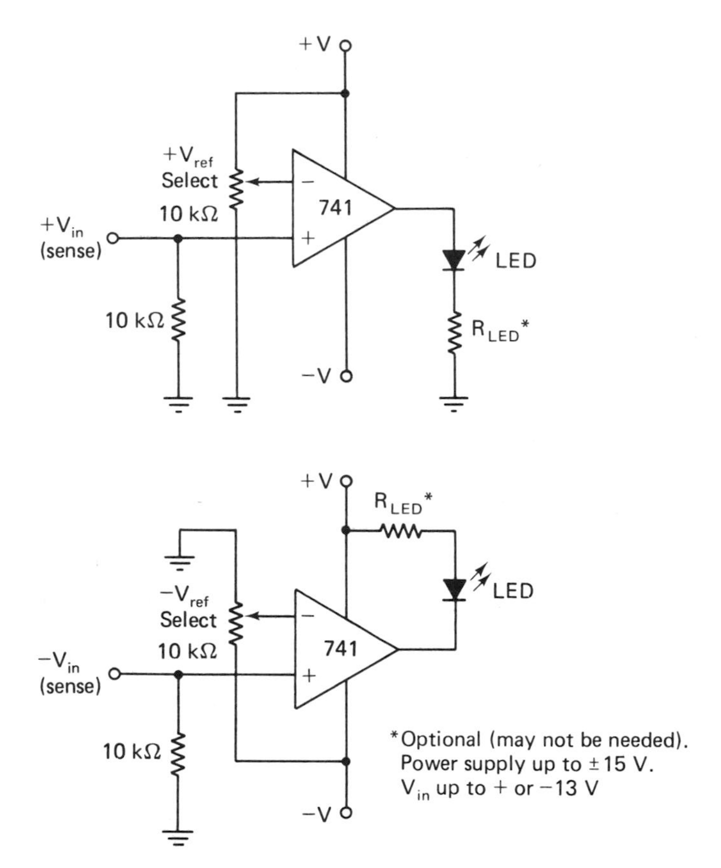

FIGURE 9-62 Voltage level detector: (a) positive indicator; (b) negative indicator.

the middle op amp. Any phase difference between these two signals is converted into a correction voltage and fed back to the input op amp. This causes the phase of the output signal to change so that it tracks the input reference signal. There is also a triangle wave available at the output of the phase comparator.

9-6-21 Voltage-Level Detector

Voltage-level detectors with LED indicators are shown in Figure 9-62. The supply voltage used can range up to ±15 V, but the input voltage sensed must be about 2 or 3 V below the maximum supply. The potentiometer can be used to set the reference voltage. When V_{in} goes beyond V_{ref}, the op-amp output changes states and the LED will light.

Appendix

Selected Manufacturers' Op-Amp Specification Sheets

μA702
WIDEBAND DC AMPLIFIER
FAIRCHILD LINEAR INTEGRATED CIRCUIT

GENERAL DESCRIPTION – The μA702 is a monolithic DC Amplifier constructed using the Fairchild Planar* epitaxial process. It is intended for use as an operational amplifier in analog computers, as a precision instrumentation amplifier, or in other applications requiring a feedback amplifier useful from dc to 30 MHz.

- **LOW OFFSET VOLTAGE**
- **LOW OFFSET VOLTAGE DRIFT**
- **WIDE BANDWIDTH – 20 MHz TYP**
- **HIGH SLEW RATE – 5 V/μs TYP**

ABSOLUTE MAXIMUM RATINGS

Voltage Between V+ and V– Terminals	21 V
Peak Output Current	50 mA
Differential Input Voltage	±5.0 V
Input Voltage	+1.5 V to –6.0 V
Internal Power Dissipation (Note)	
Metal Can	500 mW
DIP	670 mW
Flatpak	570 mW
Operating Temperature Range	
Military (μA702)	–55°C to +125°C
Commercial (μA702C)	0°C to +70°C
Storage Temperature Range	–65°C to +150°C
Lead Temperature (Soldering, 60 seconds)	300°C

NOTE
Rating applies to ambient temperature up to 70 C. Above 70°C ambient derate linearly at 6.3mW/°C for Metal Can, 8.3mW/°C for DIP and 7.1mW/°C for the Flatpak.

EQUIVALENT CIRCUIT

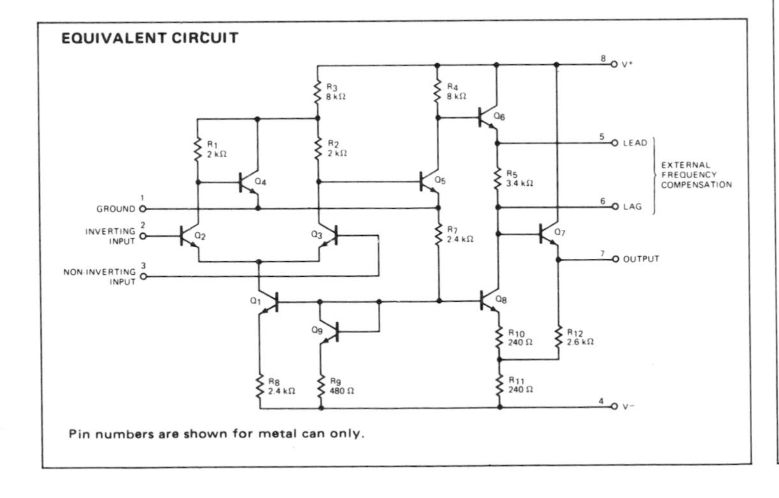

Pin numbers are shown for metal can only.

CONNECTION DIAGRAMS

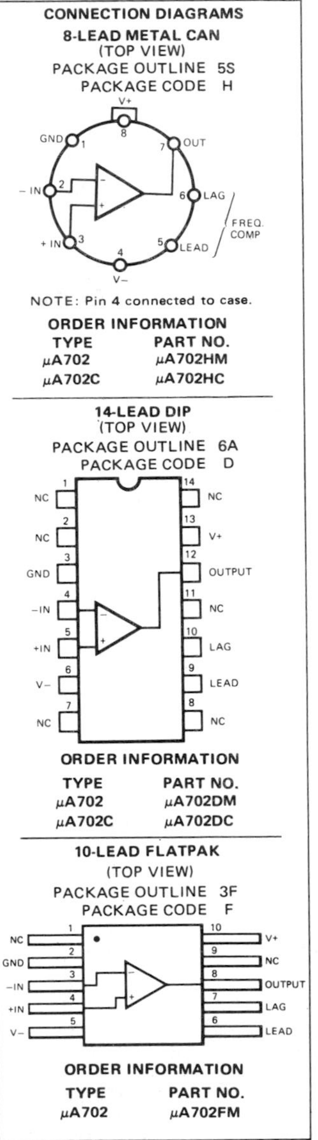

NOTE: Pin 4 connected to case.

ORDER INFORMATION (8-Lead Metal Can)

TYPE	PART NO.
μA702	μA702HM
μA702C	μA702HC

ORDER INFORMATION (14-Lead DIP)

TYPE	PART NO.
μA702	μA702DM
μA702C	μA702DC

ORDER INFORMATION (10-Lead Flatpak)

TYPE	PART NO.
μA702	μA702FM

*Planar is a patented Fairchild process.

μA702

ELECTRICAL CHARACTERISTICS (T_A = 25°C unless otherwise specified)

PARAMETER		CONDITIONS	V+ = 12.0V, V− = −6.0V			V+ = 6.0V, V− = −3.0V			UNITS
			MIN	TYP	MAX	MIN	TYP	MAX	
Input Offset Voltage		$R_S \leq 2\ k\Omega$		0.5	2.0		0.7	3.0	mV
Input Offset Current				180	500		120	500	nA
Input Bias Current				2.0	5.0		1.2	3.5	μA
Input Resistance			16	40		22	67		kΩ
Input Voltage Range			−4.0		+0.5	−1.5		+0.5	V
Common Mode Rejection Ratio		$R_S \leq 2\ k\Omega$, f ≤ 1 kHz	80	100		80	100		dB
Large Signal Voltage Gain		$R_L \geq 100\ k\Omega$, $V_{OUT} = \pm 5.0$ V	2500	3600	6000				
		$R_L \geq 100\ k\Omega$, $V_{OUT} = \pm 2.5$ V				600	900	1500	
Output Resistance				200	500		300	700	Ω
Supply Current		$V_{OUT} = 0$		5.0	6.7		2.1	3.3	mA
Power Consumption		$V_{OUT} = 0$		90	120		19	30	mW
Transient Response (unity-gain)	Rise Time	Cl = 0.01 μF, Rl = 20 Ω, $R_L \geq 100\ k\Omega$, V_{IN} = 10 mV		25	120				ns
	Overshoot	$C_L \leq 100$ pF		10	50				%
Transient Response (x100 gain)	Rise Time	C3 = 50 pF, $R_L \geq 100\ k\Omega$,		10	30				ns
	Overshoot	V_{IN} = 1 mV		20	40				%
The following specifications apply for −55°C ≤ T_A ≤ +125°C:									
Input Offset Voltage		$R_S \leq 2\ k\Omega$			3.0			4.0	mV
Average Temperature Coefficient of Input Offset Voltage		R_S = 50 Ω, T_A = 25°C to +125°C		2.5	10		3.5	15	μV/°C
		R_S = 50 Ω, T_A = 25°C to −55°C		2.0	10		3.0	15	μV/°C
Input Offset Current		T_A = +125°C		80	500		50	500	nA
		T_A = −55°C		400	1500		280	1500	nA
Average Temperature Coefficient of Input Offset Current		T_A = 25°C to +125°C		1.0	5.0		0.7	4.0	nA/°C
		T_A = 25°C to −55°C		3.0	16		2.0	13	nA/°C
Input Bias Current		T_A = −55°C		4.3	10		2.6	7.5	μA
Input Resistance			6.0			8.0			kΩ
Common Mode Rejection Ratio		$R_S \leq 2\ k\Omega$, f ≤ 1 kHz	70	95		70	95		dB
Supply Voltage Rejection Ratio		V+ = 12 V, V− = −6.0 V to V+ = 6.0 V, V− = −3.0 V $R_S \leq 2\ k\Omega$		75	200		75	200	μV/V
Large Signal Voltage Gain		$R_L \geq 100\ k\Omega$, $V_{OUT} = \pm 5.0$ V	2000		7000				
		$R_L \geq 100\ k\Omega$, $V_{OUT} = \pm 2.5$ V				500		1750	
Output Voltage Swing		$R_L \geq 100\ k\Omega$	±5.0	±5.3		±2.5	±2.7		V
		$R_L \geq 10\ k\Omega$	±3.5	±4.0		±1.5	±2.0		V
Supply Current		T_A = +125°C, $V_{OUT} = 0$		4.4	6.7		1.7	3.3	mA
		T_A = −55°C, $V_{OUT} = 0$		5.0	7.5		2.1	3.9	mA
Power Consumption		T_A = +125°C, $V_{OUT} = 0$		80	120		15	30	mW
		T_A = −55°C, $V_{OUT} = 0$		90	135		19	35	mW

TYPICAL PERFORMANCE CURVE FOR μA702

VOLTAGE TRANSFER CHARACTERISTIC

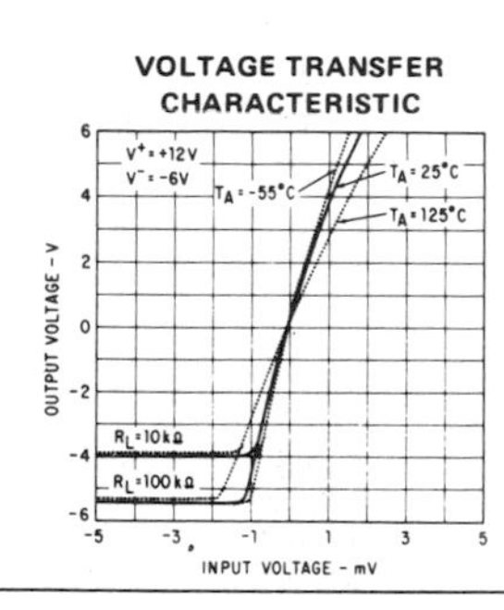

VOLTAGE TRANSFER CHARACTERISTIC

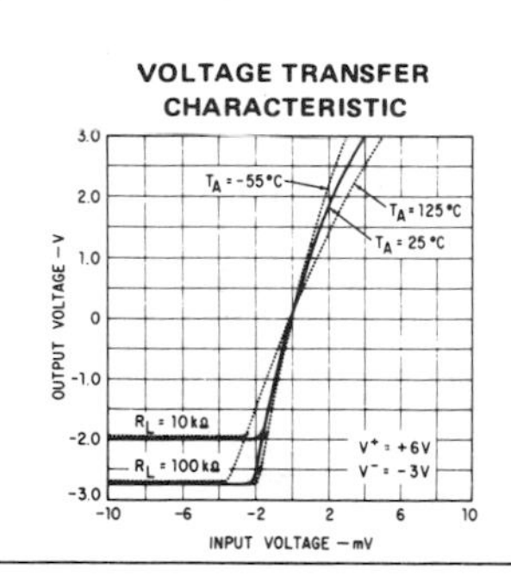

VOLTAGE GAIN AS A FUNCTION OF AMBIENT TEMPERATURE

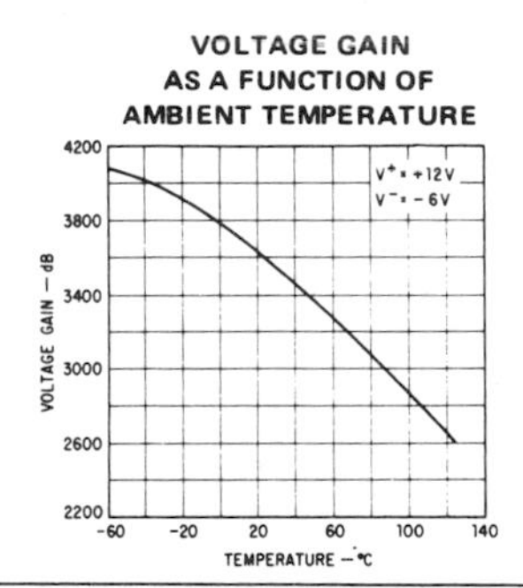

μA709

HIGH PERFORMANCE OPERATIONAL AMPLIFIER

FAIRCHILD LINEAR INTEGRATED CIRCUITS

GENERAL DESCRIPTION – The μA709 is a monolithic High Gain Operational Amplifier constructed using the Fairchild Planar* epitaxial process. It features low offset, high input impedance, large input common mode range, high output swing under load and low power consumption. The device displays exceptional temperature stability and will operate over a wide range of supply voltages with little performance degradation. The amplifier is intended for use in dc servo systems, high impedance analog computers, low level instrumentation applications and for the generation of special linear and nonlinear transfer functions.

ABSOLUTE MAXIMUM RATINGS

Supply Voltage	±18 V
Internal Power Dissipation (Note)	
Metal Can	500 mW
DIP	670 mW
Flatpak	570 mW
Differential Input Voltage	±5.0 V
Input Voltage	±10 V
Storage Temperature Range	
Metal, Hermetic DIP, and Flatpak	−65°C to +150°C
Molded DIP	−55°C to +125°C
Operating Temperature Range	
Military (μA709A and μA709)	−55°C to +125°C
Commercial (μA709C)	0°C to +70°C
Lead Temperature	
Metal Can, Hermetic DIP, and Flatpak (Soldering 60 s)	300°C
Molded DIP	260°C
Output Short Circuit Duration	5 s

NOTE:
Rating applies to ambient temperature up to 70°C. Above 70°C ambient derate linearly at 6.3mW/°C for Metal Can, 8.3mW/°C for DIP, 7.1mW/°C for the Flatpak and 5.6mW/°C for the Mini DIP.

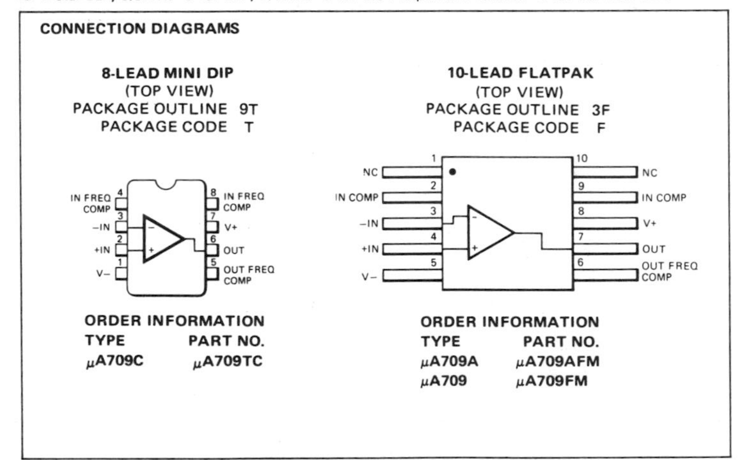

CONNECTION DIAGRAMS

8-LEAD METAL CAN
(TOP VIEW)
PACKAGE OUTLINE 5S
PACKAGE CODE H

IN FREQ COMP (8), IN FREQ COMP (1), V+ (7), −IN (2), OUT (6), +IN (3), OUT FREQ COMP (5), V− (4)

NOTE: Pin 4 connected to case

ORDER INFORMATION

TYPE	PART NO.
μA709A	μA709AHM
μA709	μA709HM
μA709C	μA709HC

14-LEAD DIP
(TOP VIEW)
PACKAGE OUTLINE 6A 9A
PACKAGE CODE D P

NC (1), NC (2), IN FREQ COMP (3), −IN (4), +IN (5), V− (6), NC (7), NC (8), OUT FREQ COMP (9), OUT (10), V+ (11), IN FREQ COMP (12), NC (13), NC (14)

ORDER INFORMATION

TYPE	PART NO.
μA709A	μA709ADM
μA709	μA709DM
μA709C	μA709DC
μA709C	μA709PC

*Planar is a patented Fairchild process.

μA709A

ELECTRICAL CHARACTERISTICS (T_A = +25°C, ±9 V ≤ V_S ≤ ±15 V unless otherwise specified)

PARAMETER (see definitions)	CONDITIONS	MIN	TYP	MAX	UNITS
Input Offset Voltage	R_S ≤ 10 kΩ		0.6	2.0	mV
Input Offset Current			10	50	nA
Input Bias Current			100	200	nA
Input Resistance		350	700		kΩ
Output Resistance			150		Ω
Supply Current	V_S = ±15 V		2.5	3.6	mA
Power Consumption	V_S = ±15 V		75	108	mW
Transient Response — Rise time	V_S = ±15 V, V_{IN} = 20 mV, R_L = 2 kΩ, C1 = 5 nF, R1 = 1.5 kΩ, C2 = 200 pF, R2 = 50Ω, C_L ≤ 100 pF			1.5	μs
Transient Response — Overshoot				30	%
The following specifications apply for −55°C ≤ T_A ≤ +125°C:					
Input Offset Voltage	R_S ≤ 10 kΩ			3.0	mV
Average Temperature Coefficient of Input Offset Voltage	R_S = 50Ω, T_A = +25°C to +125°C		1.8	10	μV/°C
	R_S = 50Ω, T_A = +25°C to −55°C		1.8	10	μV/°C
	R_S = 10 kΩ, T_A = +25°C to +125°C		2.0	15	μV/°C
	R_S = 10 kΩ, T_A = +25°C to −55°C		4.8	25	μV/°C
Input Offset Current	T_A = +125°C		3.5	50	nA
	T_A = −55°C		40	250	nA
Average Temperature Coefficient of Input Offset Current	T_A = +25°C to +125°C		0.08	0.5	nA/°C
	T_A = +25°C to −55°C		0.45	2.8	nA/°C
Input Bias Current	T_A = −55°C		300	600	nA
Input Resistance	T_A = −55°C	85	170		kΩ
Input Voltage Range	V_S = ±15 V	±8.0			V
Common Mode Rejection Ratio	R_S ≤ 10 kΩ	80	110		dB
Supply Voltage Rejection Ratio	R_S ≤ 10 kΩ		40	100	μV/V
Large Signal Voltage Gain	V_S = ±15 V, R_L ≥ 2 kΩ, V_{OUT} = ±10 V	25,000		70,000	V/V
Output Voltage Swing	V_S = ±15 V, R_L ≥ 10 kΩ	±12	±14		V
	V_S = ±15 V, R_L ≥ 2 kΩ	±10	±13		V
Supply Current	T_A = +125°C, V_S = ±15 V		2.1	3.0	mA
	T_A = −55°C, V_S = ±15 V		2.7	4.5	mA
Power Consumption	T_A = +125°C, V_S = ±15 V		[illegible]	[illegible]	mW
	T_A = −55°C, V_S = ±15 V		81	135	mW

PERFORMANCE CURVES FOR μA709A

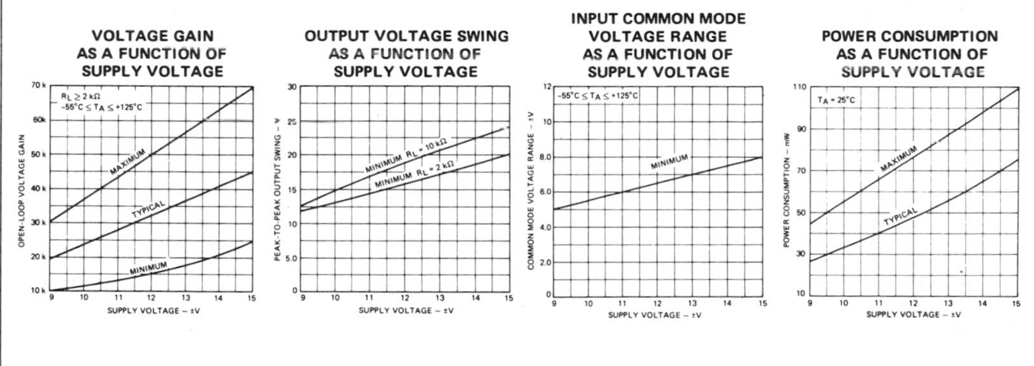

μA101 • μA201

GENERAL PURPOSE OPERATIONAL AMPLIFIERS

FAIRCHILD LINEAR INTEGRATED CIRCUITS

GENERAL DESCRIPTION – The 101 and 201 are General Purpose monolithic Operational Amplifiers constructed using the Fairchild Planar* epitaxial process. They are intended for a wide range of analog applications where tailoring of frequency characteristics is desirable. The 101 and 201 compensate easily with a single external component. High common mode voltage range and absence of "latch-up" make the 101 and 201 ideal for use as voltage followers. The high gain and wide range of operating voltages provide superior performance in integrator, summing amplifier, and general feedback applications. The 101 and 201 are short-circuit protected and have the same pin configuration as the popular μA741, μA748 and μA709.

- **SHORT-CIRCUIT PROTECTION**
- **OFFSET VOLTAGE NULL CAPABILITY**
- **LARGE COMMON-MODE AND DIFFERENTIAL VOLTAGE RANGES**
- **LOW POWER CONSUMPTION**
- **NO LATCH-UP**

CONNECTION DIAGRAMS
8-LEAD METAL CAN
(TOP VIEW)
PACKAGE OUTLINE 5S
PACKAGE CODE H

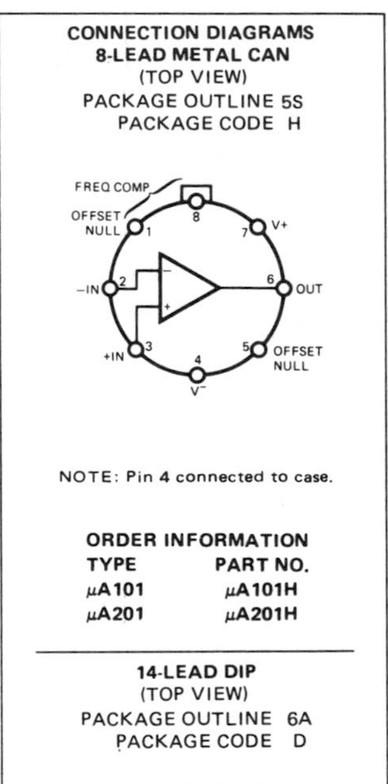

NOTE: Pin 4 connected to case.

ORDER INFORMATION

TYPE	PART NO.
μA101	μA101H
μA201	μA201H

14-LEAD DIP
(TOP VIEW)
PACKAGE OUTLINE 6A
PACKAGE CODE D

Pin		Pin	
1	NC	14	NC
2	NC	13	NC
3	OFFSET NULL (COMP)	12	FREQ COMP
4	–IN	11	V+
5	+IN	10	OUT
6	V–	9	OFFSET NULL
7	NC	8	NC

ORDER INFORMATION

TYPE	PART NO.
μA101	μA101D
μA201	μA201D

ABSOLUTE MAXIMUM RATINGS

Supply Voltage	±22V
Internal Power Dissipation (Note 1)	
Metal Can	500 mW
DIP	670 mW
Differential Input Voltage	±30V
Input Voltage (Note 2)	±15V
Storage Temperature Range	
Metal Can, DIP	–65°C to +150°C
Operating Temperature Range (Note 3)	
Military (μA101)	–55°C to +125°C
Commercial (μA201)	0°C to +70°C
Lead Temperature (Soldering, 60 seconds)	300°C

EQUIVALENT CIRCUIT

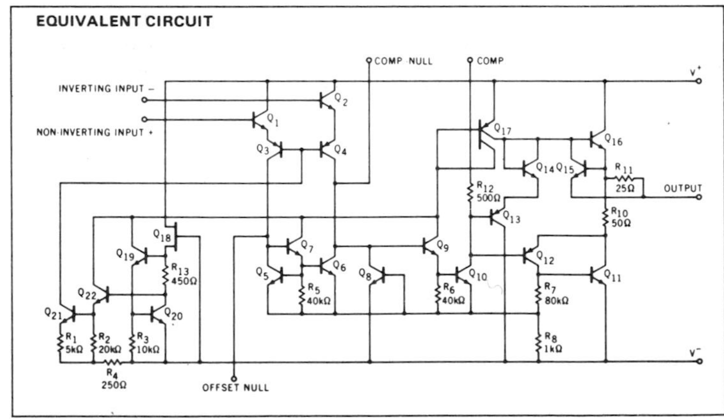

Notes on following pages

*Planar is a patented Fairchild process.

ELECTRICAL CHARACTERISTICS FOR µA101 (± 5.0 V $\leq V_S \leq \pm 20$ V, $T_A = 25°C$, C1 = 30 pF unless otherwise specified)

PARAMETER	CONDITIONS		MIN	TYP	MAX	UNITS
Input Offset Voltage	$R_S \leq 10\,k\Omega$			1.0	5.0	mV
Input Offset Current				40	200	nA
Input Bias Current				120	500	nA
Input Resistance			300	800		kΩ
Supply Current	$V_S = \pm 20V$			1.8	3.0	mA
Large Signal Voltage Gain	$V_S = \pm 15V$, $V_{OUT} = \pm 10V$, $R_L \geq 2\,k\Omega$		50	160		V/mV
The following specifications apply for $-55°C \leq T_A \leq +125°C$:						
Input Offset Voltage	$R_S \leq 10\,k\Omega$				6.0	mV
Average Temperature Coefficient of Input Offset Voltage	$R_S \leq 50\,\Omega$			3.0		µV/°C
	$R_S \leq 10\,k\Omega$			6.0		µV/°C
Input Offset Current	$T_A = +125°C$			10	200	nA
	$T_A = -55°C$			100	500	nA
Average Temperature Coefficient of Input Offset Current	$+25°C \leq T_A \leq +125°C$			0.01	0.1	nA/°C
	$-55°C \leq T_A \leq +25°C$			0.02	0.2	nA/°C
Input Bias Current	$T_A = -55°C$			0.28	1.5	µA
Supply Current	$T_A = +125°C$, $V_S = \pm 20V$			1.2	2.5	mA
Large Signal Voltage Gain	$V_S = \pm 15V$, $V_{OUT} = \pm 10V$, $R_L \geq 2\,k\Omega$		25			V/mV
Output Voltage Swing	$V_S = \pm 15V$	$R_L = 10\,k\Omega$	±12	±14		V
		$R_L = 2\,k\Omega$	±10	±13		V
Input Voltage Range	$V_S = \pm 15V$		±12			V
Common Mode Rejection Ratio	$R_S \leq 10\,k\Omega$		70	90		dB
Supply Voltage Rejection Ratio	$R_S \leq 10\,k\Omega$		70	90		dB

NOTES

1. Rating applies to ambient temperature up to 70°C. Above 70°C ambient derate linearly at 6.3mW/°C for the Metal Can and 8.3mW/°C for the DIP.
2. For supply voltages less than ±15V, the absolute maximum input voltage is equal to the supply voltage.
3. Short circuit may be to ground or either supply. The 101 ratings apply to +125°C case temperature or +75°C ambient temperature. The 201 ratings apply to case temperatures up to +70°C.

µA741

FREQUENCY-COMPENSATED OPERATIONAL AMPLIFIER

FAIRCHILD LINEAR INTEGRATED CIRCUIT

GENERAL DESCRIPTION — The µA741 is a high performance monolithic Operational Amplifier constructed using the Fairchild Planar* epitaxial process. It is intended for a wide range of analog applications. High common mode voltage range and absence of latch-up tendencies make the µA741 ideal for use as a voltage follower. The high gain and wide range of operating voltage provides superior performance in integrator, summing amplifier, and general feedback applications. Electrical characteristics of the µA741A and E are identical to MIL-M-38510/10101.

- **NO FREQUENCY COMPENSATION REQUIRED**
- **SHORT CIRCUIT PROTECTION**
- **OFFSET VOLTAGE NULL CAPABILITY**
- **LARGE COMMON MODE AND DIFFERENTIAL VOLTAGE RANGES**
- **LOW POWER CONSUMPTION**
- **NO LATCH-UP**

ABSOLUTE MAXIMUM RATINGS

Supply Voltage	
µA741A, µA741, µA741E	±22 V
µA741C	±18 V
Internal Power Dissipation (Note 1)	
Metal Can	500 mW
Molded and Hermetic DIP	670 mW
Mini DIP	310 mW
Flatpak	570 mW
Differential Input Voltage	±30 V
Input Voltage (Note 2)	±15 V
Storage Temperature Range	
Metal Can, Hermetic DIP, and Flatpak	−65°C to +150°C
Mini DIP, Molded DIP	−55°C to +125°C
Operating Temperature Range	
Military (µA741A, µA741)	−55°C to +125°C
Commercial (µA741E, µA741C)	0°C to +70°C
Lead Temperature (Soldering)	
Metal Can, Hermetic DIPs, and Flatpak (60 s)	300°C
Molded DIPs (10 s)	260°C
Output Short Circuit Duration (Note 3)	Indefinite

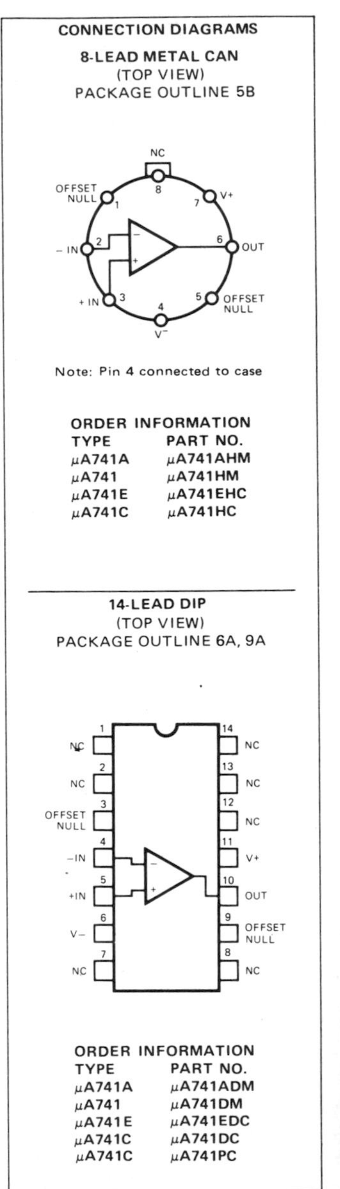

Note: Pin 4 connected to case

ORDER INFORMATION

TYPE	PART NO.
µA741A	µA741AHM
µA741	µA741HM
µA741E	µA741EHC
µA741C	µA741HC

ORDER INFORMATION

TYPE	PART NO.
µA741A	µA741ADM
µA741	µA741DM
µA741E	µA741EDC
µA741C	µA741DC
µA741C	µA741PC

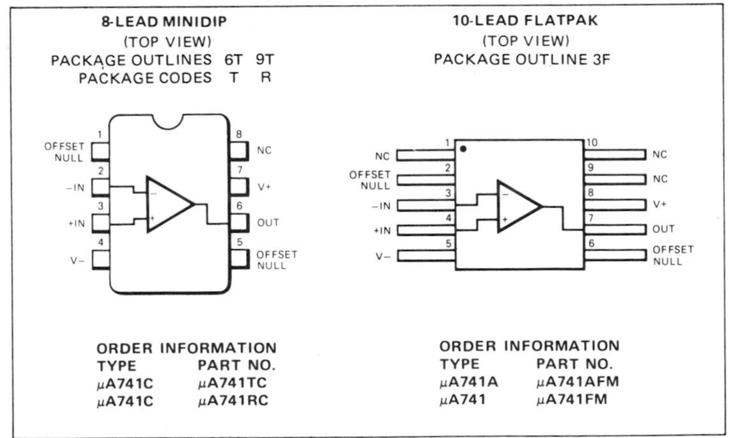

ORDER INFORMATION (8-Lead Minidip)

TYPE	PART NO.
µA741C	µA741TC
µA741C	µA741RC

ORDER INFORMATION (10-Lead Flatpak)

TYPE	PART NO.
µA741A	µA741AFM
µA741	µA741FM

Notes on following pages.

*Planar is a patented Fairchild process.

μA741A

ELECTRICAL CHARACTERISTICS ($V_S = \pm 15V$, $T_A = 25°C$ unless otherwise specified)

PARAMETERS (see definitions)	CONDITIONS	MIN	TYP	MAX	UNITS
Input Offset Voltage	$R_S \leq 50\Omega$		0.8	3.0	mV
Average Input Offset Voltage Drift				15	μV/°C
Input Offset Current			3.0	30	nA
Average Input Offset Current Drift				0.5	nA/°C
Input Bias Current			30	80	nA
Power Supply Rejection Ratio	$V_S = +10, -20$; $V_S = +20, -10V$, $R_S = 50\Omega$		15	50	μV/V
Output Short Circuit Current		10	25	35	mA
Power Dissipation	$V_S = \pm 20V$		80	150	mW
Input Impedance	$V_S = \pm 20V$	1.0	6.0		MΩ
Large Signal Voltage Gain	$V_S = \pm 20V$, $R_L = 2k\Omega$, $V_{OUT} = \pm 15V$	50			V/mV
Transient Response (Unity Gain) — Rise Time			0.25	0.8	μs
Transient Response (Unity Gain) — Overshoot			6.0	20	%
Bandwidth (Note 4)		.437	1.5		MHz
Slew Rate (Unity Gain)	$V_{IN} = \pm 10V$	0.3	0.7		V/μs
The following specifications apply for $-55°C \leq T_A \leq +125°C$					
Input Offset Voltage				4.0	mV
Input Offset Current				70	nA
Input Bias Current				210	nA
Common Mode Rejection Ratio	$V_S = \pm 20V$, $V_{IN} = \pm 15V$, $R_S = 50\Omega$	80	95		dB
Adjustment For Input Offset Voltage	$V_S = \pm 20V$	10			mV
Output Short Circuit Current		10		40	mA
Power Dissipation	$V_S = \pm 20V$, −55°C			165	mW
	$V_S = \pm 20V$, +125°C			135	mW
Input Impedance	$V_S = \pm 20V$	0.5			MΩ
Output Voltage Swing	$V_S = \pm 20V$, $R_L = 10k\Omega$	±16			V
	$V_S = \pm 20V$, $R_L = 2k\Omega$	±15			V
Large Signal Voltage Gain	$V_S = \pm 20V$, $R_L = 2k\Omega$, $V_{OUT} = \pm 15V$	32			V/mV
	$V_S = \pm 5V$, $R_L = 2k\Omega$, $V_{OUT} = \pm 2 V$	10			V/mV

NOTES

1. Rating applies to ambient temperatures up to 70°C. Above 70°C ambient derate linearly at 6.3mW/°C for the metal can, 8.3mW/°C for the DIP and 7.1mW/°C for the Flatpak.
2. For supply voltages less than ±15V, the absolute maximum input voltage is equal to the supply voltage.
3. Short circuit may be to ground or either supply. Rating applies to +125°C case temperature or 75°C ambient temperature.
4. Calculated value from: $BW(MHz) = \dfrac{0.35}{\text{Rise Time } (\mu s)}$

MC1456/1556-F,N,T

DESCRIPTION

The MC1456/1556 is an internally compensated precision monolithic operational amplifier featuring extremely low offset and bias currents and offset null capability. The MC1456/1556 is short circuit protected and its high common mode and differential input voltage range provides exceptional performance when used as an integrator, summing amplifier, and voltage follower.

PIN CONFIGURATIONS

T PACKAGE

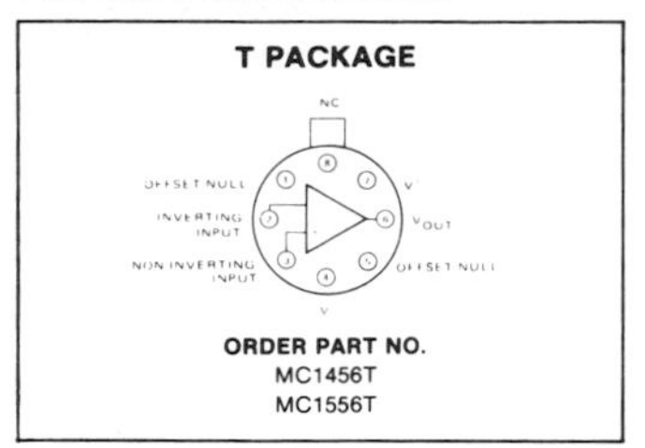

ORDER PART NO.
MC1456T
MC1556T

N PACKAGE

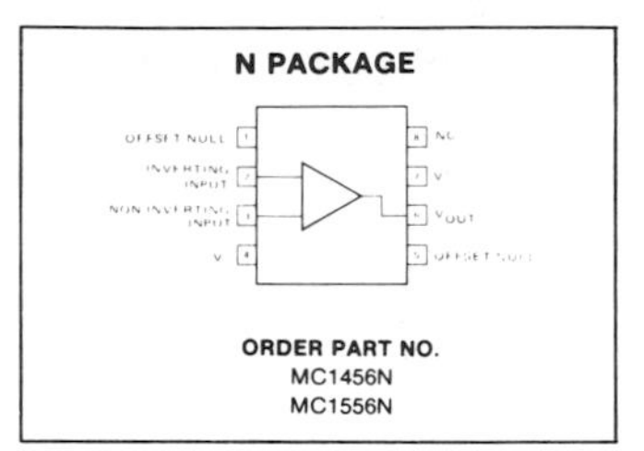

ORDER PART NO.
MC1456N
MC1556N

OFFSET ADJUST CIRCUIT

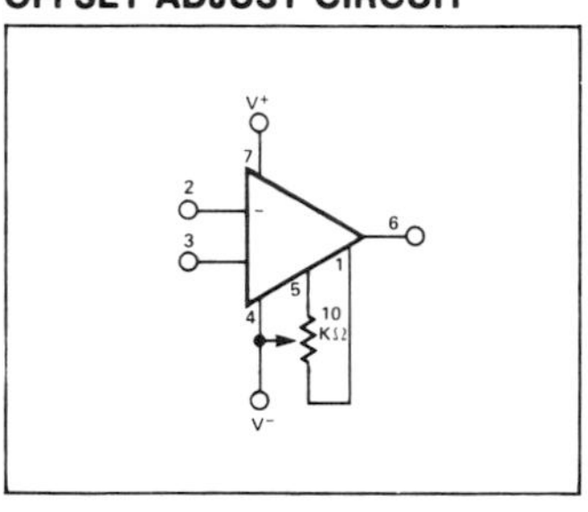

FEATURES

- Low input bias current—15nA maximum
- Low input offset current—2.0nA maximum
- Low input offset voltage—4.0mV maximum
- High slew rate—2.5V/μs typical
- Large power bandwidth—40kHz typical
- Low power consumption—45mW maximum
- Offset voltage null capability
- Output short circuit protection
- Input over-voltage protection
- Mil std 883A,B,C, available

F PACKAGE

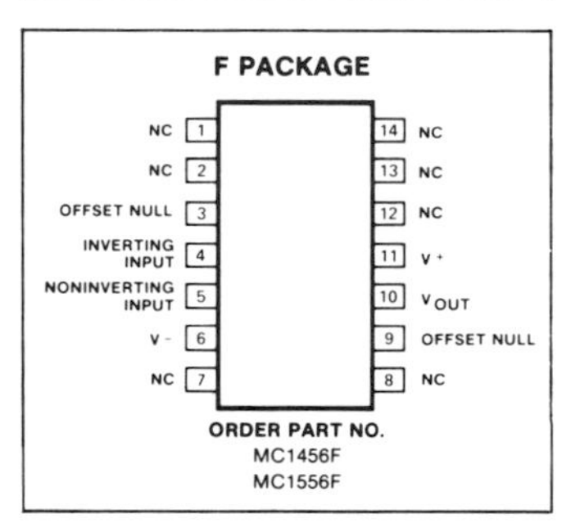

ORDER PART NO.
MC1456F
MC1556F

EQUIVALENT SCHEMATIC

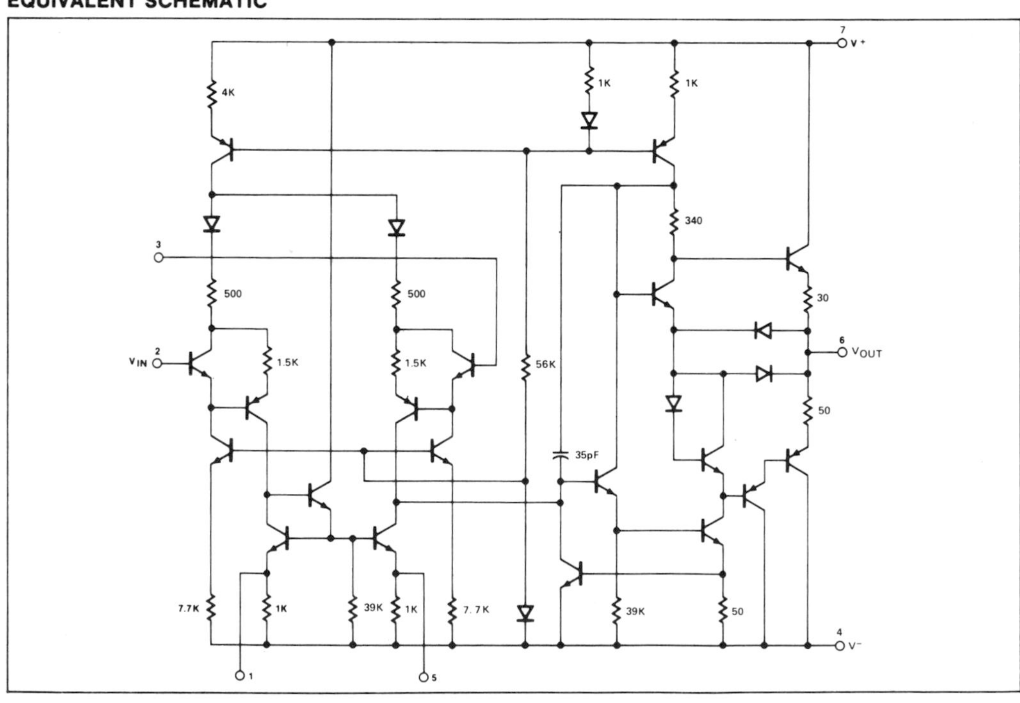

MC1456/1556-F,N,T

ABSOLUTE MAXIMUM RATINGS

PARAMETER	RATING	UNIT
Power supply voltage MC1556	±22	V
MC1456	±18	V
Differential input voltage	± V_{CC}	V
Common mode input voltage	± V_{CC}	V
Load current	20	mA
Output short circuit duration	Continuous	
Power dissipation	680	mW
Derate above T_A = 25°C	4.6	mW/°C
Operating temperature range		
MC1556	-55 to +125	°C
MC1456	0 to +70	°C
Storage temperature range	-65 to +150	°C

DC ELECTRICAL CHARACTERISTICS T_A = 25°C, V_S = ± 15V unless otherwise specified

	PARAMETER	TEST CONDITIONS	MC1556			MC1456			UNIT
			Min	Typ	Max	Min	Typ	Max	
V_{OS}	Offset voltage			2.0	4.0		5.0	10.0	mVdc
		Over temperature			6.0			14.0	mVdc
I_{OS}	Offset current			1.0	2.0		5.0	10.0	nA
		0°C ≤ T_A ≤ 70°C						14	nA
		25°C ≤ T_A ≤ 125°C			3.0				nA
		-55°C ≤ T_A ≤ 25°C			5.0				nA
I_{BIAS}	Input current			8.0	15		15.0	30.0	nA
		Over temperature			30			40	nA
V_{CM}	Common mode voltage range		+12	±13		±11	±12		V
CMRR	Common mode rejection ratio	R_S ≤ 10kΩ, T_A = 25°C, f = 100Hz	80	110		70	110		dB
Z_{IN}	Common mode input impedance	f = 20Hz		250			250		MΩ
V_{OUT}	Output voltage swing	R_L = 2kΩ	±12	±13		+11	±12		V
I_{CC}	Supply current			1.0	1.5		1.3	3.0	mA
P_D	DC quiescent power dissipation (V_O = 0)			30	45		40	90	mW
P_{SRR}	Supply voltage rejection ratio	R_S ≤ 10kΩ		50	100		75	200	μV/V
	Large signal voltage gain	R_L ≤ 2kΩ, V_{OUT} = ±10V, T_A = 25°C	100	200		70	100		V/mV
		Over temperature	40			40			V/mV

AC ELECTRICAL CHARACTERISTICS T_A = 25°C, V_S = ± 15V unless otherwise specified.

	PARAMETER	TEST CONDITIONS	MC1556			MC1456			UNIT
			Min	Typ	Max	Min	Typ	Max	
	Differential input impedance								
C_p	Parallel input capacitance	Open loop f = 20Hz		6.0			6.0		pF
r_p	Parallel input resistance			5			3		MΩ
e_n	Equivalent input noise voltage	A_V = 100, R_S = 10kΩ, f = 1.0kHz, BW = 1.0Hz		45			45		nV/$\sqrt{Hz}$
BW_p	Power bandwidth	A_V = 1, R_L = 2kΩ, THD ≤ 5% V_{OUT} = ±10V		40			40		kHz
	Phase margin (open loop, unity gain)			70			70		degrees
	Gain margin			18			18		dB
S_R	Slew rate (unity gain)			2.5			2.5		V/μsec
Z_{OUT}	Output impedance	f = 20Hz		1.0	2.0		1.0	2.5	kΩ
BW	Unity gain crossover frequency (open loop)			1.0			1.0		MHz

LF155/A/156/A/157/A, LF255/256/257,
LF355/A/356/A/357/A-T

DESCRIPTION

LF155, LF155A, LF255, LF355, LF355A (Low Supply Current)
LF156, LF156A, LF256, LF356, LF356A (Wide Band)
LF157, LF157A, LF257, LF357, LF357A (Wide Band)
The LF155, LF156, LF157 Series of operational amplifiers employ well matched, high voltage JFET input structures on the same monolithic chip as bipolar devices. These amplifiers feature low input bias and offset currents, low offset voltage and offset voltage drift, coupled with offset adjust which does not degrade drift or common mode rejection. The devices are also designed for high slew rate, wide bandwidth, extremely fast settling time and low noise.

COMMON FEATURES

(LF155A/156A/157A)

- **Low input bias current 30pA**
- **Low input offset current 3pA**
- **High input impedance $10^{12}\Omega$**
- **Low input offset voltage 1mV**
- **Low V_{OS} temperature drift $3\mu V/^{\circ}C$**
- **Low input noise current $0.01pA/\sqrt{Hz}$**

SPECIFIC FEATURES

	LF155A	LF156A
• Settling time (0.01%)	$4\mu s$	$1.5\mu s$
• High slew rate	$5v/\mu s$	$12v/\mu s$
• Wide bandwidth	2.5MHz	5MHz
• Low input noise	$20nV/\sqrt{Hz}$	$12nV/\sqrt{Hz}$

- **LF155, LF156—military qualifications pending**

	LF157A ($A_V = 5$)
• Settling time (0.01%)	$1.5\mu s$
• High slew rate	$50v/\mu s$
• Wide bandwidth	20MHz
• Low input noise	$12nV/\sqrt{Hz}$

APPLICATIONS

- **Precision high speed integrators**
- **Fast A/D, D/A converters**
- **High impedance buffers**
- **Wideband, low noise, low drift amplifier**

PIN CONFIGURATION

T PACKAGE

NC
BALANCE ① ⑧ ⑦ V+
INVERTING INPUT ② ⑥ OUTPUT
NONINVERTING INPUT ③ ④ ⑤ BALANCE
V-

ORDER PART NO.

LF155AT	LF156AT	LF157AT
LF155T	LF156T	LF157T
LF255T	LF256T	LF257T
LF355AT	LF356AT	LF357AT
LF355T	LF356T	LF357T

EQUIVALENT SCHEMATIC

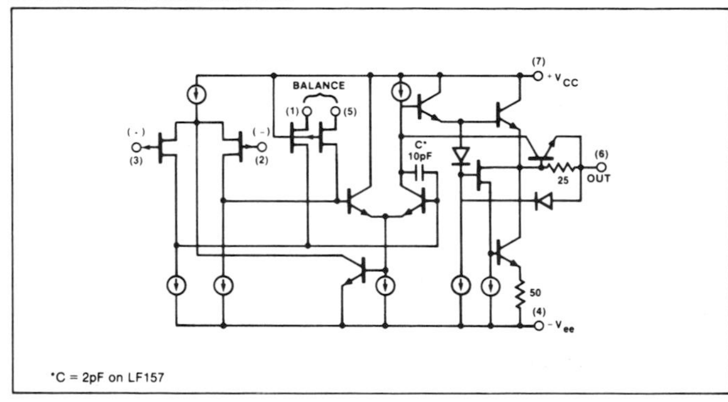

*C = 2pF on LF157

LF155/A/156/A/157/A, LF255/256/257,
LF355/A/356/A/357/A-T

ABSOLUTE MAXIMUM RATINGS

PARAMETER	RATING	UNIT
Supply voltage		
LF155A/6A/7A, LF155/6/7	±22	V
LF255/6/7	±22	V
LF355A/6A/7A, LF355/6/7	±18	V
Power dissipation[1] TO-99 (T-package)		
LF155A/6A/7A, LF155/6/7	670	mW
LF255/6/7	570	mW
LF355A/6A/7A, LF355/6/7	500	mW
Operating temperature range		
LF155A/6A/7A, LF155/6/7	-55 to +125	°C
LF255/6/7	-25 to +85	°C
LF355A/6A/7A, LF355/6/7	0 to +70	°C
T_J (Max)		
LF155A/6A/7A, LF155/6/7	150	°C
LF255/6/7	115	°C
LF355A/6A/7A, LF355/6/7	100	°C
Input voltage range[2]		
LF155A/6A/7A, LF155/6/7	±20	V
LF255/6/7	±20	V
LF355A/6A/7A, LF355/6/7	±20	V
Output short circuit duration		
LF155A/6A/7A, LF155/6/7	Continuous	
LF255/6/7	Continuous	
LF355A/6A/7A, LF355/6/7	Continuous	
Storage temperature range		
LF155A/6A/7A, LF155/6/7	-65 to +150	°C
LF255/6/7	-65 to +150	°C
LF355A/6A/7A, LF355/6/7	-65 to +150	°C
Lead temperature (soldering, 10sec.)	300	°C
LF155A/6A/7A, LF155/6/7	300	°C
LF255/6/7	300	°C
LF355A/6A/7A, LF355/6/7	300	°C

NOTES

1. The TO-99 package must be derated based on a thermal resistance of 150°C/W junction to ambient or 25°C/W junction to case.
2. Unless otherwise specified, the absolute maximum negative input voltage is equal to the negative power supply voltage.

LF155/A/156/A/157/A, LF255/256/257,
LF355/A/356/A/357/A-T

DC ELECTRICAL CHARACTERISTICS

$T_A = 25°C$ unless otherwise specified. (See notes on following page.)

PARAMETER		TEST CONDITIONS	LF155A/6A/7A			LF355A/6A/7A			UNIT
			Min	Typ	Max	Min	Typ	Max	
V_{OS}	Input offset voltage	$R_S = 50\Omega$		1	2 2.5		1	2 2.3	mV mV
$\Delta V_{OS}/\Delta T$	Avg. TC of input offset voltage	$R_S = 50\Omega$		3	5		3	5	μV/°C
$\Delta TC/\Delta V_{OS}$	Change in average TC[2] with V_{OS} adjust	$R_S = 50\Omega$		0.5			0.5		μV/°C per mV
I_{OS}	Input offset current[1,3]	$T_J = 25°C$ $T_J \leq T_{high}$		3	10 10		3	10 1	pA nA
I_B	Input bias current[1,3]	$T_J = 25°C$ $T_J \leq T_{high}$		30	50 25		30	50 5	pA nA
R_{IN}	Input resistance	$T_J = 25°C$		10^{12}			10^{12}		Ω
A_{VOL}	Large signal voltage gain	$V_S = \pm 15V$ $V_O = \pm 10V$, $R_L = 2k\Omega$ Over temp.	50 25	200		50 25	200		V/mV V/mV
V_O	Output voltage swing	$V_S = \pm 15V$, $R_L = 10k\Omega$ $V_S = \pm 15V$, $R_L = 2k\Omega$	±12 ±10	±13 ±12		±12 ±10	±13 ±12		V V
V_{CM}	Input common mode Voltage range	$V_S = \pm 15V$	±11	+15.1 -12		±11	+15.1 -12		V V V
CMRR	Common-mode rejection ratio		85	100		85	100		dB
PSRR	Supply volt. rej. ratio[4]		85	100		85	100		dB

DC ELECTRICAL CHARACTERISTICS

(Cont'd) $T_A = 25°C$ unless otherwise specified. (See notes on following page.)

PARAMETER		TEST CONDITIONS	LF155/6/7			LF255/6/7			UNIT
			Min	Typ	Max	Min	Typ	Max	
V_{OS}	Input offset voltage	$R_S = 50\Omega$		3	5 7		3	5 6.5	mV mV
$\Delta V_{OS}/\Delta T$	Avg. TC of input offset voltage	$R_S = 50\Omega$		5			5		μV/°C
$\Delta TC/\Delta V_{OS}$	Change in average TC[2] with V_{OS} adjust	$R_S = 50\Omega$		0.5			0.5		μV/°C per mV
I_{OS}	Input offset current[1,3]	$T_J = 25°C$ $T_J \leq T_{high}$		3	20 20		3	20 1	pA nA
I_B	Input bias current[1,3]	$T_J = 25°C$ $T_J \leq T_{high}$		30	100 50		30	100 5	pA nA
R_{IN}	Input resistance	$T_J = 25°C$		10^{12}			10^{12}		Ω
A_{VOL}	Large signal voltage gain	$V_S = \pm 15V$ $V_O = \pm 10V$, $R_L = 2k\Omega$ Over temp.	50 25	200		50 25	200		V/mV V/mV
V_O	Output voltage swing	$V_S = \pm 15V$, $R_L = 10k\Omega$ $V_S = \pm 15V$, $R_L = 2k\Omega$	±12 ±10	±13 ±12		±12 ±10	±13 ±12		V V
V_{CM}	Input common mode Voltage range	$V_S = \pm 15V$	±11	+15.1 -12		±11	+15.1 -12		V V V
CMRR	Common-mode rejection ratio		85	100		85	100		dB
PSRR	Supply volt. rej. ratio[4]		85	100		85	100		dB

LF155/A/156/A/157/A, LF255/256/257,
LF355/A/356/A/357/A-T

DC ELECTRICAL CHARACTERISTICS (Cont'd) $T_A = 25°C$ unless otherwise specified.

	PARAMETER	TEST CONDITIONS	LF355/6/7			UNIT
			Min	Typ	Max	
V_{OS}	Input offset voltage	$R_s = 50\Omega$		3	10 13	mV mV
$\Delta V_{OS}/\Delta T$ $\Delta TC/\Delta V_{OS}$	Avg. TC of input offset voltage Change in average TC[2] with V_{OS} adjust	$R_s = 50\Omega$ $R_s = 50\Omega$		5 0.5		$\mu V/°C$ $\mu V/°C$ per mV
I_{OS}	Input offset current[1,3]	$T_J = 25°C$ $T_J \le T_{high}$		3	50 2	pA nA
I_B	Input bias current[1,3]	$T_J = 25°C$ $T_J \le T_{high}$		30	200 8	pA nA
R_{IN} A_{VOL}	Input resistance Large signal voltage gain	$T_J = 25°C$ $V_s = \pm 15V$ $V_O = \pm 10V$, $R_L = 2k\Omega$ Over temp.	 25 15	 10^{12} 200		Ω V/mV V/mV
V_O	Output voltage swing	$V_s = \pm 15V$, $R_L = 10k\Omega$ $V_s = \pm 15V$, $R_L = 2k\Omega$	±12 ±10	±13 ±12		V V
V_{CM}	Input common mode Voltage range	$V_s = \pm 15V$	±10	+15.1 -12		V V V
CMRR PSRR	Common-mode rejection ratio Supply volt. rej. ratio[4]		80 80	100 100		dB dB

DC ELECTRICAL CHARACTERISTICS $T_A = 25°C$, $V_s = \pm 15V$ unless otherwise specified.

PARAMETER	LF155A/355A LF155/255			LF355			LF156A/LF156/256			UNIT
	Min	Typ	Max	Min	Typ	Max	Min	Typ	Max	
Supply current		2	4		2	4		5	7	mA

DC ELECTRICAL CHARACTERISTICS (Cont'd) $T_A = 25°C$, $V_s = \pm 15V$ unless otherwise specified.

PARAMETER	LF356A/LF356			LF157A/LF157/257			LF357A/LF357			UNIT
	Min	Typ	Max	Min	Typ	Max	Min	Typ	Max	
Supply current		5	10		5	7		5	10	mA

NOTES

1. These specifications apply for $\pm15V \le V_S \le \pm 20V$, $-55°C \le T_A \pm125°C$ and $T_{HIGH} = +125°C$ unless otherwise stated for the LF155A/6A/7A and the LF155/6/7. For the LF255/6/7, these specifications apply for $\pm15V \le V_S \le \pm20V$, $-25°C \le T_A \le +85°C$ and $T_{HIGH} = 85°C$ unless otherwise stated. For the LF355A/6A/7A, these specifications apply for $\pm15V \le V_S \le \pm20V$, $0°C \le T_A \le +70°C$ and $T_{HIGH} = +70°C$, and for the LF355/6/7 these specifications apply for $V_S = \pm15V$ and $0°C \le T_A \le +70°C$. V_{OS}, I_B and I_{OS} are measured at $V_{CM} = 0$.
2. The Temperature Coefficient of the adjusted input offset voltage changes only a small amount (0.5$\mu V/°C$ typically) for each mV of adjustment from its original unadjusted value. Common mode rejection and open loop voltage gain are also unaffected by offset adjustment.
3. The input bias currents are junction leakage currents which approximately double for every 10°C increase in the junction temperature, T_J. Due to limited production test time, the input bias currents measured are correlated to junction temperature. In normal operation the junction temperature rises above the ambient temperature as a result of internal power dissipation, Pd. $T_j = T_A + \theta_{jA}$ Pd where θ_{jA} is the thermal resistance from junction to ambient. Use of a heat sink is recommended if input bias current is to be kept to a minimum.
4. Supply Voltage Rejection is measured for both supply magnitudes increasing or decreasing simultaneously, in accordance with common practice.

LF155/A/156/A/157/A, LF255/256/257,
LF355/A/356/A/357/A-T

AC ELECTRICAL CHARACTERISTICS $T_A = 25°C$, $V_S = \pm 15V$ unless otherwise specified.[1]

PARAMETER		TEST CONDITIONS	LF155A/LF355A			LF156A/356A			LF157A/357A			UNIT
			Min	Typ	Max	Min	Typ	Max	Min	Typ	Max	
SR	Slew rate	LF155/156 LF155A/6A: $A_V = 1$	3	5		10	12		40	50		V/μs
GBW	Gain bandwidth product			2.5		4	4.5		15	20		MHz
t_s	Settling time[5] to 0.01%			4			1.5			1.5		μs
e_n	Equiv. input noise volt.	$R_S = 100\Omega$										
		f = 100Hz		25			15			15		nV/√Hz
		f = 1000Hz		20			12			12		nV/√Hz
i_n	Equiv. input noise current	f = 100Hz		0.01			0.01			0.01		pA/√Hz
		f = 1000Hz		0.01			0.01			0.01		pA/√Hz
C_{IN}	Input capacitance			3			3			3		pF

AC ELECTRICAL CHARACTERISTICS (Cont'd) $T_A = 25°C$, $V_S = \pm 15V$ unless otherwise specified.[1]

PARAMETER		TEST CONDITIONS	LF155/255/355			LF156/256			LF356			UNIT
			Min	Typ	Max	Min	Typ	Max	Min	Typ	Max	
SR	Slew rate	LF155/156 LF155A/6A: $A_V = 1$,		5		7.5	12			12		V/μs
GBW	Gain bandwidth product			2.5			5			5		MHz
t_s	Settling time[5] to 0.01%			4			1.5			1.5		μs
e_n	Equiv. input noise volt.	$R_S = 100\Omega$										
		f = 100Hz		25			15			15		nV/√Hz
		f = 1000Hz		20			12			12		nV/√Hz
i_n	Equiv. input noise current	f = 100Hz		0.01			0.01			0.01		pA/√Hz
		f = 1000Hz		0.01			0.01			0.01		pA/√Hz
C_{IN}	Input capacitance			3			3			3		pF

LF155/A/156/A/157/A, LF255/256/257,
LF355/A/356/A/357/A-T

AC ELECTRICAL CHARACTERISTICS (Cont'd) $T_A = 25°C$, $V_s = \pm 15V$ unless otherwise specified.[1]

PARAMETER		TEST CONDITIONS	LF157/257			LF357			UNIT
			Min	Typ	Max	Min	Typ	Max	
SR	Slew rate	LF157A/LF157: $A_V = 5$	30	50			50		V/μs
GBW	Gain bandwidth product			20			20		MHz
t_s	Settling time[5] to 0.01%			1.5			1.5		μs
e_n	Equiv. input noise volt.	$R_s = 100\Omega$							
		f = 100Hz		15			15		nV/√Hz
		f = 1000Hz		12			12		nV/√Hz
i_n	Equiv. input noise current	f = 100Hz		0.01			0.01		pA/√Hz
		f = 1000Hz		0.01			0.01		pA/√Hz
C_{IN}	Input capacitance			3			3		pF

NOTE

5. Settling time is defined here, for a unity gain inverter connection using 2kΩ resistors for the LF155/6. It is the time required for the error voltage (the voltage at the inverting input pin on the amplifier) to settle to within 0.01% of its final value from the time a 10V step input is applied to the inverter. For the LF157, $A_V = -5$, the feedback resistor from output to input is 2kΩ and the output step is 10V (See Settling Time Test Circuit).

TYPICAL DC PERFORMANCE CHARACTERISTICS (curves are for LF155, LF156 and LF157 unless otherwise specified.)

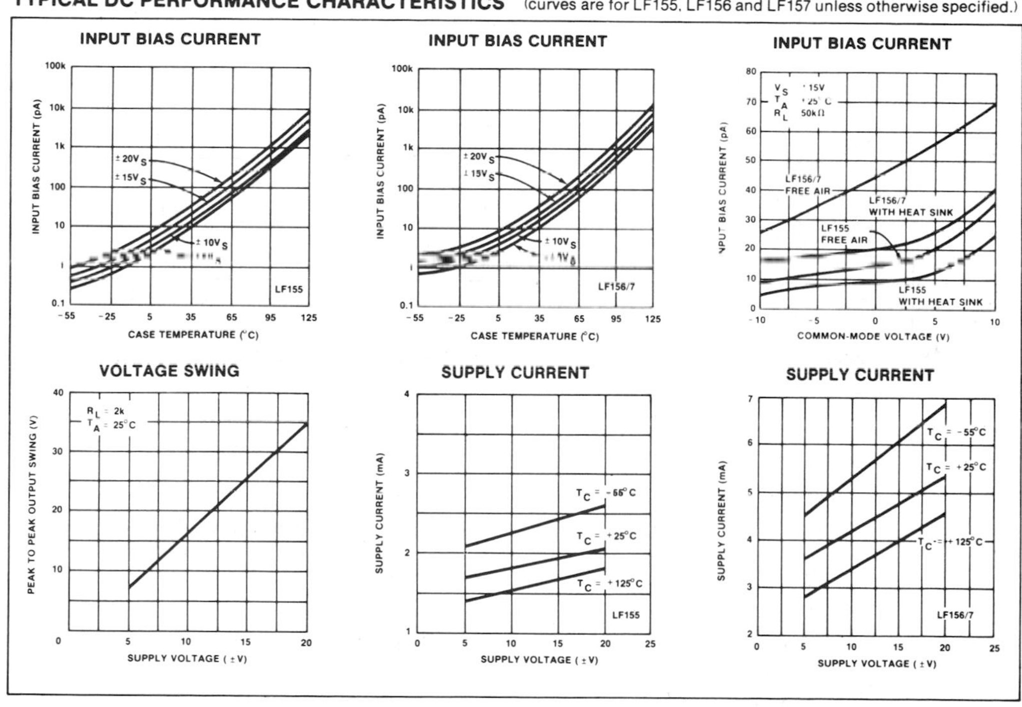

Consumer Circuits

LM377 dual 2 watt audio amplifier

general description

The LM377 is a monolithic dual power amplifier which offers high quality performance for stereo phonographs, tape players, recorders, and AM-FM stereo receivers, etc.

The LM377 will deliver 2W/channel into 8 or 16Ω loads. The amplifier is designed to operate with a minimum of external components and contains an internal bias regulator to bias each amplifier. Device overload protection consists of both internal current limit and thermal shutdown.

features

- A_{VO} typical 90 dB
- 2W per channel
- 70 dB ripple rejection
- 75 dB channel separation
- Internal stabilization
- Self centered biasing
- 3 MΩ input impedance
- 10–26V operation
- Internal current limiting
- Internal thermal protection

applications

- Multi-channel audio systems
- Tape recorders and players
- Movie projectors
- Automotive systems
- Stereo phonographs
- Bridge output stages
- AM-FM radio receivers
- Intercoms
- Servo amplifiers
- Instrument systems

schematic diagram

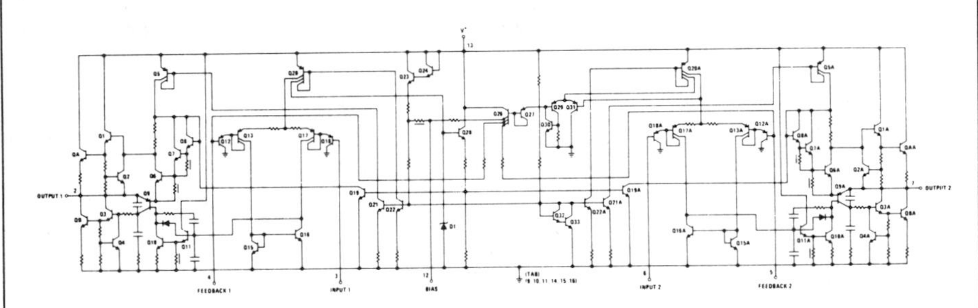

connection diagram

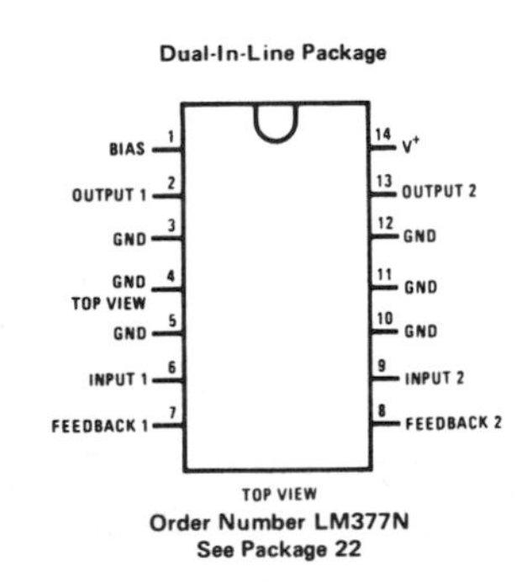

Order Number LM377N
See Package 22

typical applications

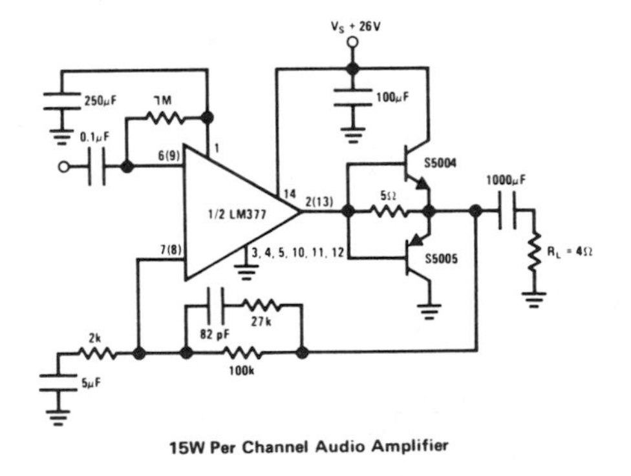

15W Per Channel Audio Amplifier

absolute maximum ratings

Supply Voltage	26V
Input Voltage	0V – V_{SUPPLY}
Operating Temperature	0°C to +70°C
Storage Temperature	–65°C to +150°C
Junction Temperature	150°C
Lead Temperature (Soldering, 10 seconds)	300°C

electrical characteristics

V_S = 20V, T_{TAB} = 25°C, R_L = 8Ω, A_V = 50 (34 dB), unless otherwise specified.

PARAMETER	CONDITIONS	MIN	TYP	MAX	UNITS
Total Supply Current	P_{OUT} = 0W		15	50	mA
	P_{OUT} = 1.5W/Channel		430	500	mA
DC Output Level			10		V
Supply Voltage		10		26	V
Output Power	T.H.D. = < 5%	2	2.5		W
T.H.D.	P_{OUT} = 0.05W/Channel, f = 1 kHz		0.25		%
	P_{OUT} = 1W/Channel, f = 1 kHz		0.07	1	%
	P_{OUT} = 2W/Channel, f = 1 kHz		0.10		%
Offset Voltage			15		mV
Input Bias Current			100		nA
Input Impedance		3			MΩ
Open Loop Gain	R_S = 0Ω	66	90		dB
Output Swing			V_S–6		V_{P-P}
Channel Separation	C_F = 250μF, f = 1 kHz	50	70		dB
Ripple Rejection	f = 120 Hz, C_F = 250μF	60	70		dB
Current Limit			1.5		A
Slew Rate			1.4		V/μs
Equivalent Input Noise Voltage	R_S = 600Ω, 100 Hz – 10 kHz		3		μVrms

Note 1: For operation at ambient temperatures greater than 25°C the LM377 must be derated based on a maximum 150°C junction temperature using a thermal resistance which depends upon device mounting techniques.

Note 2: Dissipation characteristics are shown for four mounting configurations.

a. Infinite sink – 13.4°C/W
b. P.C. board +V_7 sink – 21°C/W. P.C. board is 2 1/2 square inches. Staver V_7 sink is 0.02 inch thick copper and has a radiating surface area of 10 square inches.
c. P.C. board only – 29°C/W. Device soldered to 2 1/2 square inch P.C. board.
d. Free air – 58°C/W.

Consumer Circuits

LM378 dual 4 watt audio amplifier

general description

The LM378 is a monolithic dual power amplifier which offers high quality performance for stereo phonographs, tape players, recorders, and AM-FM stereo receivers, etc.

The LM378 will deliver 4W channel into 8 or 16Ω loads. The amplifier is designed to operate with a minimum of external components and contains an internal bias regulator to bias each amplifier. Device overload protection consists of both internal current limit and thermal shutdown.

features

- A_{VO} typical 90 dB
- 4W per channel
- 70 dB ripple rejection
- 75 dB channel separation
- Internal stabilization
- Self centered biasing
- 3 MΩ input impedance
- Internal current limiting
- Internal thermal protection

applications

- Multi-channel audio systems
- Tape recorders and players
- Movie projectors
- Automotive systems
- Stereo phonographs
- Bridge output stages
- AM-FM radio receivers
- Intercoms
- Servo amplifiers
- Instrument systems

schematic diagram

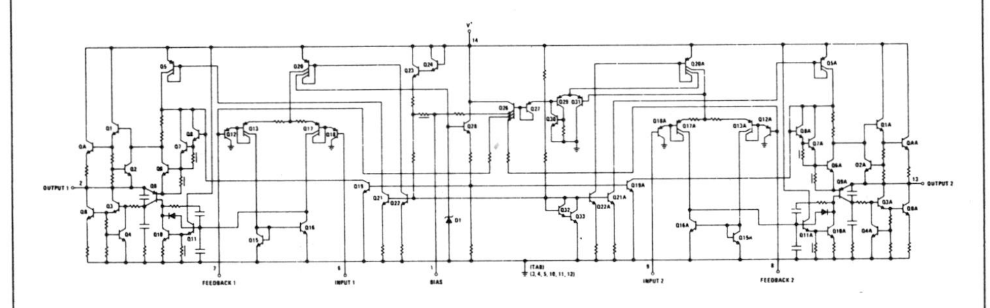

connection diagram

Dual-In-Line Package

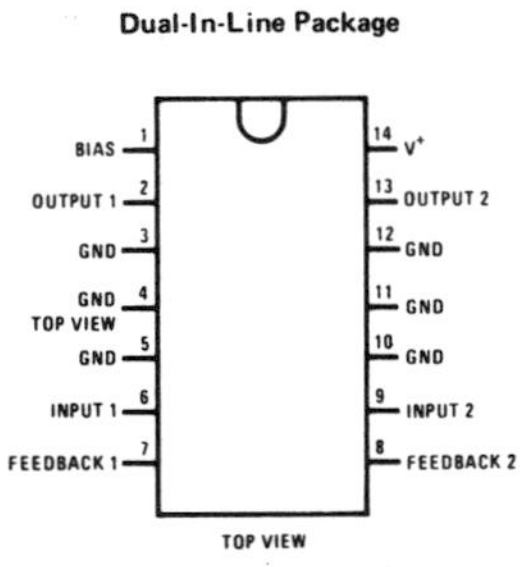

Order Number LM378N
See Package 22

typical applications

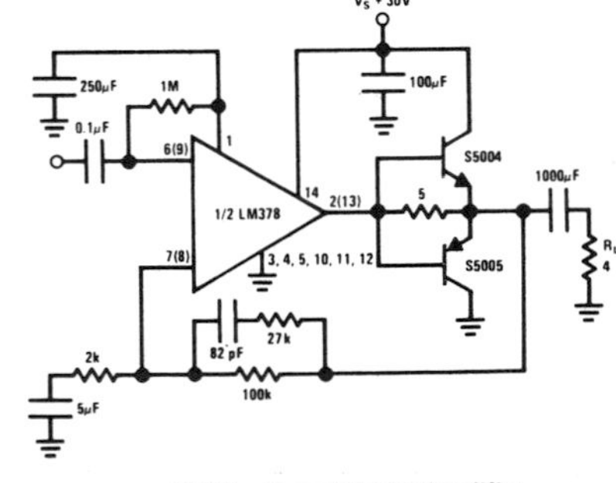

15W Per Channel Audio Amplifier

absolute maximum ratings

Supply Voltage	35V
Input Voltage	0V – V_{SUPPLY}
Operating Temperature	0°C to +70°C
Storage Temperature	−65°C to +150°C
Junction Temperature	150°C
Lead Temperature (Soldering, 10 seconds)	300°C

electrical characteristics

V_S = 24V, T_{TAB} = 25°C, R_L = 8Ω, A_V = 50 (34 dB), unless otherwise specified.

PARAMETER	CONDITIONS	MIN	TYP	MAX	UNITS
Total Supply Current	P_{OUT} = 0W		15	50	mA
	P_{OUT} = 1.5W/Channel		430	500	mA
DC Output Level			12		V
Supply Voltage		10			V
Output Power	T.H.D. = < 5%, R_L = 8Ω	4	5		W
	T.H.D. = < 5%, R_L = 16Ω	4	5		W*
T.H.D.	P_{OUT} = 0.05W/Channel, f = 1 kHz		0.25		%
	P_{OUT} = 1W/Channel, f = 1 kHz		0.07	1	%
	P_{OUT} 2W/Channel, f = 1 kHz		0.10		%
Offset Voltage			15		mV
Input Bias Current			100		nA
Input Impedance		3			MΩ
Open Loop Gain	R_S = 0Ω	66	90		dB
Channel Separation	C_F = 250μF, f = 1 kHz	50	70		dB
Ripple Rejection	f = 120 Hz, C_F = 250μF	60	70		dB
Current Limit			1.5		A
Slew Rate			1.4		V/μs
Equivalent Input Noise Voltage	R_S = 600Ω, 100 Hz – 10 kHz		3		μVrms

Note 1: For operation at ambient temperatures greater than 25°C the LM378 must be derated based on a maximum 150°C junction temperature using a thermal resistance which depends upon device mounting techniques.

Note 2: Dissipation characteristics are shown for four mounting configurations.

a. Infinite sink – 13.4°C/W

b. P.C. board +V_7 sink – 21°C/W. P.C. board is 2 1/2 square inches. Staver V_7 sink is 0.02 inch thick copper and has a radiating surface area of 10 square inches.

c. P.C. board only – 29°C/W. Device soldered to 2 1/2 square inch P.C. board.

d. Free air – 58°C/W.

*Tested at V_S = 30V.

Consumer Circuits

LM387 low noise dual preamplifier

general description

The LM387 is a dual preamplifier for the amplication of low level signals in applications requiring optimum noise performance. Each of the two amplifiers is completely independent, with an internal power supply decoupler-regulator, providing 110 dB supply rejection and 60 dB channel separation. Other outstanding features include high gain (104 dB), large output voltage swing (V_{CC}−2V)p-p, and wide power bandwidth (75 kHz, 20 Vp-p). The LM387 operates from a single supply across the wide range of 9 to 40V.

The amplifiers are internally compensated for. All gains greater than 10. The LM387 is available in an 8 lead dual-in-line package.

features

- Low noise — 0.8μV total input noise
- High gain — 104 dB open loop
- Single supply operation
- Wide supply range — 9 to 40V
- Power supply rejection — 110 dB
- Large output voltage swing (V_{CC}−2V)p-p
- Wide bandwidth 15 MHz unity gain
- Power bandwidth 75 kHz, 20 Vp-p
- Internally compensated
- Short circuit protected

schematic and connection diagrams

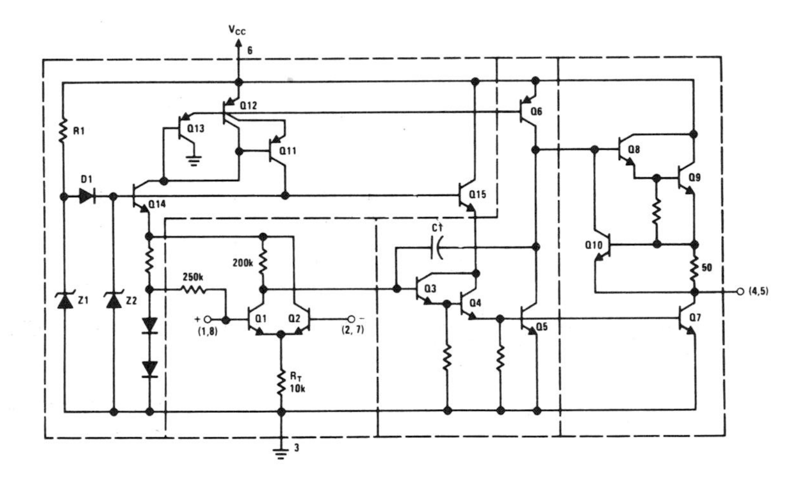

Dual-In-Line Package

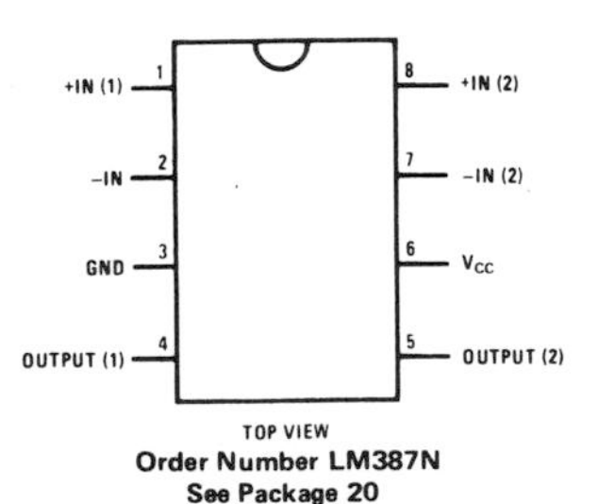

TOP VIEW

Order Number LM387N
See Package 20

absolute maximum ratings

Supply Voltage	+40V
Power Dissipation	660 mW
Operating Temperature Range	0°C to +70°C
Storage Temperature Range	−65°C to +150°C
Lead Temperature (Soldering, 10 seconds)	300°C

electrical characteristics T_A = 25°C, V_{CC} = 14V, unless otherwise stated.

PARAMETER	CONDITIONS	MIN	TYP	MAX	UNITS
Voltage Gain	Open Loop		160,000		V/V
Supply Current	V_{CC} 9 to 40V, $R_L = \infty$		10		mA
Input Resistance					
Positive Input			100		kΩ
Negative Input			200		kΩ
Input Current					
Negative Input			0.5		μA
Output Resistance	Open Loop		150		Ω
Output Current	Source		8		mA
	Sink		2		mA
Output Voltage Swing	Peak-to-Peak		$V_{CC}-2$		V
Small Signal Bandwidth			15		MHz
Power Bandwidth	20 Vp-p (V_{CC} = 24V)		75		kHz
Maximum Input Voltage	Linear Operation			300	mVrms
Supply Rejection Ratio	f = 1 kHz		110		dB
Channel Separation	f = 1 kHz		60		dB
Total Harmonic Distortion	75 dB Gain, f = 1 kHz		0.1		%
Total Equivalent Input Noise	R_S = 600Ω, 100 – 10,000 Hz		0.8	1.4	μVrms
Noise Figure	50 kΩ, 10 – 10,000 Hz		1.0		dB
	10 kΩ, 10 – 10,000 Hz		1.6		dB
	5 kΩ, 10 – 10,000 Hz		2.8		dB

Operational Amplifiers

LM747/LM747C dual operational amplifier

general description

The LM747 and the LM747C are general purpose dual operational amplifiers. The two amplifiers share a common bias network and power supply leads. Otherwise, their operation is completely independent.

features

- No frequency compensation required
- Short-circuit protection
- Wide common-mode and differential voltage ranges
- Low-power consumption
- No latch-up
- Balanced offset null

Additional features of the LM747 and LM747C are: no latch-up when input common mode range is exceeded, freedom from oscillations, and package flexibility.

The LM747C is identical to the LM747 except that the LM747C has its specifications guaranteed over the temperature range from 0°C to 70°C instead of −55°C to +125°C.

schematic diagram (each amplifier)

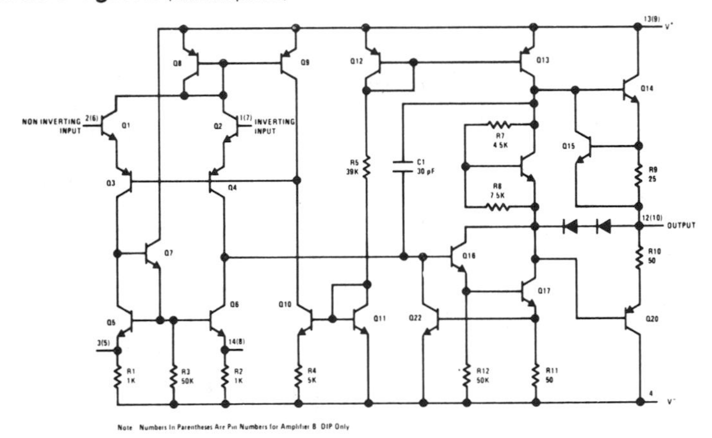

Note: Numbers In Parentheses Are Pin Numbers for Amplifier B DIP Only

connection diagrams

Metal Can Package

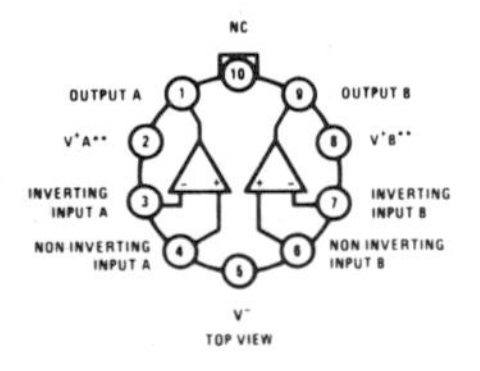

Order Number LM747H or LM747CH
See Package 14

Flat Package

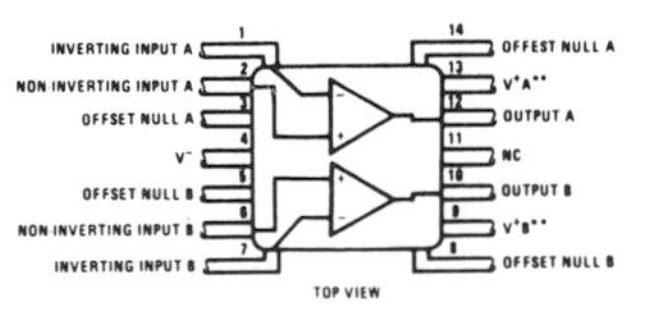

Order Number LM747F or LM747CF
See Package 4

Dual-In-Line Packages

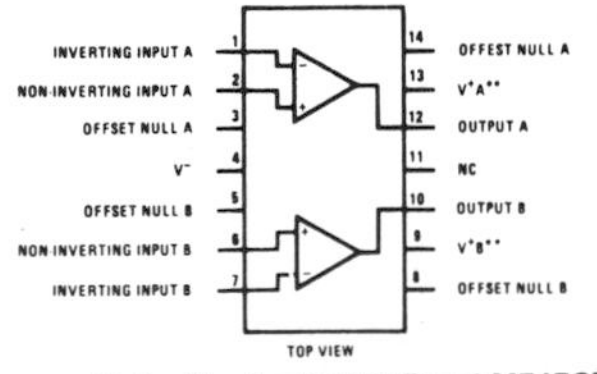

Order Number LM747D or LM747CD
See Package 1
Order Number LM747CN
See Package 22

**V+A and V+B are internally connected.

absolute maximum ratings

Supply Voltage LM747	±22V
LM747C	±18V
Power Dissipation (Note 1)	800 mW
Differential Input Voltage	±30V
Input Voltage (Note 2)	±15V
Output Short-Circuit Duration	Indefinite
Operating Temperature Range LM747	−55°C to 125°C
LM747C	0°C to 70°C
Storage Temperature Range	−65°C to 150°C
Lead Temperature (Soldering, 10 sec)	300°C

electrical characteristics (Note 3)

PARAMETER	CONDITIONS	LM747			LM747C			UNITS
		MIN	TYP	MAX	MIN	TYP	MAX	
Input Offset Voltage	$T_A = 25°C$, $R_S \leq 10\ k\Omega$		1.0	5.0		1.0	6.0	mV
Input Offset Current	$T_A = 25°C$		80	200		80	200	nA
Input Bias Current	$T_A = 25°C$		200	500		200	500	nA
Input Resistance	$T_A = 25°C$	0.3	1.0		0.3	1.0		MΩ
Supply Current Both Amplifiers	$T_A = 25°C$, $V_S = \pm 15V$		3.0	5.6		3.0	5.6	mA
Large Signal Voltage Gain	$T_A = 25°C$, $V_S = \pm 15V$ $V_{OUT} = \pm 10V$, $R_L \geq 2\ k\Omega$	50	160		50	160		V/mV
Input Offset Voltage	$R_S \leq 10\ k\Omega$			6.0			7.5	mV
Input Offset Current				500			300	nA
Input Bias Current				1.5			0.8	μA
Large Signal Voltage Gain	$V_S = \pm 15V$, $V_{OUT} = \pm 10V$ $R_L \geq 2\ k\Omega$	25			25			V/mV
Output Voltage Swing	$V_S = \pm 15V$, $R_L = 10\ k\Omega$	±12	±14		±12	±14		V
	$R_L = 2\ k\Omega$	±10	±13		±10	±13		V
Input Voltage Range	$V_S = \pm 15V$	±12			±12			V
Common Mode Rejection Ratio	$R_S \leq 10\ k\Omega$	70	90		70	90		dB
Supply Voltage Rejection Ratio	$R_S \leq 10\ k\Omega$	77	96		77	96		dB

Note 1: The maximum junction temperature of the LM747 is 150°C, while that of the LM747C is 100°C. For operating at elevated temperatures, devices in the TO-5 package must be derated based on a thermal resistance of 150°C/W, junction to ambient, or 45°C/W, junction to case. For the flat package, the derating is based on a thermal resistance of 185°C/W when mounted on a 1/16-inch-thick epoxy glass board with ten, 0.03-inch-wide, 2-ounce copper conductors. The thermal resistance of the dual-in-line package is 100°C/W, junction to ambient.

Note 2: For supply voltages less than ±15V, the absolute maximum input voltage is equal to the supply voltage.

Note 3: These specifications apply for $V_S = \pm 15V$ and $-55°C \leq T_A \leq 125°C$, unless otherwise specified. With the LM747C, however, all specifications are limited to $0°C \leq T_A \leq 70°C$ $V_S = \pm 15V$.

Operational Amplifiers

LM3900 quad amplifier

general description

The LM3900 consists of four independent, dual input, internally compensated amplifiers which were designed specifically to operate off of a single power supply voltage and to provide a large output voltage swing. These amplifiers make use of a current mirror to achieve the non-inverting input function. Application areas include: AC amplifiers, RC active filters; low frequency triangle, squarewave and pulse waveform generation circuits, tachometers and low speed, high voltage digital logic gates.

features

- Wide single supply voltage range — 4 V_{DC} to 36 V_{DC}; or dual supplies — ±2 V_{DC} to ±18 V_{DC}
- Supply current drain independent of supply voltage
- Low input biasing current — 30 nA
- High open-loop gain — 70 dB
- Wide bandwidth — 2.5 MHz (Unity Gain)
- Large output voltage swing — $(V^+ - 1)$ $V_{p\text{-}p}$
- Internally frequency compensated for unity gain
- Output short-circuit protection

schematic and connection diagrams

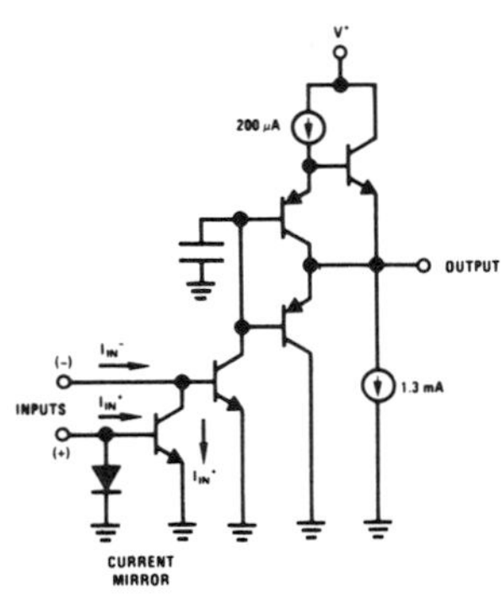

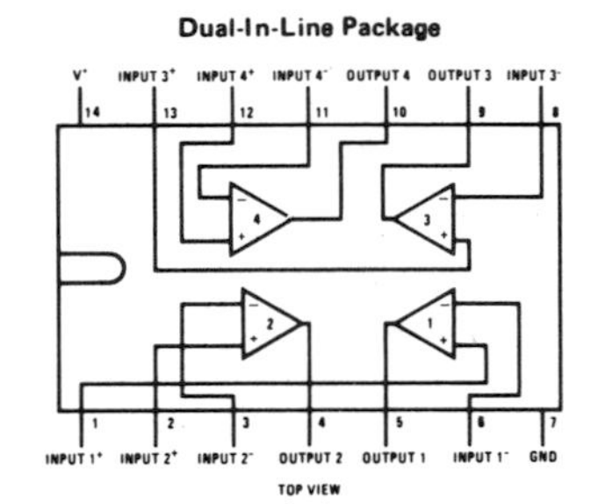

Order Number LM3900N
See Package 22

typical applications ($V^+ = 15V_{DC}$)

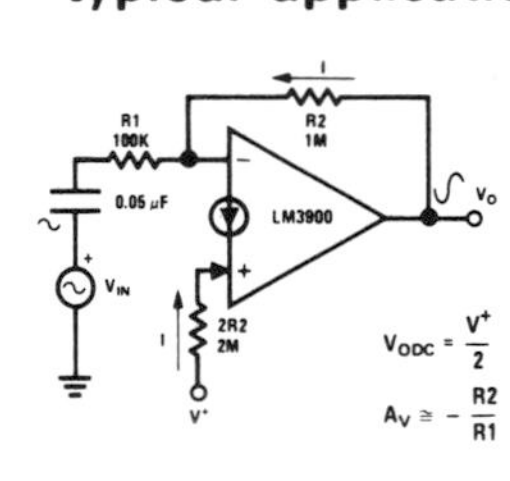

Inverting Amplifier

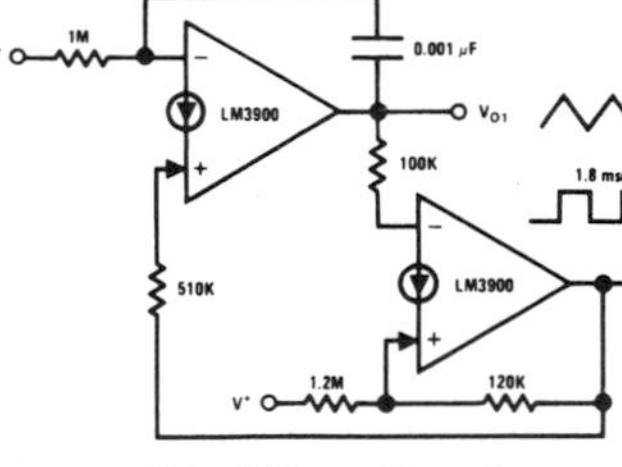

Triangle/Square Generator

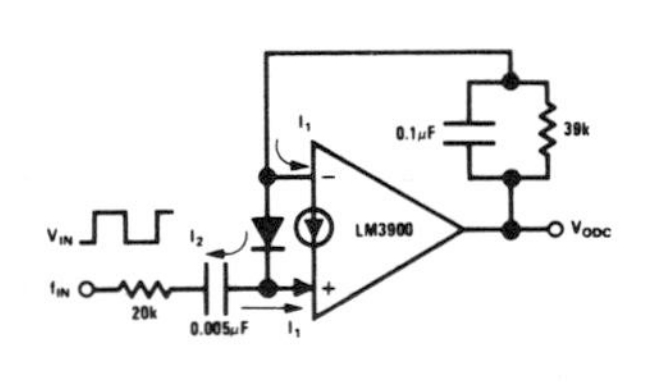

Frequency-Doubling Tachometer

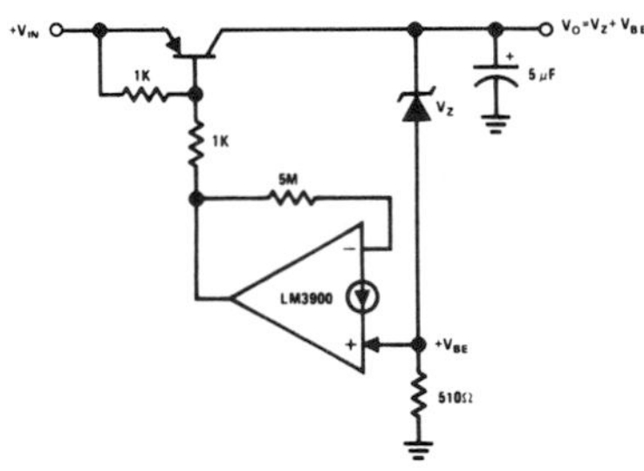

Low V_{IN}–V_{OUT} Voltage Regulator

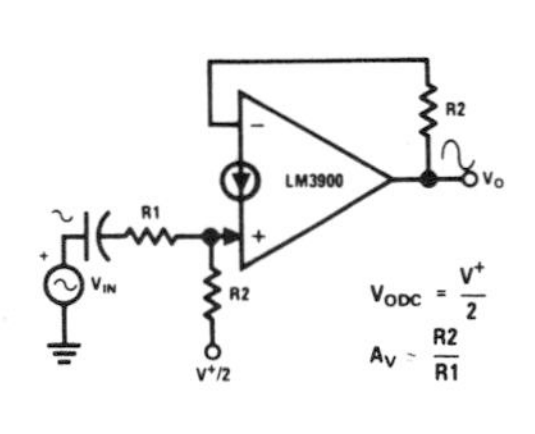

Non-Inverting Amplifier

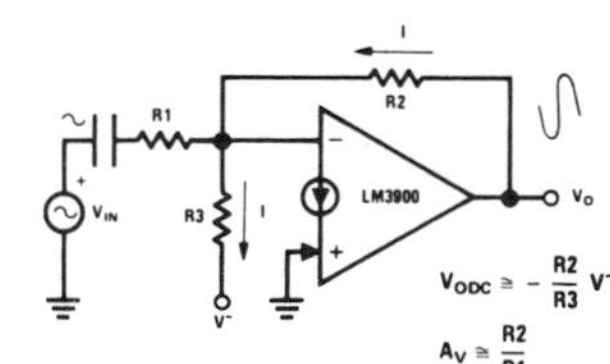

Negative Supply Biasing

absolute maximum ratings

Supply Voltage	+32 VDC ±18 VDC
Power Dissipation (T_A = 25°C) (Note 1)	570 mW
Input Currents, $I_{IN}+$ or $I_{IN}-$	20 mA DC
Output Short Circuit Duration – One Amplifier T_A = 25°C (See Application Hints)	Continuous
Operating Temperature Range	0°C to +70°C
Storage Temperature Range	-65°C to +150°C
Lead Temperature (Soldering, 10 seconds)	300°C

electrical characteristics (V^+ = +15 VDC and T_A = 25°C unless otherwise noted)

PARAMETER	CONDITIONS	MIN	TYP	MAX	UNITS
Open Loop					
Voltage Gain	f = 100 Hz	1200	2800		V/V
Input Resistance	Inverting Input		1		MΩ
Output Resistance			8		kΩ
Unity Gain Bandwidth	Inverting Input		2.5		MHz
Input Bias Current	Inverting Input		30	200	nA
Slew Rate	Positive Output Swing		0.5		V/µs
	Negative Output Swing		20		V/µs
Supply Current	$R_L = \infty$ On All Amplifiers		6.2	10	mA DC
Output Voltage Swing	R_L = 5.1k				
V_{OUT} High	$I_{IN}- = 0$, $I_{IN}+ = 0$	13.5	14.2		VDC
V_{OUT} Low	$I_{IN}- = 10\ \mu A$, $I_{IN}+ = 0$		0.09	0.2	VDC
Output Current Capability					
Source		3	18		mA DC
Sink	(Note 2)	0.5	1.3		mA DC
Power Supply Rejection	f = 100 Hz		70		dB
Mirror Gain	$I_{IN}+ = 200\ \mu A$ (Note 3)	0.9	1	1.1	µA/µA
Mirror Current	(Note 4)		10	500	µA DC
Negative Input Current	(Note 5)		1.0		mA DC

Note 1: For operating at high temperatures, the device must be derated based on a 125°C maximum junction temperature and a thermal resistance of 175°C/W which applies for the device soldered in a printed circuit board, operating in a still air ambient.

Note 2: The output current sink capability can be increased for large signal conditions by overdriving the inverting input. This is shown in the section on Typical Characteristics.

Note 3: This spec indicates the current gain of the current mirror which is used as the non-inverting input.

Note 4: Input V_{BE} match between the non-inverting and the inverting inputs occurs for a mirror current (non-inverting input current) of approximately 10 µA. This is therefore a typical design center for many of the application circuits.

Note 5: Clamp transistors are included on the IC to prevent the input voltages from swinging below ground more than approximately –0.3 VDC. The negative input currents which may result from large signal overdrive with capacitance input coupling need to be externally limited to values of approximately 1 mA. Negative input currents in excess of 4 mA will cause the output voltage to drop to a low voltage. This maximum current applies to any one of the input terminals. If more than one of the input terminals are simultaneously driven negative smaller maximum currents are allowed. Common-mode current biasing can be used to prevent negative input voltages; see for example the "Differentiator Circuit" in the applications section.

Answers to Self-Checking Quizzes

Chapter 1

1.	c, l	11.	False
2.	f, k	12.	True
3.	a	13.	False
4.	i	14.	True
5.	h	15.	True
6.	e	16.	True
7.	d	17.	True
8.	j	18.	False
9.	g	19.	True
10.	b	20.	True

Chapter 2

1.	f	11.	b
2.	d	12.	a
3.	e	13.	c
4.	a	14.	f
5.	c	15.	b
6.	b	16.	a
7.	a	17.	b
8.	c	18.	c
8.	a	19.	b
10.	d	20.	d

Chapter 3

1.	e	11.	b
2.	f	12.	f
3.	b	13.	True
4.	c	14.	False
5.	a	15.	False
6.	d	16.	True
7.	e	17.	True
8.	c	18.	True
9.	d	19.	False
10.	a	20.	True

Chapter 4

1.	c	6.	True
2.	a	7.	True
3.	e	8.	False
4.	b	9.	True
5.	d	10.	False

Chapter 5

1.	OP-3	6.	T.C.-2
2.	OP-4	7.	b
3.	T.C.-1	8.	c
4.	R_{11}	9.	a
5.	OP-1	10.	d

Chapter 6

1.	b	6.	e
2.	g	7.	a
3.	d	8.	d
4.	c	9.	b
5.	f	10.	d

Index